Routledge Revivals

Charles Darwin

This biography of Charles Darwin, first published in 1937, re-lives Darwin's life year by year, allowing the reader to share his experiences. The book displays Darwin's ideas and how they developed and grew over time. This title will be of great interest to students of the history of science and philosophy.

Charles Darwin
The Fragmentary Man

Geoffrey West

First published in 1937
by George Routledge & Sons, Ltd.

This edition first published in 2018 by Routledge
2 Park Square, Milton Park, Abingdon, Oxon, OX14 4RN
and by Routledge
711 Third Avenue, New York, NY 10017

Routledge is an imprint of the Taylor & Francis Group, an informa business

© 1937 Geoffrey West

All rights reserved. No part of this book may be reprinted or reproduced or utilised in any form or by any electronic, mechanical, or other means, now known or hereafter invented, including photocopying and recording, or in any information storage or retrieval system, without permission in writing from the publishers.

Publisher's Note
The publisher has gone to great lengths to ensure the quality of this reprint but points out that some imperfections in the original copies may be apparent.

Disclaimer
The publisher has made every effort to trace copyright holders and welcomes correspondence from those they have been unable to contact.

A Library of Congress record exists under LCCN: 38001703

ISBN 13: 978-1-138-49655-2 (hbk)
ISBN 13: 978-1-351-02130-2 (ebk)
ISBN 13: 978-1-138-49656-9 (pbk)

CHARLES DARWIN IN 1881

[*frontispiece*

CHARLES DARWIN
THE FRAGMENTARY MAN

BY

GEOFFREY WEST

Illustrated

LONDON
GEORGE ROUTLEDGE & SONS, LTD.
BROADWAY HOUSE: CARTER LANE, E.C.

First published . . . 1937

Printed in Great Britain by T. and A. CONSTABLE LTD.
at the University Press, Edinburgh

To
HELEN REBECCA WELLS
MY MOTHER

CONTENTS

	PAGE
PREFACE	xi

BOOK ONE: FAMILY OVERTURE

CHAP.
| I. | DARWIN-WEDGWOOD . . . . . . | 3 |
| II. | FROM ONE GENERATION . . . . . | 21 |

BOOK TWO: THE MAKING OF THE MAN

III.	SHREWSBURY DAYS	43
IV.	DOCTOR, CLERGYMAN	59
V.	OFFER OF ADVENTURE	81
VI.	THE ADVENTURE BEGINS	96
VII.	SHIP AND SHORE	108
VIII.	THE WORLD BEFORE US	124
IX.	LONDON: THE FIRST FRUITS	145
X.	LONDON: MACAW COTTAGE	161

BOOK THREE: THE MAN MAKING

XI.	EARLY DAYS AT DOWNE	177
XII.	BARNACLES AND MIDDLE AGE	197
XIII.	THE GREAT WORK	216
XIV.	CLIMACTIC	229
XV.	RECEPTION OF A BOOK	241
XVI.	THE GREAT GOD PAN	257

CHAP.		PAGE
XVII.	THE QUESTION OF QUESTIONS	272
XVIII.	BOTANY BAY	283
XIX.	THE ULTIMATE HAVEN	299

BOOK FOUR: COMMENTARY

XX.	THE FRAGMENTARY MAN	315
BIBLIOGRAPHY		333
INDEX		343

LIST OF ILLUSTRATIONS

Charles Darwin in 1881	*frontispiece*
Dr. Erasmus Darwin	*facing p.* 16
Josiah Wedgwood and his Family at Etruria Hall in 1780	,, 32
Robert W. Darwin	,, 96
Charles Darwin in 1840	,, 144
Emma Darwin in 1840	,, 162
Charles Darwin in 1854	,, 208
Robert FitzRoy	,, 252

PREFACE

IF there is anything to be said for the present work, it is possibly best implied in two brief quotations from the Darwin memorial notices which appeared in *Nature* in 1882. The first, by G. J. Romanes, applies to its subject: "It is in Charles Darwin's case particularly and pre-eminently true that the first duty of biographers will be to render some idea, not of what he did, but of what he was." The second, by W. Thistleton Dyer, bears more upon the author: "From not being, till he took up any point, familiar with the literature bearing on it, his mind was absolutely free from any prepossession." If I had any relevant views at all, when I first approached the matter six years ago, they were dominantly a taking of Darwinism for granted joined to a faint distaste (not uncommon, I find) for Darwin himself as a rather more than usually "typical Victorian." My inclination to-day is to take Darwinism very much less for granted—as will appear in due course—and to feel a considerable liking, even an affection, for the man Darwin and for many of his family and friends, who seem to me to represent, if "Victorianism" still, some of the best elements of English life in their period. (Actually this is, as I have told it, a more than nineteenth-century story, beginning in the eighteenth century with Darwin's grandfathers as young men.)

So far as my account has any originality, it probably resides in the fact that I have, I believe, tried to keep closer to Romanes's dictum than any other biographer, to re-live Darwin's life imaginatively year by year, almost day by day, sharing his experience and displaying his ideas as growing out of it. I have done that primarily in the belief that the validity of any man's ideas depends ultimately upon what he is or was—that in the last resort a sow's ear cannot even *think* a silk purse! It was a desire to test that faith in a field previously foreign to me that really impelled me to undertake this study, and on the whole it has been confirmed.

Throughout I have striven wherever possible for complete accuracy in factual detail. Frankly, I have found the airy assumptions and unconfirmed borrowings of some Darwin biographers distinctly disturbing. When one against all the evidence can make Emma Wedgwood Charles Darwin's "sweetheart" in 1825, and another on no evidence at all adopt the acknowledged speculations of a psycho-analytical theorist as established fact, there seems no reason why one should not resort to fairy-tales direct.

Perhaps if any part of this book needs apology it is the last chapter, which for adequate treatment should have been much longer. But the work, despite drastic cutting throughout the manuscript, was already long enough, and I have had to content myself with deliberately free and various quotation to show that the views suggested there are actually part of a wide and strong current of contemporary thought.

Little unpublished material has been draw upon, though a considerable amount—the withheld parts of Darwin's autobiography, personal letters, manuscripts and notebooks—must be in existence somewhere. I do not think this omission matters much, for I cannot believe that any secrets remain to be revealed. The Darwin the world knows is the whole Darwin.

On the other hand I have done my best to utilize printed sources as widely as possible, though I have quoted only as seemed absolutely necessary. All such citations, save for the briefest phrases, are acknowledged in footnotes, which in conjunction with the bibliography at the end will, I hope, make identification easy. (I should perhaps say here that the abbreviations *LL.* and *ML.* refer to the *Life and Letters of Charles Darwin* and the *More Letters of Charles Darwin* respectively; in other cases the name of the subject is added, as *LL. Hooker, LL. Huxley*, etc.) References are to first editions, save when otherwise dated.

For very kind permission to make my more extensive copyright quotations I have to thank especially John Murray (as publisher of the *Life and Letters*, the *More Letters*, and that work so essential to any biographer of Charles Darwin and so delightful for its own sake, the *Emma Darwin: A Century of*

Family Letters); the Cambridge University Press (as publishers of *The Foundations of the Origin of Species, Charles Darwin's Diary of the Voyage of H.M.S. "Beagle,"* and *Darwin and Modern Science*); Messrs. Williams & Norgate, Ltd. (*Life: Outlines of Modern Biology*, J. A. Thomson and P. Geddes); as also the authors, editors and publishers of works of which I have made more passing use, all, I trust, well within the limit of what the Copyright Act terms "reasonable quotation."

It was at one time my intention to include a formal bibliography of Darwinian writings in this volume, but I abandoned that idea on discovering that Mr. Paul Victorius, of 30 Museum Street, W.C.1, whom I have to thank for various kindnesses, had had such a work some while in preparation; it is, I understand, shortly to appear, but meanwhile he would welcome any further information on unusual bibliographical points connected with works on the Theory of Evolution generally. I must also thank Mr. J. B. Oldham, Librarian of Shrewsbury School Library, for allowing me to see certain Darwin items in possession of the Library; Mrs. Nora Barlow, the Rev. J. E. Auden, and Mr. H. E. Forrest for various courtesies; my wife for her assistance; and my publishers for their patience.

G. W.

# BOOK ONE

FAMILY OVERTURE

CHAPTER I

DARWIN-WEDGWOOD

I

CHARLES DARWIN began to be born almost exactly one hundred years before the publication of *The Origin of Species*. The occasion was strictly masculine, the meeting, as doctor and patient, of Erasmus Darwin and Josiah Wedgwood, who were to be, though neither would live to know it, Charles Darwin's grandfathers. The date cannot be precisely fixed. It cannot have been earlier than 1759, or later than 1764—1760 or 1761 seems most likely—and the place was doubtless Ivy House in Burslem, Wedgwood's home.

The two men were well matched, almost of an age (Wedgwood born 1730, Darwin seventeen months later in 1731), both already prospering, both men of integrity, intelligence, and character, purposeful too, and with keen intellectual interests extending far beyond their purely professional preoccupations. Science for its own sake, and as basis for philosophical postulation, attracted them both, and if Dr. Darwin ran the faster, Wedgwood was admirable complement, giving the needed ballast to keep on even keel and within sight of earth the soaring speculations of his friend.

For friends they became at once, and close friends they remained for more than thirty years, till Wedgwood's death. There was more than their obvious mental congeniality to make and keep them so. They were, in deeper sense, beings of the same—a new—kind, forward-looking men who sensed, if they did not consciously know, the changes which were to come upon the world. Darwin in his theorizing, Wedgwood not a little in practice, were among the pioneers of that industrial process of which this eighteenth century saw the effective beginning, the nineteenth the maturity, the twentieth the culmination and crisis.

The eighteenth century was a period of transition from a

medieval to a modern economy. Though in many ways seemingly static, English society from top to bottom was instinct with forces working for fundamental change. The reigns of the first two Georges, from 1714 to 1760, saw parliamentary government established to a point of security from which not all the autocratic ambitions of the third George could oust it. These years witnessed the end of the agricultural revolution, the initiation of the industrial. The long story of land enclosure was come at last to conclusion, bringing for the while high prosperity to farmers and rural landlords, to market towns and county centres, and creating so decisive a demand for rapid transport of agricultural products as to cause a sudden improvement in the state of the English roads such as had hardly been known since the days of the Romans. Canals too were projected to the same end, though it was usurping industrial necessity which eventually brought most of them into being. There was peace, even social peace, throughout the country after the one futile Jacobite incursion of 1745, though abroad the King's soldiers waged war all but continuously in three continents, principally against the prime enemy France in India, in America, and in Europe; the British Empire was in vigorous foundation. In that island-peace, mechanical invention was proceeding rapidly, bringing out of the bowels of the earth convenient coal for the manipulation of metals and presently, with the steam-engine, the direct production of power. Iron, coal, textiles—these were the bases of the industrial development which grew steadily throughout the middle and later decades of the century.

II

The two friends watched that development with attention. Wedgwood actively promoted both turnpike roads and canals, and Dr. Darwin shared his interest, while among their most intimate acquaintances were James Watt and Matthew Boulton, constructors of the first adequately effective steam-engine, grandparents of the Power Age, one might say, as these of its philosopher. Wedgwood certainly was drawn to these matters by self-interest, though also by more than that. He was a manufacturing tradesman, a potter, by both ancestry and upbringing, and his advantage was involved both directly and potentially. His

Staffordshire family had been farmers and minor gentry a century earlier, but had turned to pottery and produced, in immediately previous generations, several skilled craftsmen whose special talent, if little else, he inherited.

He had left school at the age of nine, on his father's death, to take a place in his brother's pottery—a small room in the modest family home—working there until, when he was twelve, an affection of the right knee led to lifelong lameness, and finally, nearly thirty years later, amputation. This interrupted the apprenticeship to his brother, but he persevered, and in due course entered into two successive partnerships, the second, in 1753, with Thomas Whieldon, one of the cleverest master-potters of the day. He learned all that Whieldon had to teach, but already he was experimenting with his own private processes, and these he put into immediate operation when at last, in 1759, he was able to start in business for himself, a fellow-workman with his few but ever-increasing employees.

Almost immediately the improved quality of his ware gave him a ready market, and within three years he was well on the way to fortune; nevertheless his name cannot have been widely known, or he by any impersonal evidence more than one among a score of other small manufacturers, when he and Dr. Darwin first came face to face.

III

Erasmus Darwin had, in that respect, the advantage. Though a year-and-a-half younger, he had been more than two years longer independently established in his profession, and his success had been even more immediate. Presumably Wedgwood sent to Lichfield for him because already his reputation had travelled the intervening thirty miles. There were many physicians nearer, and Wedgwood can scarcely yet have been in a position to afford what must have seemed, obvious advantage apart, an unnecessary expense.

Darwin's success, no less than Wedgwood's, was a personal achievement. His family might be socially the more elevated, it might own estates in Nottinghamshire and Lincolnshire and even have a country living in its gift, but all were relatively small, and the name of Darwin was in fact rather respectable than impressively respected, at any rate so far away as Lichfield in Staffordshire.

That was in consonance with its past, whose records went no further back than the mid-sixteenth century to the rather unilluminating fact of a Darwin father and son, both probably named William, both yeomen of Marton in Lincolnshire, and both dead before 1542. The surname belonged to the north-country, and Cumberland, Yorkshire and Derbyshire were each suggested as the family's earlier home, on the rather slender ground of a presumed derivation from one of the rivers Derwent common to these counties. The second William begat a third, and he a Richard, whose son, William the fourth (c. 1573–1644), holding the Marton property at the close of Elizabeth's long reign, prospered and enlarged the original modest estate by both purchase and judicious marriage. King James I appointed him to the post of Yeoman of the Royal Armory at Greenwich, a mainly honorary place he held for over thirty years until his death during the Civil War, in which his son and namesake (1620–1675) fought for the King as Captain-Lieutenant in Sir William Pelham's troop of horse. The son was for this loyalty heavily fined by the victorious Commonwealth, but managed to retain Marton, and at the Restoration was made Recorder of the City of Lincoln. Himself a barrister in the intervals of these excitements, he introduced a familiar name into the family by marrying a daughter of Erasmus Earle, sergeant-of-law, of which union was born yet another William (1655–1682) who, like his grandfather, wedded a woman of property, a daughter of Robert Waring, who brought her husband the manor and hall of Elston, near Newark, a handsome Elizabethan building complete with castellated doorway. This, inherited by Robert, the last William's younger son, became in due course the birthplace of Erasmus Darwin of Lichfield. The link was but precariously joined, for this William died young, and Robert (1682–1754) was born but sixteen days before his decease. Robert was a barrister, but of such indifferent ambition that upon his marriage to the energetic, capable and scholarly Elizabeth Hill he retired from his profession to live pleasantly but unenterprisingly as a country gentleman on the Elston estate, following desultorily those mild scientific or perhaps only antiquarian interests which made him an occasional correspondent of William Stukeley and a member of the Gentleman's Society—

"for improvement in the liberal sciences and in polite learning" —of Spalding in Lincolnshire. His great-grandson Charles thought him, in his portrait which showed him wearing heavy wig and bands, less like a lawyer than some "dignified doctor of divinity," and he was indeed a Darwin alike in his rather sombre dignity and in his freedom from that sin whereby the angels fell.

He indulged not only in science but also, apparently, in verse, though his one surviving specimen of the latter activity might really have been conceived and delivered impromptu by any sporting squire on the morning after a Hunt Supper! These inclinations were to find more evident and fruitful embodiment in two of his four sons—the eldest, Robert Waring (1724–1816), and the youngest, Erasmus (1731–1802). Robert on the one hand had "a taste for poetry," and on the other actually published, when in his sixties, a *Principia Botanica: or, A Concise and Easy Introduction to the Sexual Botany of Linneus*, of which three editions appeared and which, for all its unambitious aim (to instruct youth in "the harmony of creation, as nothing more strongly evinces the existence of a Supreme Cause"), was praised by Charles Darwin as containing "many curious notes on biology" at a time when the subject was almost wholly neglected in England.

IV

Erasmus, seventh child of barrister Robert and learned Elizabeth, appears as indubitably the most dynamic personality produced by the Darwins up to his time or since it. He was, from early years, robust and boisterous, compact of purpose and mental and physical vitality. His range of interest and activity was always wide. Even as a boy he distributed his attentions almost equally between poetry, mechanisms, and crude electrical experiments. If he had, even then, a distaste for all field sports and similar rural diversions, and shrank from most varieties of bodily exercise, it was not from lack of intrinsic energy.

It has been said that a consecutive narrative of Erasmus's life and development is impossible in the absence of fuller records, but that matters less than it might since wherever we can approach to him we find essentially the same versatile and attractive if at times rather portentous figure. Of his childhood we know nothing.

Most of his years from the ages of ten to nineteen he was a pupil at Chesterfield School, and seems to have been happy there. His letters to his brother Bob, now in London, mingled verses with school gossip, and at sixteen he was writing to his sister Susanna, two years older, in a way suggesting that he held the jester's acknowledged licence, for answering her report of Lenten abstinence he declared that for his own part he ate freely of beef, mutton, goose, and whatever else might come his way, for "all flesh is grass!" following this with a "panegyrick upon abstinence," and ending abruptly: "Excuse haste, supper being called, very hungry." [1]

From Chesterfield he went to St. John's College, Cambridge, fortunately winning a scholarship of an annual value of £16, for two brothers were already at the university, and their father was finding their expense, together with that of three girls at home, as much as his means could carry. Erasmus studied classics, mathematics, and medicine, the writing of poetry remaining his active hobby. He resided some twelve terms, but with a break to attend the anatomical, or more likely obstetrical, lectures in London of Dr. William Hunter. In 1754 he took his B.A. examination, and went to Edinburgh University for a period of specialized study. There he met James Keir the chemist, his lifelong friend, and there too he "sacrificed to both Bacchus and Venus; but he soon discovered that he could not continue his devotions to both these deities without destroying his health and constitution. He therefore resolved to relinquish Bacchus, but his affection for Venus was retained to the last period of his life." [2] He was a perpetual advocate in after-life of abstinence from alcohol, but whether he ever publicly commended it on this particular ground is not recorded.

While he was at Edinburgh his father died, and in this connexion he wrote to a friend declaring the existence of a God, creator of the world and of mankind, but adding: "That He influences things by a particular providence, is not so evident. The probability, according to my notion, is against it, since general laws seem sufficient for that end. Shall we say," he asked, with evident Newtonian reference, "no particular providence is necessary to roll this planet round the sun, and yet affirm it necessary in turning up *cinque* and *quatorze* while shaking a box of

[1] *Erasmus Darwin*, 10–11.   [2] q. *Dr. Darwin*, Pearson, 5.

dies, or giving each his daily bread?"¹ These lines have been cited to show that he was already an evolutionist, but really they display no more than the inclination of his mind towards the conception of a universe of law rather than of special creation and divine interference, a simple following of the general line of Newtonian thought in the direction of its clear conclusions. He was in this only in accord with the mental movement of the day, and if its logical end was its application to man's own being, growth, and origin, still the evolutionary explanation was, if a solution, by no means an implicit or even necessary one.

What one would like to know is his attitude at this time to the medical teaching of the day, which at Edinburgh as elsewhere regarded the human body mainly as a hydraulic mechanism, a system of pipes wherein illnesses arose principally from fermentation of essential fluids or from obstruction. When, Keir wondered, writing fifty years later, after his friend's death, did the mind of Erasmus move from this narrow view to his "more enlarged consideration of man as a living being, which affects the phenomena of health and disease more than his merely mechanical and chemical properties?"² If Keir was in doubt, we can scarcely be otherwise to-day, or suppose more than that, while this tendency too was of the period, and while he may have owed something to the vitalist teachings of John Hunter, Erasmus was by way of being a pioneer himself, finding his path ahead of his fellows. From quite early days he was known as an experimentalist in treatment—a dangerous one, his rivals said, but, even they could not deny, often effective. And in his experimenting he did not neglect the psychological, the organic, aspect.

He spent two years at Edinburgh, returning there after taking his medical degree at Cambridge in 1755. Not until September 1756 did he seek to set up in practice. He would have liked to have done so in London itself, but the odds were too heavy, and he went to Nottingham. It is recorded only that he had no success there, but possibly he did not like the place, or the opposition was plainly too strong, for he gave it but brief trial, unable to divert himself even by his current amusements of learning shorthand, writing letters in Latin, composing verse, and speculating upon the electric qualities of the human soul. Two months or less, and he had packed up, to arrive in Lichfield in early November

¹ *Erasmus Darwin*, 15. ² *Ibid.*, 13-14.

with letters of introduction to Lady Gresley and the learned and literary Canon Seward, who gracefully occupied the fine seventeenth-century Bishop's Palace. Here fortune smiled, for a case, given up by all the local physicians, was brought to him in despair, and he, by a "novel course of treatment," effected a speedy cure. The patient being "a young gentleman of family," Erasmus's reputation was made. As Anna Seward was to relate, he "soon eclipsed the hopes of an ingenious rival, who resigned the contest," and thereafter he reigned supreme.

Perhaps it was not so immediate and easy as all that. Anna was at the time only fifteen years old, and such victories sharpen their outlines in the perspective of passing days. Still, victory it was, and quickly he became what he remained to his life's end—the premier physician not only of Lichfield but of the Midlands. Nearly twenty-five years he was to live in Lichfield, and those years were the most active of his life. He had his own way to make, for his father's death had brought him little fortune. In 1757 his earnings were under £200, in 1758 over £300; from 1760 they never fell below £500 a year, and after 1770 never below £1000, a considerable income in those days. His patients included the highest and the humblest, peers and beggars. The poor, and the Cathedral clergy and their families, he tended for trifling fees or none; to the former, in real need, he would often give food.

He also became, and not merely professionally, a notable figure in the city. Striking he always was, if not, at first sight, especially attractive. Even at twenty-five he was already physically unwieldy, tall, broad, fat, a great lump of a man who bore himself clumsily, his florid, pitted, plum-pudding face staring out beneath a heavy full-bottomed wig which, even when not set carelessly awry, made him look twice his age. He was soon to lose his teeth, too, for though Anna Seward denied that he commonly walked with his tongue hanging between his lips she could not but admit the disclosure of certain "ravages of time," though "by no means in any offensive degree." His very expression, in repose, tended to the gloomy, even—said some—to the saturnine.

But he reposed seldom, and if the exterior repelled the manner was quick to charm. His smile and his greeting were infectious, and his conversation, despite a sometimes extreme stammer—a defect he characteristically converted to a quality—was an entertaining compound of wit, thought, and knowledge. (Late in life he was to write of a stuttering apothecary that he did not think "a little impediment in his speech would at all injure him, but rather the contrary by attracting notice," adding that he had found it no obstacle in his own case.) He talked frankly of what interested himself, but always so as to engage his company. People might come to scorn but remained to listen, and to come, or beg him to come, again. They did so as patients, they did so as friends; some, like Wedgwood, beginning as one, quickly found themselves the other.

v

The lessons learnt at Edinburgh were not forgotten. He preached temperance and wedded quickly—within fourteen months of his arrival in Lichfield. His bride was a girl of seventeen, the intelligent and attractive but delicate Mary Howard, daughter of a local resident and friend of the Sewards. They were married on December 28th, 1757, as unostentatiously as possible, he writing from Darlaston on Christmas Eve plans by which they might have the ceremony over early in the morning, "before anybody in Lichfield can know almost of my being come home." The earliest part of this love-letter deserves quotation, exhibiting as it does with all its eighteenth-century artificiality not a little of that individual and tender charm and humour which evidently, for more than one woman, offset his indifferent exterior. "Dear Polly," it begins, "As I was turning over some old mouldy volumes, that were laid upon a Shelf in a Closet of my Bedchamber; one I found, after blowing the Dust from it with a Pair of Bellows, to be a Receipt Book, formerly, no doubt, belonging to some good old Lady of the Family. The Title Page (so much of it as the Rats had left) told us it was 'a Bouk off verry monny muckle vallyed Receipts bouth in Kookery and Physicks.' Upon one Page was 'To make Pye-Crust,'—in another 'To make Wall-Crust,'—'To make Tarts,'—and at length 'To make Love.' 'This Receipt,' says I, 'must be curious, I'll send it to Miss Howard next Post, let the way of making it be what it will.' . . . At the Top of the next Page, begins 'To make an honest Man.' 'This is no new dish to me,' says I, 'besides it is now quite old Fashioned; I won't read it.' Then followed 'To make a good

Wife.' 'Pshaw,' continued I, 'an acquaintance of mine, a young Lady of Lichfield, knows how to make this dish better than any other Person in the World, and she has promised to treat me with it sometime,' and thus in a Pett threw doun the Book, and would not read any more at that Time. If I should open it again tomorrow, whatever curious and useful receipts I shall meet with, my dear Polly may expect an account of them in another Letter." He ended: "Nothing about death in this letter, Polly."[1] Charles Darwin endorsed Anna Seward's poor opinion of his grandfather as a letter-writer; possibly he underestimated him in this as in other respects.

Erasmus and his Polly seem to have been mutually happy together, but she remains an almost unknown figure not much illuminated by the resplendent ancestry ascribed to her through her mother Penelope Foley, scaling the ages backward to Scottish, Frank, Saxon, Norman and French royal sources, and laying Charlemagne, Alfred the Great and William the Conqueror all under incidental contribution. Dubious honours, perhaps, for somewhere there crept in a strain of weakness not to be balanced even by the sound bourgeois blood of the Foleys themselves, and Mary's physical frailty could not sustain what Anna Seward termed "the frequency of her maternal situation," a matter of five children—two dying in infancy—in nine years. She was an invalid for some while before her death in 1770.

In her earlier married years she appeared, however, a hostess of wit and perception, and had full scope for her talents. Some time in 1758 Erasmus purchased the house where he was to spend the rest of his Lichfield days, an old half-timbered building cut off from Beacon Street by a narrow dell which he terraced and planted. A picturesque bridge was set across it, and the house itself re-fronted and enlarged. He was soon keeping four horses, and a groom, gardener, and other servants.

Here his children were born—Charles in 1758, Erasmus the second in 1759, Robert Waring in 1766—and his visitors entertained: patients, relatives, friends. The social position of doctors in those days was equivocal; more than half a century later it was still customary for the prosperous country physician to be received in the housekeeper's quarters rather than the drawing-rooms of larger houses, while Lichfield etiquette in 1780 allowed

[1] *Erasmus Darwin*, 21-24.

surgeons to be admitted to dances but not to card assemblies. But Erasmus made his own rules, and the man of whom King George III declared that if he would but come to London he should be the royal physician, was no company for the servants even of the nobility.

Once he had wanted to go to London. Now he went only under protest. He had found his own kingdom, and neither royal favour nor increased income could tempt him away. Moneymaking was far from being his principal purpose in life, and for the rest he liked better to hold first place in his own court. His and Dr. Johnson's mutual antipathy was inevitable, for they were in many respects too much alike, each apt to resent more than a stimulating modicum of opposition or rivalry. But while an essential melancholy underlay Johnson's domineering manner, Erasmus was basically a happy man. Depression was foreign to him; energy of body and mind gave him success, and in turn success gave him energy of body and mind. Always it was intrinsic vitality rather than ill-nature, one feels, which evoked his apparent irascibility, bringing sarcasm or irony easily to his lips; at heart he was sympathetic and amiable. In the bulk of the stories that multiplied about him he appears as a man of bluff but chuckling commonsense, his vivid imagination for ever bubbling with notions grave and extravagant, profound and improbable, many of them duly entered in his omnivorous commonplace book, as many probably spent on an immediate audience and long since lost in thin air.

Most of those ideas which achieved record were first set down to while away the hours as he travelled from patient to patient in his two-wheeled carriage of that type termed a "sulky" because it unsociably held one person only. He had had it fitted to his own design for writing and reading, with an adaptable skylight set in the roof and holders for books, pen, paper, and ink. He carried also a hamper of fruit and sweetmeats for himself, hay and oats for his horses, and a bucket for their water, always starting out in fact fully prepared for what the day, or night, would bring. Such precautions were necessary, for at any hour he might find himself far afield and on roads sometimes muddy to impassibility. For such occasions, when the worst occurred, he had a horse, named, with odd humour, "Doctor," which he would mount and so proceed, leaving the carriage and its horses to await his own or some rescuing groom's return.

The astonishing thing is that, jolting from country house to cottage, from village to village and town to town, along pre-macadam highways with surfaces the modern traveller would deem disgraceful in the veriest country lane, he was able to write, or even read, at all. But idleness would have been the harder part. His inexhaustible, ebullient vitality would not be denied, and his busy mind, quivering with ideas like milk upon the boil, spilled over the blotted pages to mingle diagrams of projected machines, descriptions of diseases and other medical observations, notes on wind and weather, sleep, lunacy, sanitation, science, phonetics, language, slavery, and a thousand and one such topics. For him a fool was "a man who never tried an experiment in his life," and he had views on every subject under the sun. Some of them looked far ahead:

> Soon shall thy arm, Unconquer'd Steam! afar
> Drag the slow barge, or drive the rapid car;
> Or on wide-waving wings expanded bear
> The flying chariot through the fields of air.
> —Fair crews triumphant, leaning from above,
> Shall wave their fluttering kerchiefs as they move;
> Or warrior bands alarm the gaping crowd,
> And armies shrink beneath the shadowy cloud.[1]

The topical lines are almost too well known for quotation, as is his pre-vision of the submarine, but the future was to endorse him no less faithfully if less spectacularly in his advocacy of isolated cemeteries, removal of sewage, and pure water supplies. These came perhaps within his professional purview. Mechanical devices remained his hobby. He installed a speaking tube in his house, and scared a servant almost out of her wits. Before Watt's advent he suggested to Matthew Boulton a mutual partnership for the construction of a locomotive steam engine he had conceived and wished to patent. He sank an artesian well in the garden. He devised a manifold writer, a talking machine which said "mamma" and the like, a canal lock, a rotary pump, a "very singular carriage" which somewhat signally and disastrously failed to carry, and a horizontal windmill actually used by Wedgwood for grinding flints.

[1] *Economy of Vegetation*, canto i, ll. 289-96.

VI

Such matters were the consolation of his enforced solitudes—he was so avid of life that he could not bear to waste the most passing moment—but he was no natural recluse, rather a man with a love amounting to a genius for friendships. In his Lichfield years he drew about him a circle of congenial acquaintances—Wedgwood of course; Matthew Boulton, the Birmingham manufacturer; James Watt, Boulton's partner, whose improvements to the steam-engine have caused him to be looked upon almost as its inventor; Dr. William Small, fellow physician and philosopher; Thomas Day, the author of *Sandford and Merton*; Richard Lovell Edgeworth; Samuel Galton; Joseph Priestley, nonconformist as minister and chemist alike; James Keir, retired from military life to resume the Edinburgh intimacy—these and more. They were all men of much the same outlook: fellow-followers of Rousseau; intensely interested in science and invention; politically radical, rejoicing in the War of Independence, and hailing the French Revolution as a nearer dawn of freedom; most of them, in religion, theists who held atheism to be folly and approved the maxims of Jesus if not the claims of Christ.

Boulton and Small he met early, through attending professionally the Robinsons of Lichfield, a family related to Boulton by marriage, and these three were among the founders, somewhere in the middle 'sixties, of the celebrated Lunar Society, a dining club of personal friends who met monthly at each other's houses near the time of the full moon, that those members who had farthest to go might have illumination on their homeward way. Most of Erasmus's friends were "Lunatics," except Wedgwood, and he was a frequent guest. At the meetings literature and science were freely discussed, views exchanged, the results of experiments communicated. "Lord," Erasmus wrote once when unable to attend, "what inventions, what wit, what rhetoric, metaphysical, mechanical, and pyrotechnical, will be on the wing, bandied like a shuttlecock from one to another of your troupe of philosophers! while poor I, I by myself I, imprisoned in a post-chaise, am joggled, and jostled, and bumped, and bruised along the King's highroad to make war upon a stomach-ache or fever."[1] Undoubtedly echoes of many of these discussions appear in the footnotes to, even the very verses of, his poetical writings, for he took his wealth where he found it, and here it was scattered broadcast in abundance.

Edgeworth came early on the scene, introducing himself as a fellow contriver of improved carriages, and it was he who made Day and Erasmus known to each other, with the result that the former took up residence in Lichfield for some years and long remained the other's correspondent. Erasmus had this power of attaching the most varied individuals to himself, and of retaining their affection. Watt he came to know by way of Boulton, in 1767; Galton was a Birmingham resident, his house a favourite meeting-place for the Society; Keir was living from about 1770 at West Bromwich, and later at Stourbridge; Priestley joined the circle last of all, in 1780, when Small was some years dead, Day and Edgeworth long since far removed, and Erasmus himself about to withdraw.

These were his closer companions, but he met and carried on correspondence with other even more distinguished men. One of these was James Hutton, the geologist, another Benjamin Franklin, another the great Rousseau, whom he met in the flesh when the Frenchman was staying at Wootton Hall in Staffordshire in 1766. Rousseau was at that time in Diogenes mood, absorbed in writing his *Confessions* and resentful of intrusion, but Erasmus drew him into conversation by walking slowly past the Hall, peering so intently at plants or the like that the recluse was enticed to speak, curious to know what this casual passer-by could be so eagerly examining. Erasmus had little mind for sport, but he knew more than one way of catching a fish.

VII

Mary died on June 30th 1770, after four years of ill-health. Erasmus was deeply grieved, but also no man to be overwhelmed by grief, and life went on, though it is worth remarking that it was in the three or four years following her death that he brought together, very much in the form of its publication twenty years later, the matter of his most substantial work, the *Zoonomia*. Exertion was, he always held, the only antidote to loss. There

[1] *Scientific Correspondence of Priestley*, 199.

DR. ERASMUS DARWIN

From *The Life, Letters and Labours of Francis Galton*. By courtesy of the Cambridge University Press

were all the while work, friendships, his wide intellectual interests, the education of his sons, to attend to. A sister came to keep house for him, and some time in this period, from hygienic or more tender considerations, he took to himself an unidentified mistress—a woman of no social status—who bore him two daughters, the Miss Parkers.

If he paid tribute to Venus, the goddess was not unkind to him. He grew, with the passing years, no handsomer—teeth gone, huge body, coarse Darwin nose, broad jowl and fleshy chins, heavy head sunk low on short thick neck between rounded shoulders. Also, from 1768, the year incidentally of Wedgwood's amputation, he limped, one of his knees having been permanently injured in a fall from that "very singular" carriage. But none of these things seems vitally to have diminished his fascination.

Anna Seward at any rate thought not. She had been still a child when he first came to Lichfield and her father's house. Now in these middle 'seventies she was well into her thirties, the long tunnel of perpetual spinsterhood closing upon her. From the beginning he had shown interest in her immature talents, had been the first to acclaim her verses, and shared with her the pleasure of her early triumphs. Undoubtedly she admired him, and there were those in Lichfield who believed that she wanted nothing so much as to succeed to poor dead Mary's place in his household and his affections. If he would but ask her! Perhaps he was too well aware of her hopes, and in deliberate self-protection against the dumb appeal of that long solemn face and constrained mouth adopted that crushing sarcasm with which he at times withered her tenderest confidences. Erasmus might be poetically inclined, but he stood no nonsense from his muses; if any inquired he would declare, quite uncharacteristically, that he wrote only for money. Anna undoubtedly wanted to write exclusively with a golden pen on tablets whiter than a star; a hymning angel was exactly how she saw herself. He would play with poetry and with poetesses, but had little wish to become seriously entangled with either.

Yet in 1777 Anna may have felt herself well on the road to success. Erasmus was turning to botany with more than usual interest. He brought into being the Lichfield Botanical Society, which in the next decade issued well-received translations of Linneus's *System of Vegetables* (1781) and *Family of Plants* (1787), though it consisted of three persons only, and brought posthumously to birth the second of its twin fruits. The members were Erasmus, "dissipated" Sir Broke Boothby, and Jackson, a low Cathedral proctor who was, according to Anna, a mocker at Christianity, "a would-be philosopher, a turgid and solemn coxcomb, whose morals were not of the best," but whom his betters found a useful drudge, he doing the donkey-work of translation and they supplying the polish. In the absence of further recruits (alas that women were not admitted, or there would have been one) these three sustained the Society, dignifying with its authority occasional observations conveyed to appropriate periodicals. Anna experienced a certain acid amusement when scientifically inclined travellers asked for particulars of this learned and exclusive body.

Pursuing this diversion Erasmus in 1777 purchased a small property on the outskirts of Lichfield. A century before a well-known local physician had established there a picturesque grotto and bathing pool, but it had fallen into neglect, and now Erasmus proposed to restore it as a botanical garden. Anna of course was soon upon the spot, and immediately composed a poem which (she said) so delighted Erasmus that he declared it "ought to form the exordium of a great work. The Linnean System is unexplored poetic ground, and happy subject for the Muse. It affords fine scope for poetic landscape; it suggests metamorphoses of the Ovidian kind, though reversed." Forthwith he outlined such a work, for which she was to write the verses while he supplied the notes, "which must be scientific." Modestly she protested that even for the verse the subject was better suited to "the efflorescence of his own fancy," and when he urged "the professional danger of coming forward as an acknowledged poet," piled compliment on compliment by asserting his reputation to be above such injury, "especially since the subject of the poetry, and still more the notes, would be connected with pathology."[1]

For twenty years, he had stated two years earlier, he had neglected the Muses, and meant to write no more verse to the end of his days, caring only to improve the prose *Zoonomia* for posthumous publication. But inspiration, and Anna, would not be denied, and soon he was engaged upon both verse and notes, encouraged also by Edgeworth, who only regretted that his fine

[1] *Dr. Darwin*, Seward, 130–131.

images could not be released from their vegetable birth. As a compliment to Anna, he included her poem in its proper place, unfortunately without any acknowledgment of its authorship. This was to cause later ill-feeling, but no rift appeared as yet. It was pleasant to walk the garden with him in innocent (if hopeful) enjoyment of the higher pleasures of the mind.

Too soon her hopes were dashed. That next spring of 1778 Elizabeth Chandos Pole, the wife of Colonel Pole of Radburn Hall, near Derby, brought her ailing children to Lichfield, staying some weeks in the Doctor's home. When at last she returned husband-wards with "her renovated little ones," she took with her also Erasmus's heart. Shortly afterwards he sent her a specially designed tea-vase decorated with love-emblems and inscribed "To the Fairest," which seems so to have affected the Colonel that the would-be lover was soon to be found lamenting his exile from "yon friendly towers" where "Avarice dwells in Love's polluted fane."

> Farewell! a long farewell!—your shades among
> No more these eyes shall drink Eliza's charms;
> No more these ears the music of her tongue!—
> O! doom'd for ever to another's arms! [1]

It was not quite as final as that. Mrs. Pole, that autumn, fell seriously ill. Erasmus was sent for, permitted to see her and to prescribe for her, but not requested, despite his distance from home, to stay the night. He must go forth! Forth he went, but to return, watching the night through beneath her bedroom window and executing to pass the anxious hours a paraphrase of Petrarch in his best elegiac manner. One hopes it kept fine for him.

Eliza was all too evidently another's; but that did not aid Anna's suit.

Here was tragi-comedy coming at moments close to farce. A deeper and unrelieved sorrow came in the spring of the year from Charles's sudden death. Not yet twenty, he had been a brilliant boy and his father's favourite. Like Erasmus he stammered, and though he wrote verses his primary interest too was in the sciences—mechanics, chemistry, the collection of fossils. He was barely sixteen when he went to Christ Church College,

[1] *Dr. Darwin*, Seward, 107.

Oxford, but "his mind languished in the pursuit of classical elegance like Hercules at the distaff," [1] and he begged to be sent without further ado to the medical school at Edinburgh. There he had three years of hard work, and gave promise of a brilliant future, receiving the first gold medal of the Æsculapian Society for a paper describing his experiments. He died from blood-poisoning contracted during a dissection. Erasmus's distress is evident even in the conventional language of his brief published obituary of the son he had set his hopes upon. The second boy, Erasmus, was always something of an ugly duckling, gifted, but retiring to the point of inanition. Robert was still a child, yet to him his father's interest seems to have shifted. Erasmus was plainly an affectionate parent, yet he aroused in each of these younger sons of his a fondness ever mixed with fear. As he said of his own father, he was "very tender to his children, but still kept them at an awful kind of distance." That this was probably due mainly to his stress of activity appears in the close attachment to him of the children of his second marriage, who grew up when he had much more leisure. In those later years his attitude to Robert also softened, though the early impression was never overcome in his son's mind.

To draw the story short, that second marriage had not long to wait. In 1780 Colonel Pole died. At once there were several suitors in the field, younger than Erasmus and also wealthier—not to say handsomer. Eliza, now thirty-three, had had one experience of a husband twice her age, and found him peevish and suspicious. She was healthy, wealthy, and if not especially wise at least lively, lovely and fascinating; she might have made almost any match. She preferred Erasmus, with all his physical defects. They were married on March 6th, 1781.

The one change meant another. Eliza did not like Lichfield. Perhaps people talked too much. Perhaps Anna was the obstacle. Since she would not come thither, he went to Radburn Hall. But if she could not settle comfortably in her second husband's home, neither could he in her first's; before two years were over they had moved into the town of Derby.[2]

---

[1] *Experiments, etc.*, C. Darwin, 131.
[2] Margaret Ashmun, in *The Singing Swan*, rejects the idea of Anna's wish to marry Erasmus on the ground of her affection for John Saville, but offers no irresistible evidence that her love for the latter, as distinct from her friendship for him, really preceded Erasmus's meeting with Mrs Pole.

CHAPTER II

FROM ONE GENERATION

I

ANNA, and Lichfield too, now belonged to the past, save for a most occasional visit. So the Lunar Society also, though its gatherings continued for two decades longer, the poorer for Erasmus's absence, until the dwindling of its membership brought it to a melancholy close. He missed those jovial and informative meetings—"I am here cut off from the milk of science, which flows in such redundant streams from your learned Lunatics" [1]— and the Derby Philosophical Society, which he founded in 1783 to console himself, had no such galaxy of remarkable minds to offer. He made the best of correspondence.

There was, though, one personal contact no change of place or circumstance could sever—that with Josiah Wedgwood, now in the world of pottery a European figure. His rise, in the years following their first meeting, had continued with unchecked rapidity. A cream ware service, presented to Queen Charlotte, won him the position and title of Royal Potter. Soon he was England's foremost manufacturer of both useful and ornamental ware; not much longer, and orders were coming from all parts of the Continent, where his name and work were widely advertised by the dinner-service, bearing some 1300 views of British scenery, made especially to the command of the Russian Empress Catherine. He created in fact an international demand for English earthenware, and gave Staffordshire pre-eminence in the world-market. Twice in the 'sixties he was forced to move to new and more extensive quarters, in the second case—on his partnership with Thomas Bentley, who took charge of the London warehouse and showrooms—to the famous Etruria Works and village, whose name, with that of the new family home, Etruria Hall, was suggested by Erasmus. Throughout the 'seventies trade was

[1] *Dr. Darwin*, Pearson, 97.

rapidly expanding both at home and abroad, and though it was of his own inclination that he gave more and more time and attention to the creation of costly "artistic pottery"—vases and medallions of and bearing classical designs—his work in this field also found favour and served to increase further his reputation and general sales.

He appears in all accounts as individually a very admirable figure, of complete integrity alike in his personal and commercial relationships. He was no democrat, and well aware of class distinctions, but his attitude to his workers was benevolent if self-interested; even if he did on one or two occasions address them more like an employer than an honest man, his general care for them, in an age of evil working conditions, was notable. Apt to be patronized by the local gentry, despite much genuine friendship, he stood independent in thought and action, a Whig tending to radicalism in politics, in religion a Unitarian (a featherbed, Erasmus mocked, to catch a falling Christian). He did not drink, or bet, or hunt. Instead he pursued on one hand schemes of country development (he was treasurer of the Trent and Mersey Canal project), and on the other antiquarian and artistic interests, collecting books and prints. He has been termed a great artist, but a truer estimate would assess him as an artist a good potter, and as a potter a good artist; he often worked upon fine classical models, but his tastes were imitative, not originative.

Like Erasmus, he hated slavery, and publicly expressed his detestation of it by making and circulating several hundred copies of a cameo depicting a black slave kneeling in chains and bearing the words: "Am I not a man and a brother?" He applauded the American bid for independence, and the Wedgwoods as a family were sympathetic to revolutionary France, even to the point, later, of making something of a hero of Napoleon even when Britain was most bitterly at war with him.

This independence of outlook doubtless influenced the family's intense reserve outside its own circle. Josiah established what became almost a family tradition when he married, in 1764, his distant cousin Sarah, the daughter of Richard Wedgwood of Spen Green in Cheshire, a wealthy retired cheese factor turned country gentleman. She was a strong-featured woman of education and ability, an admirable wife and mother, stern perhaps but always with an underlying sympathy, one who had, even with much ill-health, a mental energy to cope with all situations. From 1765 to 1778 she bore her husband eight children, of whom the third died in infancy, the last in childhood.

Because of her indispositions the children--three boys and the three elder girls—were early sent away to boarding schools, but these were never regarded as wholly satisfactory, and in 1779, when young Josiah had been ill and Tom's constitutional weakness was already apparent, their father decided that they must be taught at home. He did so in no small part on Erasmus's advice, sharing the latter's preference for "modern" over "classical" subjects, at any rate for boys not specifically intended for "the professions." The features of "the Etruscan school" were regular exercise, short lessons, good food and absence of corporal punishment. Tutors were engaged—a chemist from Birmingham who was also a scholar, and a French prisoner-of-war—and Josiah himself actively supervised. Young Robert Darwin, in intervals of Manchester and Bolton schooling, came to share the lessons with John, Tom, young Josiah, and Susannah, the eldest of all, and when he was ill the Wedgwood lads sometimes went to Lichfield to have lessons there with him. The younger generation of "Etruscans" and "Darwinians," as they called each other, were thus intimate from the schoolroom.

But this arrangement lasted only a year or two. Robert was soon off to Edinburgh, to follow his father's profession. Then Bentley died, and the whole burden of the business fell upon Josiah. In 1782 both John and young Josiah also went north to Edinburgh, first to the High School and then the University, and Tom followed a little later.

II

Erasmus, after the move to Derby, carried on his work as doctor still, but somewhat less busily. Patients now came to him, not only from all over the United Kingdom, but from many parts of the Continent. His income more than sufficed him, and Eliza stood in no need. He told Robert, when the latter was setting out to make his way in the world, that "as to fees, if your business pays you well on the whole, I would not be uneasy about making absolutely the most of it; to live comfortably all one's life is better than to make a very large fortune towards the end of it,"[1] and what he preached he practised. He felt now that he had better work to do than the daily round of doctoring he had sustained so long. There was his elaborate botanical poem to complete, and he began to wonder whether after all he might not have the satisfaction of seeing that prose work, by which he set much greater store, appear in his own lifetime. What could his reputation suffer now? At the worst it could scarcely affect him commercially, and, for the rest, he had always the courage of his conviction. It was as early as 1784 that he decided to publish even if it damned him, but *The Loves of the Plants*, Part Two of *The Botanic Garden*, did not appear till 1789.

The whole was then somewhile since written, but he deferred the printing of the first part, *The Economy of Vegetation*, because he thought the second would make the more immediate appeal. Certainly *The Loves* sold well; he received a thousand pounds for it, and as much, on the strength of that, for *The Economy*.

To-day the once popular *Botanic Garden*, like its posthumous successor *The Temple of Nature*, is read only by students, and the world at large loses nothing by its abstinence. Erasmus had a restricted conception of poetry, expounded in three Interludes thrust into *The Loves*, seeing "the essential difference between Poetry and Prose" as existing in that "the Poet writes principally to the eye, the Prose-writer uses more abstracted terms." Ornament was proper to poetry much more than to "graver works, where we expect to be instructed rather than amused." These were didactic poems, "nevertheless Science is best delivered in Prose, as its mode of reasoning is from stricter analogies than metaphors or similes," and accordingly the real substance of the "poetical works" is found invariably in the notes, which far exceed the verse in bulk. Of logical meaning his verse seldom fails—it is as plain as a pikestaff, and about as appropriate to the occasion. The title of George Canning's famous parody, *The Loves of the Triangles*, was apt, for it does of itself acutely criticize a poetic method as apposite—or inapposite—to geometry as to botany. He never began to be a poet; he was no idle versifier but a man of scientific interests and "literary accomplishments" who used the latter as sugar-coating to the former.

[1] *Erasmus Darwin*, 65.

He could attain eloquence, and rose to it not infrequently and sometimes with dignity, especially in *The Temple of Nature*, but on the whole preferred preciosity:

> Weak with nice sense, the chaste Mimosa stands,
> From each rude touch withdraws her timid hands;
> Oft as light clouds o'erpass the summer glade,
> Alarm'd she trembles at the moving shade;
> And feels, alive through all her tender form,
> The whisper'd murmurs of the gathering storm.
> Shuts her sweet eye-lids to approaching night;
> And hails with freshen'd charms the rising light.
> Veil'd with gay decency and modest pride,
> Slow to the mosque she moves, an eastern bride;
> There her soft vows unceasing love record,
> Queen of the bright seraglio of her Lord.[1]

From which one gathers that the Mimosa is extremely sensitive to touch and to light, and is "of the class Polygamy, one house."

The *Zoonomia*, though, published in 1794 and 1796, was the sun of all his literary creation, and in truth the first-born from which the other planetary works came forth, shining by the reflected light of its ideas. Yet it was not in form, or in the main, a philosophical essay at all, but a medical text-book, its primary purpose to "unravel the theory of diseases" and thereby aid their treatment. The first volume opened with an involved and careful attempt to account for organic phenomena in terms of a "sensorial fluid" variously existent in states of irritation, sensation, volition and association, acting, reacting, and interacting the one upon the others — a view in which the unity of body and mind, mutually interacting, is implicit — but the bulk of the work deals with specific functions and diseases and remedies. Larger issues occur almost incidentally, this being true even of the evolutionary outline, appropriately found in the long section on Generation. As an outline it was, considering all things, remarkably complete, owing something perhaps to Buffon's half-hidden views on mutability as directly due to changing conditions and the struggle for existence, but containing new and original elements.

The evolutionary idea, that of the natural development of complex from simple organic forms by transmutation and survival

[1] *Loves of the Plants*, canto i, ll. 247-58.

of the fitter, was of course very ancient, and yet in a sense new. Both Greeks and Romans, notably Empedocles, Aristotle and Lucretius, adumbrated it. Yet if they guessed at the truth they could do little more, lacking any real basis of observation, and their speculations remained aerial. The long centuries of the early Christian and medieval periods went by, and when questioning began again in the seventeenth and more vitally the eighteenth centuries it took almost nothing from the past. Linneus and Cuvier were not evolutionists, but they provided much of the material for the speculation which followed. Erasmus, despite these others, despite Buffon even, was an authentic pioneer.

His account was succinct. He pointed to the practically universal organic capacity for change, to the "great similarity of structure in all the warm-blooded animals," suggested changes as produced in part at least by the animals' own "perpetual endeavour" to supply their desires; premised transmission of these acquired forms or propensities to their posterity, which were not "new animals" but "branches or elongations of the parent," and able therefore to "retain some of the habits of the parent system"; and declared lust, hunger and security to be the three main impelling desires, shaping the beasts to fight for their mates, to win food, and to sustain safety by battle or escape. "From thus meditating on the great similarity of the structure of the warm-blooded animals, and at the same time of the great changes they undergo both before and after their nativity; and by considering in how minute a portion of time many of the changes of animals above described have been produced; would it be too bold to imagine, that in the great length of time, since the earth began to exist, perhaps millions of ages before the commencement of the history of mankind, would it be too bold to imagine, that all warm-blooded animals have arisen from one living filament, which THE GREAT FIRST CAUSE endued with animality, with the power of acquiring new parts, attended with new propensities, directed by irritations, sensations, volitions and associations; and thus possessing the faculty of continuing to improve by its own inherent activity, and of delivering down those improvements by generation to its posterity, world without end!"[1]—and not only these but the cold-blooded creatures too,

[1] *Zoonomia*, i. 505.

the vegetables even, all from the same single source! Approving Hume's conclusion that "the world itself might have been generated, rather than created; that is, it might have been gradually produced from very small beginnings, increasing by the activity of its inherent principles, rather than by a sudden evolution of the whole by the Almighty fiat," he broke into a higher applause, of "THE GREAT ARCHITECT! THE CAUSE OF CAUSES!" Himself, and forthwith linked to the development theory that ghost which has dogged it ever since, the conception of a perpetual Progress: ". . . the improving excellence observable in every part of the creation; such as in the progressive increase of the solid or habitable parts of the earth from water; and in the progressive increase of the wisdom and happiness of its inhabitants; . . . our present situation being a state of probation, which by our exertions we may improve, and are consequently responsible for our actions." [1]

This was the formal statement of the conception which underlay all of Erasmus's more than occasional writings. The *Phytologia: or the Philosophy of Agriculture and Gardening* was avowedly a prose supplement to the *Zoonomia*, intended to establish "a true theory of vegetation, so much wanted to collect the various facts in the memory, to appreciate their value, and to compare them with each other," [2] and plainly classifying vegetables as "inferior animals" despite their lack of digestive and locomotive organs. *The Botanic Garden*, for all its large beginning with the creation of the universe, provides little scope for wider matters, being an account mainly of the structure and sexual habits of plants, nevertheless the first canto of *The Loves of the Plants* had a note referring to rudimentary organs in animals as "marks of having in the long process of time undergone changes in some parts of their bodies, which may have been effected to accommodate them to new ways of procuring their food," the query being added: "Perhaps all the productions of nature are in their progress to greater perfection?" [3] Similarly, in notes to *The Economy of Vegetation*, a series of speculative suggestions ended with the question: "Or do some animals change their forms gradually and become new genera?" [4]

It was, however, his last poetic work, *The Temple of Nature*, which put his evolutionary views into verse, describing the Origin

[1] *Zoonomia*, i. 509.   [2] p. viii.   [3] p. 181.   [4] p. 120.

of Society from the Production of Life by spontaneous generation, "improvement" by reproduction—

> These, as successive generations bloom,
> New powers acquire, and larger limbs assume;
> Whence countless groups of vegetation spring,
> And breathing realms of fin, and feet, and wing—

the "emigration of animals from the sea," the developing means of reproduction culminating in the sexual, and so leading forward to the Progress of the Mind and the knowledge of Good and Evil. In a note he declared passingly the benefits of cross- over self-fertilization, and of "occasionally intermixing seeds from different situations together," and not only did he recognize Man's unity with other organic life, but apparently approved the view of Buffon and Helvetius "that mankind arose from one family of monkeys on the banks of the Mediterranean" who "accidentally had learned to use" the thumb.

Both in *The Temple* and *Phytologia* he stressed the universal struggle for existence:

> From Hunger's arms the shafts of death are hurl'd,
> And one great Slaughter-house the warring world!

The first law of organic nature was, he said, "Eat or be eaten!"—a slogan he not only urged upon his more finicking patients, but to which he was at the dinner-table almost indelicately attentive. Even the plants, he declared, were at "vegetable war." It seemed dreadful, yet it was necessary:

> War, and pestilence, disease, and dearth,
> Sweep the superfluous myriads from the earth;

and in the last resort he saw the process as making for actual increase of happiness, eliminating the old and sickly—those who enjoyed life least—and making the more room for the young and strong whose pleasure was greater. In thus approaching the idea of a natural selection by survival of the fitter Erasmus showed how near a man may come to a fundamental notion without grasping it, without realizing perhaps that it is there to be grasped. He asserted the unity of organic life and indicated its process of development, yet never really faced the problem of means. So he left it to Lamarck to develop one solution, and to his grandson Charles to develop yet another.

Necessarily but one aspect of his work has been mentioned.

His wider range of subjects was remarkable and refreshing; the reader never knew what might come next. Ninon de L'Enclos, Arkwright's cotton-mills, Moses, Matlock, Montgolfier, meteors, the primary colours, Memnon's lyre, the steam-engine, fairy-rings, the Portland Vase, coal, shell-fish, vegetable perspiration, the coldness of mountain heights, twilight, explosions—these were but a handful of the topics of which his omnivorous notes made free, while on medicine, botany, education, and horticulture he spoke as an assured authority!

His books were on the whole extraordinarily well received, the poems with an acclaim, of hard cash as well as soft words, which has never ceased to astonish later generations. Even the *Zoonomia*, which he withheld so long, went into three editions in seven years, and was expeditiously translated into French, German and Italian. It succeeded more perhaps as a text-book (as which Robert thought its influence considerable) than as scientific speculation, for commentators upon his evolutionary theory found his evidence inadequate, amounting to scarcely more than a few loose analogies, and his fears of losing reputation were justified to the point that for a period "Darwinizing" became a cant term for wild hypothesis. Nevertheless it had, and continued to have, its admirers.[1]

III

Robert and young Erasmus were now full-grown men, living their own lives, but throughout these years of active authorship a new family was growing up, for Eliza bore him seven more children, one of them Violetta, who married a Samuel Galton (grandson of the "Lunatic") and was the mother of Francis Galton. There were the Pole children too—and the Parker girls to be educated and established as proprietors of a school at Ashbourne in Derbyshire, to aid and advertise which the Doctor wrote and published his *Plan for the Conduct of Female Education in Boarding Schools* (1797), in which he advocated the instruction of girls not only in the social graces but to develop "internal strength and activity of mind, capable to transact the business or combat the evils of life," for then and then only "the female character becomes complete, excites our love, and commands

---

[1] It has been suggested as the source of Lamarck's speculations, but the evidence, despite Samuel Butler, is against such conjecture.

our admiration." He also promised his free medical attention for pupils. Violetta and a sister attended the school.

Etruscans and Darwinians were friendly as ever. The Wedgwood youths every February sent gay valentines to the Pole girls, and all Etruscans journeying to London were expected to "make Derby a resting-place, and recruit yourselves and your camels for a few days, after having travelled over the burning sands of Cheadle and Uttoxeter."[1] The Wedgwood girls—Susannah, Catharine, Sarah—were frequent visitors; Susannah was Erasmus's particular favourite. Gifts passed easily between the families: "Your medallion I have not yet seen, it is covered over with so many strata of caps and ruffles, and Miss Wedgwood is whirled off to a card-party—seen and vanished like a shooting star."[2]

The younger generation, it seemed, were inheriting the earth. Robert had completed his education at Edinburgh, taken his medical degree at Leyden, and was successfully practising at Shrewsbury. He and Susannah were in love, and soon to be married. Josiah's health was failing, and in 1790 he had formally retired, giving the business over to his sons. He went often, on his old friend's advice, to Blackpool or Buxton to rest and recruit his strength, and at Buxton Erasmus, troubled by occasional attacks of what seemed to be *angina pectoris* and taking life more easily, would often join him. He died on January 3rd, 1795. He had started with no more than his own intelligence and industry; he left a fortune. To each of his daughters he bequeathed £25,000, to each of his sons more, his namesake receiving the bulk of the great business from which the other two had earlier withdrawn.

But illness and death came not only to age. Tom Wedgwood was already an invalid, and quite suddenly, though apparently after a period of considerable mental depression, the younger Erasmus in December 1799 deliberately drowned himself in the river Derwent which flowed past the end of his garden. His had been an unsatisfying, unsatisfactory life. He exhibited some of his father's faculties, but always in diminuendo. He wrote a little poetry. He collected a little, it was said coins. He was desultorily interested in genealogies and statistics. Some said he wished to enter the Church, but feared his father's sarcasms.

---

[1] *Dr. Darwin*, Pearson, 214. [2] *Ibid.*

He became instead a lawyer, in Derby, but was too dilatory, too indifferent, to attain success. He shrank from decisions and from people (though Boulton and Day were among his sympathetic friends), preferring to hide his over-sensitiveness in idleness. He was aloof from women as from men. The cause of his suicide was, it was said, the oppression of postponed work. His father, despite Anna Seward's embittered accusation of callous indifference, suffered a natural distress, but not such as he had felt for Charles two decades earlier. He had never understood his second son, and the whole manner of this death was alien to him. He might have written a philosophical treatise on suicide as a rational act—he would never have committed it. He had not only a zest for life—for him, he said, a "sense of pleasure" was "annexed to the mere feeling of existence"—but a quality of stubbornness in that zest.

His own end, however, was not far distant. There is little record of these final years. The *Phytologia*, last of his works to appear in his lifetime, was issued in his seventieth year. Still he practised and still he wrote, bleeding himself to relieve his heart-attacks. After severe pleurisy in 1801 his health failed visibly. He grew pale and weak; his family and friends feared the worst.

But inactive he could not be. Early in 1802 he and all his household moved to Breadsall Priory, three miles from Derby, a pleasant, well-sheltered house in an extensive garden. Immediately he started planning improvements. On April 10th he was feverish, but recovered, and on the 17th—it was a Saturday evening—his wife remarked how well he was looking. "I generally look well before I become ill," he told her, and added that he would never live to see the contemplated changes to the house and grounds. Next morning he was up early, and was writing a letter to Edgeworth—a description of the new home—when the fever suddenly returned, bringing on a worse condition. His surgeon, Hadley, husband of one of the Parker girls, was sent for from Derby, but before he could arrive the old man had died, between eight and nine o'clock.

He had lived always with a certain boisterousness; he died calmly, soothing his wife's distress. Whatever his expectation of another life, or its absence, it was not in his nature to be fearful. As a young man he had concluded that "the light of Nature affords us not a single argument for a future state; this is the only one, that it is possible with God, since He who made us out of nothing can surely re-create us; and that He will do so is what we humbly hope," [1] and there is no evidence that he ever essentially changed that view. In *The Temple of Nature* he expressed his conviction of "the benevolence of the Deity," finding its proofs in the world's history, and though Anna declared him "not famous for holding religious subjects in veneration," it was not in his nature to be blasphemous; he was definitely a theist.

He was buried in Breadsall Church. The papers of the day lamented his death as that of a great physician and philosopher whose reputation as poet was already in decline; they numbered his many children, including his two natural daughters.

It is a matter for regret that his biography was never formally written, that he had no Boswell—only an Anna Seward. For he was a man of both deep insight and exceptional character. Immediately after his death Robert was contemplating "a biographic sketch, to be prefixed to his writings at some future time," and it was to aid this work that Anna's reminiscences were begun. But that idea was soon abandoned and Edgeworth approached by the widow and children to undertake the task. He evidently agreed, but delayed until the occasion and the impulse had passed. Another champion, Erasmus's pupil and friend, Dewhurst Bilsborrow was also in the field, but though his work was, even by June 1802, well in progress it never achieved publication, and so for all practical purposes Anna remains the only contemporary annalist—unfortunately a not very reliable one. Even more lamentably, Robert burned the bulk of his father's private papers.

So Dr. Erasmus Darwin, of Lichfield and Derby, fades into the past, glimpsed rather than squarely seen, though men have, if they will seek them, the fruits of his thinking in the forms in which he chose to present them.

IV

With the deaths first of Josiah and then of Erasmus it is possible that but for Robert's marriage to Susannah the links between Darwins and Wedgwoods might have been attenuated to breaking point. For a period the Wedgwoods, practically speaking, left

[1] *Erasmus Darwin*, 15.

JOSIAH WEDGWOOD AND HIS FAMILY AT ETRURIA HALL IN 1780
(From a painting by George Stubbs, R.A.)

Mary Anne   Tom   Catharine   Susannah   Josiah   John   Mrs. Sarah Wedgwood
   Sarah   Josiah Wedgwood

From *Emma Darwin: A Century of Family Letters*. By the courtesy of Mrs. Cecil Wedgwood and John Murray

Staffordshire altogether. As early as 1793 the rather commonplace John had withdrawn to become a banker in London, the handsome and intelligent Tom, whose memory as "the first photographer" does less than justice to a brilliant and attractive character, in vain quest of health. The younger Josiah, the best business man of the three, also moved to Surrey, convenient to the London showrooms, on or soon after his marriage to Elizabeth Allen in 1792.

Following their father's death there was a further disintegration. Etruria Hall was deserted—even apparently put up for sale, though unsuccessfully—the widowed Mrs. Wedgwood and her two unmarrying daughters staying with John in London or young Josiah at Chobham. In 1799 and 1800 Tom and Josiah both settled at Tarrant Gunville in Dorset, Josiah at Gunville Park, Tom at Eastbury Park, where his mother and sisters, Kitty and Sally, came to live with him.

Now indeed were Darwins and Wedgwoods well apart, save as friendship and Susannah's marriage in 1796 had bound them, causing them still to visit each other in Shrewsbury and in Dorset, and to spend holidays together in London and Bath. In these years even Josiah lived very remote from Staffordshire, save for business visits. His Dorset estate absorbed his attention—that and sustaining his position among a local gentry which persisted in regarding him as, after all, a tradesman. He farmed extensively and experimentally, breeding sheep with especial interest. He hunted, shot, and no doubt fished. When he travelled it was in state, his carriage-and-four attended by postillions in scarlet liveries. In 1803 he became Sheriff for the county. Many distinguished visitors came to Gunville—Sydney Smith and other bright stars of the Holland House company, Wordsworth, Coleridge, Sir James Mackintosh who married Josiah's sister-in-law Catherine Allen, and Thomas Poole of Nether Stowey, who wanted to marry Catharine Wedgwood but was abruptly rebuffed by the shocked Tom and Josiah as presuming above his station.

But the new century was still new when the necessity of an active return to the business became obvious. The Continental wars, with their tariffs, embargoes, and the seizing of cargoes on the high seas, were seriously disturbing trade; there were complaints too that the quality of the goods was declining. None of the sons had their father's ability or interest in the works, but something had to be done. John was the first to return, in 1800, but only temporarily. Two years later Josiah, borrowing from Robert, bought Maer Hall, seven miles from the Etruria Works and off the main road between Newcastle-under-Lyme and Market Drayton, and began to prepare it for occupation; though not till after the consumptive Tom's death in 1805 was the Dorset property sold and a general move made northward, Mrs. Wedgwood and her daughters to Parkfields, at Barlaston beyond Longton, and Josiah to settle at Maer in 1807. He wrote to Mackintosh: "My principal inducement for this change is, that I may look to my business, and next, that I may diminish a very heavy load of debt which I have incurred in making purchases quite disproportionate to my means."[1] Mackintosh applauded the return: "A Wedgwood living out of Staffordshire must lose something of his proper importance."[2]

Wedgwoods in Staffordshire again, Darwins at Shrewsbury—the scene was setting for the appearance of the hero!

V

Shrewsbury in 1787, despite the rivalry of three doctors, six surgeons, and "divers apothecaries," was an admirable choice for a capable and energetic young physician. In that day when bad roads and slow travelling made London almost as remote as Warsaw or Moscow now, it was, with its population of only 14,000, the "metropolis" of a considerable rural area. A wealthy aristocracy and comfortable squirearchy, inhabiting the country halls and granges of the lovely Welsh border, made Shrewsbury their meeting-place and, more, their winter quarters, taking town houses there for the months when too many paths and lanes were passable only upon horseback. It was a prosperous society, too, especially in the war years which followed from 1793 to 1815, when rising corn-prices produced higher rents and tithes.

The town then lay almost entirely within the main loop of the deep, swift-flowing Severn, and largely inside the limits of its medieval walls, now stripped of their battlements and towers and reduced in height, but still standing to provide pleasant

---

[1] *Group of Englishmen*, Meteyard, 297.   [2] *Ibid.*, 301.

walks through the outskirts. At the neck of the loop the old red stone castle, crumbling but picturesque, looked down from its elevation upon the houses and the narrow cobbled, muddy, ill-pavemented streets. Many of the larger timbered houses dated from the sixteenth century, but between their overhanging casemented fronts appeared newer and more fashionable erections with pillared façades, or mean tumbledown cottages of thatch and plaster and shops with small dark windows. Improvement was active. The new Welsh Bridge leading to Frankwell across the river was built in 1792–5, and early in the new century agitation began for "better paving, lighting, watching, cleaning, watering and regulating" the town as a whole, and for widening the streets and squares.

Inevitably Shrewsbury society was snobbish and bigoted, but it was also gay. There was at least in the winter a continual round of balls, suppers, oyster-feasts, dinner-parties at six o'clock, and evening tea-drinkings to follow. There was hunting for the sportsmen, and occasional visits of strolling players to the assembly hall for the less strenuously inclined. Hostesses strove to excel each other's hospitality. With much chatter, some good talk was not lacking. Speakers and preachers visited the town. It was perhaps the last blossoming of the flower. From the last decade of the eighteenth century London was becoming ever more accessible, with better roads, faster coach services, and presently the railway. But the change was too gradual to disturb seriously Robert's progress.

It is mainly as a man in middle and late life that he appears in the reminiscences of his son, for he was over forty when Charles was born, but clearly he was in younger life too a strikingly forceful character. The only time to ask for advice, he would say, was when your mind was made up. Then good advice encouraged you and bad advice was easily rejected. He seldom asked for advice, and his mind was always made up! The degree of his worldly ambition is a moot point, on which even Charles seemed in doubt, but whether or not he had as a young man his scientific interests and hopes stifled by the Royal Society's rejection of a paper he submitted to it, so that always afterwards he scoffed at the like ambitions of others, one feels that no more than Erasmus was he a man to accept second place in his own immediate field. Probably he was content enough to be a physician, if a successful one! Actually, when he did become a Fellow of the Royal Society in 1788—on the strength of a paper on ocular spectra which his father is said to have helped to write—he was presented to the Society one evening, visited the president's office next morning, and set off for Shrewsbury in the afternoon with a celerity which hardly suggested undue concern for the business in hand. Charles's final view was that Robert's mind was not scientific in that, though he readily formed theories, he made little attempt to systematize his knowledge. "I cannot tell why my father's mind did not appear to me fitted for advancing science; for he was fond of theorizing, and was incomparably the most acute observer whom I ever knew. But his powers in this direction were exercised almost wholly in the practice of medicine, and in the observation of human character. He intuitively recognized the disposition or character, and even read the thoughts, of those with whom he came into contact with extraordinary acuteness."[1] His attention, and one must therefore suppose his interest, turned less easily to principles than persons, and viewing his life as a whole there is no real evidence that he was not satisfied by his Shrewsbury existence, at any rate once he had settled down to it.

At first there was undoubtedly some interior conflict. He could never conquer his distaste for the sight of blood, and as a young man so hated his profession that, he said, he would never have been a doctor at all but for Erasmus's decision and his own inability to find any other way of earning a living. He was, if a Darwin unmistakably, a Howard too. Tender- and tough-mindedness fought in him. If tender-mindedness inclined him to non-surgical treatment, it was tough-mindedness that carried him through.

He had to face, in the beginning, prejudice as well as opposition. He was young. He held liberal views. He was the son of the heretical Erasmus—and Shrewsbury was rigidly orthodox and conservative. Nevertheless he made such headway as to have little need of the £1400 which came to him about this time from his long-dead mother and his aunt Susanna Darwin; his first year's fees paid for the upkeep of two horses and a servant, and his practice widened rapidly.

One early matter brought him useful publicity. He had, like

[1] *Erasmus Darwin*, 84-85.

his father before him, the good fortune to cure a "hopeless" case, whereupon the most eminent of local visiting physicians, Dr. Withering of Birmingham, a Lunar Society member in its later days, exclaimed that he "had never known a patient, labouring under her complaint, recover, after being so erroneously treated." Robert accused him of slander, and breach of medical etiquette. Acrimonious letters were exchanged, and Robert at last impatiently published these in a pamphlet that had a wide local circulation. He himself wrote throughout the correspondence warmly, pugnaciously, with the resentment and even rudeness of a young man conscious of his elder's jealousy. Withering was more dignified, in the rôle of the man of experience contemptuously exposing an ignoramus. But Robert, on the firm ground of his patient's recovery, was not abashed, and gleefully held up to scorn his rival's boastings of the world's and his own good opinion of himself. "Your composition calls to my mind the wonderful fountain described by Pomponius Mela, who says that whoever drank of it died with laughter!"[1]

VI

The controversy and cure established him, despite inevitable gossip. He quickly became and for a half-century remained Shrewsbury's leading physician, with a widespread practice and an income reputed to exceed that of any doctor outside London.

For nearly a decade he lived a bachelor on St. John's Hill. Then, on April 18th, 1796, he was married to Susannah Wedgwood at St. Marylebone Church in London. They had a house for a while in the Crescent, and there their first child was born in 1798. He might, with his rapidly increasing income, have taken one of the larger residences in the town. Instead, he chose to build a house just outside it, and not on the more fashionable English side of the river, but across the Welsh Bridge beyond the rather dingy district of Frankwell. It was an uninspired three-storey Georgian house of red brick, square, substantial and ugly, its principal charm the view from its green lawns between clustering trees over the curving Severn below to the rolling pastures and woods beyond, with no more than a glimpse of the town away to

[1] *Appeal to the Faculty*, 45.

the right, and lowly Frankwell tucked out of sight behind thick shrubberies. The garden, with its greenhouse and orchard, was Robert's and Susannah's particular pleasure, and The Mount pigeons were known to all the town.

He was an impressive figure, even at a glance—above six feet in height, and big and broad in proportion. In later years he grew enormous, until the rickety stairs and rotten floors of some of his poorer patients were a positive danger to him, and had to be tested by his coachman before he could venture upon them. He had a heavy, rather dour face of ruddy complexion, the full nose and firm-set lips very like those of Charles; his arched eyebrows seemed fixed in a perpetual, quizzical questioning. His clothes were sombre, the dark coats cut with large lapels and deep cuffs, his only ornament a heavy gold chain across his lappeted waistcoat. Out-of-doors he wore habitually a broad-brimmed shadowing hat. His small squeaky voice oddly made one think of a needle in a haystack.

His character was, no less than that of Erasmus, a mingling of the friendly and the portentous. His moral quality was high—his integrity unquestioned, his sympathy, generosity, benevolence, and general kindliness and affectionate feeling everywhere recognized. His insight was instant, and friends and patients alike readily gave him both their confidence and their confidences. He remained always sensitive to physical pain and mental distress in others, but retained an independence in his views, and could be sometimes crushingly peremptory in expressing a divergent opinion. His spirits were high too, and he liked to see people in his company enjoying themselves, but his was a Jovian laughter, holding the threat of thunderbolts, a large dominating cheerfulness that was apt to diminish the cheer of those about him, and not least in his own family circle. In other moods he could be irascible and exacting, even overbearing.

For long years he was a notability in Shrewsbury, he, Dr. Butler of the Grammar School, and Mr. Blakeway the town historian ranking as the three "local encyclopedias." His small yellow chaise, drawn by a pair of sleek horses, was as familiar a sight about the district as once Erasmus's around Lichfield. But unlike Erasmus, Robert never read in his carriage, but "always sat as though carved in stone, as though books were unnecessary to one whose treasury of knowledge and thought was alike inexhaustible." [1] He was welcomed everywhere, in the homes of the richest as the poorest. His interest in the town life and people was continuous. He was long visiting physician to the Shrewsbury Infirmary, and in 1823 he established at his own expense the first infant school in Frankwell, in accordance with the Rousseauistic ideas of Pestalozzi. Not only was he ever ready to render medical or financial aid to his own intimate friends, but gave loans to local tradesmen in unexpected distress. He was called, gratefully, the Father of Frankwell.

VII

In all these things his wife, during her lifetime, played her part. Not much, though, can be told of her. Physically she had the pleasing Wedgwood features, regular and well proportioned, intelligent if over-strong for beauty. Her gaze was direct, and humour or at least good-nature lurked in her lips. Her curling abundant hair, her dark brows, gave distinction. Everyone liked her, found her agreeable. Before her marriage she had pleased Erasmus by attention to her studies and his discourses, and she entered at once into her husband's interests and labours, doing much writing for him. She had indeed every capacity to hold her own, as her husband's wife and in Shrewsbury society, save that of strength. The "frequency of *her* maternal situation" was perhaps too much for her. Marianne was born in 1798, Caroline in 1800, Susan in 1803, and Erasmus in 1804. Her time, her attention, what energy she had, she gave to her children, fearing, it appears, she would not be with them long. "Everyone seems young but me," she lamented one June day in 1807, the cry of an ailing if not ill woman. It was just twelve months later, in the early summer of 1808, that she found herself pregnant again, and on February 12th, 1809, Charles Robert Darwin was finally physically born. It was Sunday: in the parish church, St. Chad's, the preacher took for his text: "My people are destroyed for lack of knowledge." [2]

When he was christened on the following November 17th at St. Chad's, she was already expecting another—her last—child, Catherine, born on May 10th, 1810. These successive pregnancies,

---

[1] *Group of Englishmen*, Meteyard, 263.
[2] *Shrewsbury Chronicle*, Feb. 17th, 1809.

in her forty-fourth and forty-fifth years, only too probably aided that "decline" which ended in her death seven years later. "The Lord gave, and the Lord hath taken away; blessed be the name of the Lord."

She was buried at Montford Church, a few miles from Shrewsbury, beside the river.

BOOK TWO

THE MAKING OF THE MAN

CHAPTER III

SHREWSBURY DAYS

I

OUT of the seeming darkness of pure isolated sensation the child-mind moves progressively towards that awareness of mutual relationship and relative value which is memory and self-consciousness. Charles Darwin spoke in agreement not only with modern psychology but also common sense when, asked as an old man which years of a child's life he considered the most "subject to incubative impressions," he asserted: "Without doubt, the first three," going on to explain, in the words of his interrogator, "that the brain at that period is entirely formed—it is a virgin brain adapted to receive impressions, and though unable to formulate or memorize these, they none the less remain and can affect the whole future life of the child-recipient." [1]

It is unfortunate then, if inevitable, that this should always be the time concerning which least is recoverable—in Charles's case, nothing at all. His own earliest memory went back barely beyond his fourth birthday, to an incident of sitting on his sister Caroline's knee in the drawing-room at home, and being so startled by a cow unexpectedly running past the window that he jumped and cut himself on the knife Caroline was using to peel an orange. Immediately subsequent recollections were similarly trivial. For the most part the rooms and lawns of The Mount were his world, wherein father and mother, brother and sisters, maids and nurse were figures so primordially familiar as scarcely to be apprehended, but the fresher experience of a holiday by the sea, at Abergele near Rhyl, in the summer of 1813, imprinted itself as a series of detached pictures. A house dimly seen. A small shop, and the shopkeeper who gave him a fig—which delightfully turned out to be two figs—as fee for kissing his nurse-maid. A walk to a well, past a cottage where

[1] *Richmond Papers*, 101.

a white-haired recluse was forthcoming with damsons. A broad ford, crossed in a carriage about whose wheels white foaming water washed alarmingly. But he was not sure whether it was he or Catherine who had been shut in a room for naughtiness, and in anger tried to break the windows. From visits to his Aunts Kitty and Sally at Parkfields he brought back little more than terror at the tales of a servant, Betty Harvey, who pictured to him, no doubt as a cautionary story, the dreadful deaths of little boys who ventured beside the canal, got on the wrong side of the towing-rope, and so were swept irrevocably to a watery end. Death could be but a word to him, but she communicated perhaps her own fear in her dramatic narrative.

All Europe was astir in those days, gathered for the last desperate struggle with Napoleon. The Peninsular War, Borodino, the retreat from Moscow, Elba, the Congress of Vienna, the Hundred Days—one event followed hot-foot on another, yet meaning little more to Charles than the amusement of watching the local militia exercising in a field near home. Excitement in Shrewsbury as over all the Continent rose to its height in the June and July of 1815, as word came daily of Quatre Bras, Waterloo, the Allies' entry into Paris, the restoration of Louis XVIII, Napoleon's surrender to the English. It was fun then to race across the cobbled courtyard to the gates with the others, the maids even, to see the evening coach fling out Papa's mail-package as it laboured past gay with triumphant laurel and fluttering ribbons. But opening the letters and reading aloud from the close-printed columns of the thin newspapers were dull adult matters, as was the discussion that followed of what should be the Emperor's fate, now that his acceptance of defeat was at last assured in unimpeachable black-and-white. Papa was decisive that St. Helena was the very place for him, and Aunt Elizabeth, visiting from Maer, thought that the best suggestion she had heard; all cried out upon the "abominable papers" which would have the poor man hanged. But Charles's attention was for the guns firing in the town, the bells ringing in every church-tower, the staying up late, the illuminations, the quivering sense of celebration.

Even these made but a secondary impression. He was evidently a "difficult" child, hard to arouse, slow to learn, seldom eager, a self-centred day-dreamer. He invented and narrated fictions, to draw attention to himself. He gathered fruit secretly from the orchard, hid it in the shrubbery, and then ran excitedly indoors to declare his discovery of a thief's hoard. He returned from solitary wanderings to tell of marvels glimpsed, usually strange birds, a choice he later remarked as denoting an interest even so early in natural history—as undoubtedly it did, though it might most plausibly be regarded as a purely derivative interest, reflecting an already traditional family habit of mind.

The inclination is generally ascribed to his mother, who is seen, as out of sustained regard for her deceased father-in-law and his evolutionary theories (which may have meant nothing whatever to her), piously inspiring her younger son to take up his lifelong questioning of nature's profoundest riddles. It makes a very pretty picture, but the evidence is slender. Susannah is the least figure in Charles's early memories. Later he could remember scarcely anything about her, declaring her to have died during his "infancy"—when he was eight years old! Her black gown and her work-table were clearer to him than anything of herself. He recollected that once she had asked him to do something for his own good, but what it was he had forgotten. Only by hearsay could he attest her "very agreeable in conversation." Even her death made far less impression upon him than the local burial of a dragoon which he witnessed about the same time, and certain details of which—"the horse with the man's empty boots and carbine suspended to the saddle, and the firing over the grave"[1]—remained vivid after sixty years, though of the other, and one would have supposed much more moving, event he could recall no more than being sent for, going into his dying or dead mother's room and finding his father there (again the memory was of anyone *but* his mother), and a certain amount of sympathetic weeping. His younger sister Catherine, on the other hand, could remember almost everything. There probably lies the clue. Susannah was a loving mother, but in these last years her strength was failing, and there was always, from Charles's quite early infancy, a younger child to take the lion's share of attention. The story of her especial devotion to him is, as far as facts go, plain myth.

Actually, those romantic writers who insist upon her influence, asserting that she it was who introduced him to "science" by way of botany, are forced to rest their case upon little more

[1] *LL.*, i. 30.

than a schoolfellow's recollection of Charles "bringing a flower to school and saying that his mother had taught him how by looking at the inside of the blossom the name of the plant could be discovered,"[1] an incident which seems quite at one with and about as important as his stories of stolen fruit and his other statement, made to the same friend, that he could make flowers change colour apparently by urinating upon them, a "monstrous fable" he denied in his autobiography having even attempted to put into practice. One psycho-analytical writer boldly asserts that "whether or not Darwin's mother actually propounded her enchanting riddle to her boy is not quite so important as the fact that he said she did, showing how keenly his wishes relished the fancy that she had revealed to him the one secret of life that fascinated her—the secret which, if read, would reveal the origin and creation of life and—himself,"[2] all of which would be true and to the point were there only some ponderable reason to suppose that any such idea had ever occurred to Charles.

Not his mother but his father was the effective influence. To this timid, stammering, nervous boy Robert was more than human—god-like in his omniscience as in physical bulk.[3] His equitable refusal to be either impressed or perturbed by Charles's most remarkable inventions uncomfortably suggested not only his awareness of their nature but also a superiority indifferent even to that knowledge; he ascended thereby to the height of the deity who does not so much as condescend to strike with his lightnings the petty mortal who defies him. Unforgettable was the occasion when Charles had done something childishly naughty. No one, he was sure, could know of it, yet as he sought to still his conscience beneath an unusual display of filial affection his father turned on him, ordering him to confess his wrongdoing. Charles never forgot his utter astonishment, his terror of the god who reads one's deepest thoughts.

Robert has already been drawn as in many respects an awesome figure, and in Charles's later childhood circumstance made him especially so. Susannah's death on July 15th, 1817, must have stricken him. There is every evidence that he loved his wife with all that enduring affection of which the Darwins were

---

[1] *LL.*, i. 28 n. [2] *Psychopathology*, Kempf, 214.
[3] The boys at the Grammar School, of which he was a trustee, thought him the biggest man ever seen outside a show.

capable. Unlike his more resilient father, he never married again, nor does any woman enter the story of his remaining thirty-one years. A wife's decease can bring a man closer to his children, but it is more likely, and especially in the case of so intrinsically dominant a personality, to remove him, to enhance his awesomeness by flinging about it the cloak of a living sorrow. Death—even of a mother—goes by for the young, while sensed as still present for the widower-father, and Robert was a man who lived much in the pangs and pains of the past, refusing as an old man to go driving because every road out of Shrewsbury was associated in his mind with some unhappy event. Charles grew up never quite at ease with him, awed and fearful—a fear evoking the feelings and expressions of reverence which, as with passing time the son moved more and more beyond the boundaries of his father's probable approval, became more and more extravagant.

He passed, nevertheless, many happy hours with him, when Robert on his professional rounds would take the boy riding in his chaise, pointing out the wild birds and creatures seen in the fields and along the country roads. "I was born a naturalist," Charles once said, but if the interest was inherent here we have the first effective means of its development. The collecting of objects, which his mother may or may not have initiated, is a trait common enough to almost all children, and he began with the very obvious objects of coins, seals and franks. It was a schoolfellow's suggestion or example which later turned his attention to small stones and minerals—he thought he would like to know "something about every pebble in front of the hall door," as also about that "inexplicable" geological mystery, the Shrewsbury "bell-stone"—and thence to shells. His brother Erasmus was already a collector of plants, which he would dry and preserve, and in such a household—Robert a keen gardener, Susannah a lover of flowers and birds—there seems nothing remarkable in the gathering of natural history objects becoming the one persistent pleasurable activity of Charles's early years. The remarkable thing would be had it been otherwise. To put it on the lowest level, was not a new pebble brought home in the hand well worth two imaginary if unwonted birds reported in the shrubbery? Even Papa could not gainsay the concrete object, and the odds are that he outspokenly applauded it.

Charles was, in sum, a sufficiently commonplace child. His collecting habits were no more than those normal to his age and given direction by home influence; the special significance read into them in the light of after-event is mainly fictitious. As a young schoolboy too he was quite unexceptional, if among the rather more timid, the more gullible of his fellows. He readily formed friendships with other boys of similar type and tastes, feeling a naïve affection for them, and attempting to impress them too with tales of his astonishing discoveries and personal powers. But he swallowed whole the fictions of others, and was easily frightened, alike by stray dogs in the town streets, and by fights at school. He readily reacted to others' pain. He loved fishing, but shrank from impaling the living worm upon the hook. Having once beaten a puppy, pleased by the sense of power it gave him, he felt afterwards such remorse that he never forgot the incident or the spot where it had happened. And when collecting birds' eggs, he would never take more than one from any nest, except once in a deliberate spirit of bravado.

Sensitive he most clearly was, but in these days much less certainly imaginative.

II

He began his formal education in the spring of 1817, a bare few months before his mother's death, at the day school kept by the Rev. G. Case, minister of the Unitarian Chapel in High Street which Susannah rather irregularly attended with her children when she did not take them to St. Chad's. He was there for more than a year, but later could remember nothing of the teaching, either manner or matter, or of the place itself. Even a holiday visit to Liverpool in July 1818 left only the vaguest impressions of a good dinner, fear that the coach might upset, ships seen at a distance. He was a sturdy bullet-headed boy, with an obstinate mouth and slightly sullen grey eyes under a heavy forehead.

August 10th of 1818 he became a pupil at the Shrewsbury Grammar School, where Erasmus had been attending three years now. Founded by Edward VI more than two hundred and fifty years before, the school had had a changeful history. Towards the close of the sixteenth century it claimed to be the largest public school in all England, but a succession of inadequate headmasters brought about headlong decline, until by the end of the eighteenth century boarders were almost unobtainable, and the local townsmen refused to let their sons waste time there even as free foundation scholars. It was at its last gasp when, soon after Robert's arrival in Shrewsbury, reform was set afoot, local trustees appointed, the revenues reorganized, and St. John's College, Cambridge, which held the gift of the headmastership, beseeched to exercise discretion. The Rev. Samuel Butler, product of Rugby and the nominating college, was suggested, and though but twenty-four years of age was warmly welcomed by the new Governors upon his brilliant academic record.

That was in 1798, and the grandfather of the author of *The Way of All Flesh* held his post for the next thirty-eight years, becoming in that period holder successively of the living of Kenilworth, a prebendal stall in Lichfield Cathedral, and an archdeaconry at Derby, but retiring only when, having made a fortune from the school boarders, he accepted in 1836 the "ill-paid bishopric" of Lichfield and Coventry. He was an impressive if unattractive personality, one of those incredibly *worthy* men of whom the early and middle nineteenth century had so much more than its share, a spiritual cousin of Thomas Arnold, if less completely evolved. His extraordinary rigidity of character appears in his relation to his second master at the school. He speedily quarrelled with this undismissable assistant, and thereafter for thirty-seven years communicated with him only by letter, though all the while they carried on duties side by side; Butler consented to reconciliation only on the other's death-bed, when they partook of Holy Communion together.

Partly because of this sustained quarrel Butler had to contend with much local opposition, but he succeeded in his task, restoring discipline, reviving prestige, and making his charge once again one of the first public schools in the country, from all parts of which other headmasters came or wrote to consult him. He was in no essential sense an innovator, however, despite some minor originalities, but a man of legal type of mind—scrupulous, methodical, exact and unimaginative. Competitive examinations, "merit money," and the birch were his incentives to learning. He did not flog with enthusiasm—but he flogged. His aim was to educate the sons of provincial gentlemen in gentlemanly attainments, his method instruction in the classics almost without relief and certainly without alternative. He would not even argue the point. "While your son remains here he will always be exercised in Latin and Greek composition both in prose and verse, and the higher he gets in the school the more he will have of it."[1] In the matter of educating "the Poorer Classes" he took as strong a line, impatiently protesting that he lived "in a time unexampled for morbid sensibility." Since the poor (unhappy creatures) could never attain to "real learning," why then create false hopes—and waste good men's time into the bargain? Alternatively, if that seemed over harsh, let us recall that "mighty difficulties make mighty minds"—was it, after all, true charity to make the road too smooth?[2] One would like to know in what category he set his foundation scholars, who accounted for but a fraction of his fixed income and gave him, of course, no boarders' profits at all. At least it is noteworthy that Robert Darwin preferred to send both his sons to the school as boarders, near by as their home was, thus allowing them what Charles later termed, not altogether consistently, "the great advantage of living the life of a true schoolboy."

A dubious advantage it appears to the outside view. The life was rough, for the boys were left very much to themselves outside the classrooms. Doubtless they learnt self-reliance, but they may have learnt other things too; the cultural possibilities of sleeping adolescents two in a bed were not insisted upon in the curriculum, though it was noted that for a small extra fee (all grist to the headmaster's mill) a boy might have a bed to himself. This in fact was generally done, but Charles did not pass through his boarding-school days wholly unaware of "wickedness."

Still, if the considerable leisure was devoted largely to fighting, bullying, practical joking and the like, there was opportunity too for study, miscellaneous reading, and country walking and wandering. Poaching might for some add a needed spice to these last activities, but young Charles, whose collecting instinct was moving now in botanical and even entomological channels, demanded no such adventitious aid.

These must have been the aspects in which he found virtue, for assuredly he did not discover it on the academic side. Nothing, he said in his autobiography, could have been worse

[1] *LL. Butler*, i. 369.   [2] *Ibid.*, i. 311–13.

for the development of his mind. Neither mathematics nor modern languages were taught. There was nothing to stimulate observation. Except for a little ancient geography and history, it was classics, classics all the way. "The school as a means of education to me was simply a blank."[1] His studies in no way engaged him; they remained strictly a matter of learning by heart what he would completely forget by the day after to-morrow, of working wholly and solely for the minimum necessary satisfaction of his tutors. He was weak at spelling, bad (as all his life) at languages, and worse at versifying; he cribbed whenever possible. Butler almost certainly thought him a fool, and had no hesitation in making him appear so. Late in life Charles would recall with "fearful distinctness" how the headmaster would "hum" prolongedly while contemplating his classical verses, an alarming sound expressing all too evident suspicion of their illicit origin. Nor could Butler conceive that a mind unable to respond to his own instruction might nevertheless achieve virtue in another sphere, for hearing that Charles was helping Erasmus with chemical experiments in an improvised laboratory in the back garden of The Mount, he went out of his way to jibe in a manner setting him beside his pupils who baited the boy with the nickname "Gas."

Yet Charles was plainly no fool, however unforthcoming and even apathetic in some respects. When his interest was aroused, he could be keen, eager for understanding, anxious to learn. He liked reading, and was encouraged in it by Erasmus, who pressed books upon him. He found pleasure in the poetry of Scott, Byron and Thomson, but it was Shakespeare's historical plays which held him engrossed hour after hour upon a wide schoolwindow-ledge above and aside from the loud hubbub of the mass of playing boys. The logical proofs of Euclid enthralled him when presented by a private tutor. He still collected pebbles and the like, if, as he said, quite unscientifically and with more care for numbers and novelties than varieties or series; and from 1819 he was preserving all the dead insects he could find (not thinking it right to kill them), an interest wakened during a summer visit to the Welsh coast at Towyn, in Cardigan Bay, not far from Barmouth. Reading White's *Natural History of Selborne*, he was soon watching birds, and recording his observations with such enthusiasm that he could not conceive why this

[1] *LL.*, i. 32.

was not the occupation of every man's leisure. Even wider curiosities about the natural objects of the earth's surfaces, heights, and depths were roused by a schoolfellow's copy of *The Wonders of the World*, whose entrancing pages he and sympathetic class-mates would read over and over, disputing among themselves whether these astounding things could possibly be true, and wishing that they might travel to the most distant countries to see for themselves. Rugged Welsh scenery made early appeal to him; at ten-and-a-half he could feel an almost Wordsworthian pleasure "in the evening on a blowy day walking along the beach by myself and seeing the gulls and cormorants wending their way home in a wild and irregular course." [1] The attraction of his brother's tool-shed experiments —which belonged to the later period of his schooling, after 1822, when Erasmus had left the Grammar School for Christ's College, Cambridge—was more than superficial and sensational; the two youths worked really hard together, and Charles read a number of books on the subject with care and attention. No school experience, he declared, was more educative, for here he learned in practical terms the reality of experimental science, and was deeply impressed by it.

In almost all this was evidence of material calling for active development or at least encouragement. From Dr. Butler it could, all too clearly, hope for neither. More surprising is Robert's failure to discern and aid it. Charles stood between god-omniscient master having no understanding, and god-omniscient father apparently withholding it; necessarily he was forced inward upon his own resources.

Externally as well as mentally these years at the Grammar School were comparatively uneventful for Charles. His first term saw a school rebellion in protest against the boarders' food, Butler being threatened with violence, and stones flung through the sacred study window; but Charles was almost certainly absent, ill with scarlet fever.

Lessons apart, he was not unhappy at the school. If some of the boys agreed with his masters in finding him dull and apathetic, much less evidently gifted indeed than Erasmus, to others he proved, despite an occasional show of angry temper, a pleasing and cheerful companion. His friendships were warm

[1] *ML.*, i. 5.

if most of them transitory. He had the makings of a competent athlete, and was good at bat-fives. But he preferred reading and solitary rambles. While still at school he read his grandfather's *Zoonomia*—no light task for a boy, and one carried through more out of pride in a distinguished relative than of burning interest in the subject, for he told a schoolfellow that he thought medicine "a beastly profession." [1]

The book was read at home, and it seems that right through this school period it was his home life which, after all, gave him most. Boarder as he was, he was never cut off from it. The Mount was barely a mile from the school, and often in the evening interval he would hasten there by way of the Roushill Walls to the Welsh Bridge. In earlier years, small legs had to hurry fast, and on the return journey, having delayed till the last moment and fearing to be late, he would, as he panted up steep Pride Hill, pray to the magical God of childhood whose miracles are hourly to lend him speed, "and I well remember that I attributed my success to the prayers and not to my quick running, and marvelled how generally I was aided." [2] On more leisurely occasions he was apt to lose himself in his thoughts, once so completely that he stepped right off the wall, falling more than a man's—even a Darwin's!—height to the ground below, fortunately without injury. On other lonely walks his day-dreams sufficed for company. What they were about he could never afterwards remember, but one can scarcely doubt that they were the usual timid boy's compensation-fantasies— of how by this or that extraordinary means he would astonish his chums, his teachers, his sisters, even his father or that fat ass Butler; how he would travel all the world over and see with his own eyes the marvels that other men had seen, and greater marvels he would discover for the first time and write about in books for stay-at-homes to read and wonder at; how he would know everything, the names of all the flowers and birds and beasts without having to learn them either, and other people listen when he deigned to disclose his knowledge; how he would become a Fellow of the Royal Society, and greater, much more important even than his dull old grandfather; how the local newspaper would at last—the highest honour—acclaim him "our deserving fellow-townsman."

[1] *Salopian Shreds*, April 2nd, 1884.   [2] *LL.*, i. 31.

Such dreams for such youth are endless; in their ignorant freedom they scale heaven, lacking the offence of presumption only because they are so innocent.

III

As the shadow of Susannah's death passed, and the elder children came to ages of relative independence, The Mount was probably a gayer place than ever before. The four sisters and two brothers—but the former especially—had the high spirits of youth joined to deep mutual affection. They were both intelligent and lively. They all read books, and discussed them together, sometimes, as in the case of the witty and brilliant *Letters of Madame de Sévigné*, applying the names they found therein to their best or least liked friends and neighbours.

Marianne and Caroline, respectively nineteen and seventeen when their mother died, naturally took charge of their younger brothers and sisters and the household generally. A discreet period past, they attended with their father, of formal necessity but also with informal pleasure, the frequent local balls and tea-drinkings. Both were tall, striking girls, for though Caroline was deemed not regularly handsome her brilliant eyes and bright colouring, her bang of jet-black hair over a wide forehead, drew instant admiring attention. As quite a young girl she had given lessons to Charles and then to Catherine; later she took charge of the infant school Robert had founded, and gave no less care to the pale, grubby children who attended it. Caroline and Susan—the real beauty and general favourite of the family—were especially remembered for their vivacity and feeling; they were a happy pair, taking the pleasure of the passing moment, yet never too lost in it to have instant thought for others.

There would be, sometimes, parties of young people gathered at The Mount, as in the winter of 1820 when the Darwin girls, their Wedgwood cousins Charlotte and Elizabeth, and their friends Sarah and Fanny Owen from Woodhouse, had singing lessons together from the energetic Mr. Sor, who quite fascinated Elizabeth despite his eccentricity of interrupting the proceedings to dash exercisingly to and fro about the garden paths. Sunday, though, was a graver day: "We dined at half-past one, drest afterwards, and sat about three hours expecting the tide to come in about dark, and rather stiff and awful the evening was." [1]

The "tide" obviously was the head of the household, and the metaphor is revealing. Robert was a kindly man, but apt to prove oppressive, especially to the young. He believed in continuous entertainment of guests, and when present saw to it personally. Everything had to go on under his immediate eye and in his immediate ear ("Hm, hm, what is Emma saying?"[2]), and on all matters he must make his comment, which turned the conversation into a good deal of a monologue, for not only had he always much to say—and said it endlessly though so fast that he became incomprehensible—but also fixed ideas and a heavy tongue for who dared to disagree. His demand for mental as well as material tidiness shrivelled the free wings of youthful fancy; restraint came into the room with his entry, and waited upon his exit. Young Harry Wedgwood was thought very brave because he actually jested with his uncle, returning one day from a sale to tell Robert of an item in the catalogue he should have bought—"a 'ditto to correspond,' for you know how much you hate writing letters." [3] The others looked on and giggled, but they would never have dared say it themselves.

The younger Charles was probably happy at home rather in spite of his father. Summer and other holidays away were times of a larger freedom, a spreading of the wings of imagination such as at home he achieved only in his solitary walks, or well out of sight with Erasmus at the end of the garden. It was on holiday in 1819 that he was first impelled to the collection of insects. It was on riding tours with Erasmus among the mountains of North Wales in 1820 and 1822 that his pleasure in scenery was stimulated. It was on a visit to his step-uncle Samuel Galton, at The Larches near Birmingham, when he was about fifteen, that he conceived his delight in marksmanship, and that passion for shooting birds and beasts which possessed him for a number of years. Uncle Samuel took this "very pleasant lad" out with a gun one day, and on their return reported that the birds "sat upon the tree and laughed at him." Charles laughed too, but the thrust went home to such effect that when Galton next visited The Mount he was triumphantly called into the garden to watch the boy toss a glove into the air and hit it every time.

[1] *Emma Darwin*, i. 139.   [2] *Ibid.*, 140.   [3] *Ibid.*, 55.

At Woodhouse too there was shooting. Charles got on famously with the peppery old squire, and Sarah and Fanny were always ready to join him in any field-sport. Fanny was Susan's special friend, almost her age; the Darwin girls were oftener at Woodhouse than the boys, and Susan made plain its appeal in telling Caroline: "What a delightful visit I have had. I never enjoyed anything like it—so gay—we never talked a word of common sense all day."[1] (*Hm, hm, what is Susan saying?*) Parkfields also gave the children a happy welcome until Aunt Sarah left it in 1825, her mother, and sister Catharine, having died in 1815 and 1823 respectively.

Above all there was Maer Hall, deserted for Etruria 1812 to 1819, but now once more the settled home through the 'twenties and 'thirties, the years when it meant most to Charles, of Uncle Josiah and Aunt Elizabeth and a whole host of Wedgwood aunts and cousins. It was an easy drive of just over twenty miles from The Mount, and visits both ways were frequent.

IV

The abiding affection felt by Charles for Maer and all its associations is easily understood. Here was no tide to come in about dark, no stiff and awful evening to look forward to. It might have been otherwise, for a strong strain of taciturnity was evident in each generation of the male Wedgwoods, and Uncle Josiah had all his share of it. Sydney Smith said of him: "Wedgwood is an excellent man; it is a pity he hates his friends,"[2] and one catches an unuttered reservation in a sister-in-law's comment that "Daddy Jos is always right, always just, and always generous."[3] Like Robert, he inspired awe no less than respect. The Darwin girls were frankly afraid of him, and even his wife, with all her devotion and his indulgence, was never quite at her ease; typically, he kept her wholly ignorant of his financial position. But he was never worse than solemn, despite "the Smith's" epigram, and delighted in the happy home his wife created. He did not insist, like Robert, upon any assent to his own views, and there was always in his house complete freedom of speech and much vigorous discussion. Moreover, Charles had the good fortune to be his favourite nephew, to

[1] *Emma Darwin*, i. 226.   [2] *Ibid.*, i. 7.   [3] *Ibid.*

whom he would talk frankly and intimately as to few. Caroline was Aunt Elizabeth's favourite too—"I should like her for my daughter more than anybody I know"—and accordingly the Darwin children were always especially welcome and at ease.

For this frank and happy household atmosphere it was Aunt Elizabeth who was principally responsible. Though her father, John Bartlett Allen of Cresselly in Pembrokeshire, was a far grimmer figure than either Robert or Josiah began to be, she herself stands as one of the most charming of the many attractive personalities in this Darwin-Wedgwood gallery. Allen's wife had died in 1790 after extensive child-bearing, and thenceforward, as one suspects previously, his "melancholy disposition and arbitrary temper" combined to make Cresselly more of a prison than a home, from which his nine daughters and two sons eagerly escaped into marriage or, failing that, frequent visits to more fortunate brothers and sisters. One of the girls found a husband as bad as her father; the rest were happier. Louisa Jane married John Wedgwood, Elizabeth Josiah, Catherine became the second wife of Sir James Mackintosh, and Jessie, as late as 1819, of Sismondi the Swiss historian. All were exceptional conversationalists, a gift Robert ascribed to their father's making them regularly perform under penalty of a horsewhipping. They were good-looking, witty, cultured and affectionate, equally at home in their own intimate circle or in the most distinguished social and political society. It was through Mackintosh that the Darwins and Wedgwoods came to know both Sydney Smith and the Carlyles. Smith greatly admired all the Allen sisters, but none higher than Elizabeth.

Maer was her daily task and joy, and she instilled into it her spirit of charm and beauty, sympathy, gaiety and generosity. In itself it was a delightful place, a large Tudor house with picturesque portico looking across the flower garden to the small clear lake, beyond which tall trees clustered in a rising wood, a barrier of enclosing greenery. The grounds had been planned by "Capability Brown," the landscape gardener, and were pleasantly if not completely "natural." There was a sand-walk about the lake, where boating was a summer, skating a winter amusement. There were shooting and fishing too, and horses for riding on the heath. Life was, at least for the visitor, holiday in a perpetual paradise.

Visitors came and went continually, young as well as old, the children often so numerous as to create their own independent world. There was a typical large party of young cousins in 1824, following the confirmation of Emma, then just sixteen. Marianne (soon Mrs. Henry Parker) and Caroline had been there, but returned to The Mount to make place for Susan and Catherine. There were, Emma recorded, "wicked times," "revels" which culminated in a performance of scenes from *The Merry Wives of Windsor*. It was all very noisy, with "the Tag Rag company" in full cry up and down the stairs and in and out of doors; Mackintosh, convalescing there, rallied his hostess on "the gentlest mistress in England having the noisiest household," [1] while she herself was sometimes driven to write to her sisters that "a little calm will be very agreeable," [2] or even: "I shall not be sorry to have our party lessened. There is very little pleasure in what the young ones call a row." [3] But in neither case was there the least indication of restraint to diminish anyone's enjoyment.

No wonder that Charles was happy there. Doors and windows seemed always wide open. You could do just what you liked. There were books for reading, cousins to talk to, someone always ready for a game or a walk. There was rat-catching with the dogs. There was the riding. The shooting. Wilcox the gamekeeper was friendly enough to this youth with so evident an enthusiasm for a gun and so obvious a respect for professional experience. And to end the day there were "the enchanting evenings" when all the family gathered in free informal talk, not just personal gossip but discussion of new developments in politics and thought, the latest editions of Lord Byron or Mr. Wordsworth, Mr. Coleridge's newest pamphlet, matters retailed by letter or visitor from London or abroad—and yet talk in which all, young or old, might join without fear of rebuke. There would be music too, and to the end of his life he would never forget "how delightfully" Charlotte—his boyhood favourite—sang. On warm summer evenings the gathering would be upon the porch, the younger ones on the steps watching the bright view fade into shadow and dusk and darkness, while the talk passed to and fro behind and among them, and music sounded from inside, and the beauty of song and earth and evening sky and human affection fused in a perfect placid relationship.

[1] *Emma Darwin*, i. 208.   [2] *Ibid.*, i. 161.   [3] *Ibid.*, i. 210.

CHAPTER IV

DOCTOR, CLERGYMAN

I

It might be wondered how much Robert really approved of the Maer influence in some of its aspects. Erasmus had remained at the Grammar School till he was eighteen; Charles was removed suddenly at little more than sixteen, his father telling him angrily that he cared "for nothing but shooting, dogs, and rat-catching," and would grow up "a disgrace to yourself and all your family."[1] It was partly true, partly quite false. Charles doubtless did care more for these things than for any academic subject, but clearly they did not contain all his interest, as his attention to chemistry, poetry, pebbles, insects, plants and birds, however unsystematic and undeveloped, had shown. His curiosities were awakening, as in the case of the Shrewsbury "bell-stone." Definitely he wanted to know more about it, and would have responded eagerly to any explanation.

It was not for the moment forthcoming, from his father or any other. Robert's decision to send him to Edinburgh arose from no understanding of his needs but from the decision first to make him a doctor, and second that he was wasting time where he was. Erasmus, after three years at Cambridge, was now going to Edinburgh; let Charles accompany him.

As preparation he attended, under supervision, his father's free patients. Robert thought well of the boy's bedside manner, and tempered his earlier outburst with the assurance that Charles was sure to be a successful doctor, having the essential gift of creating confidence. Charles was not completely convinced, but he found it all very interesting, and a decided improvement upon waiting in a class-room for Dr. Butler to register condemnation. He was glad to go to Edinburgh. Robert's reproaches had deeply distressed him. He wanted to dispel them, to "make good," to become a credit if not to his teachers at any rate to himself.

[1] *LL.*, i. 32.

II

It was October 1825 when he and Erasmus made their long journey northward, putting up at the Star Hotel in Princes Street. They entered their names for the medical classes and obtained tickets for the University Library and the Royal Infirmary. On Sunday, at church, they were relieved to escape with no traditional Scots sermon but a mere twenty-minute discourse. Enthusiastically they explored "the Athens of the North," marvelling at its monuments and looking for lodgings, found at last—two bedrooms and a sitting-room—at 11 Lothian Street, one of a row of plain, tall, rather dingy houses. The rooms were at the top of four flights of stairs, but airy and light when one attained them.

Before the end of the month lectures were in full swing. This was no lackadaisical English university, and they were expected to work. The first, compulsory classes began at eight in the morning, and thereafter the student passed rapidly from room to room through most of the day. Charles plunged joyfully in, but his enthusiasm suffered a sudden chill; for he found most of the lectures "intolerably dull." Dr. Andrew Duncan might speak to him "with the warmest affection" of his uncle and namesake who had died in Edinburgh nearly half a century ago and been buried in the Duncan family vault, but that personal interest did not keep the materia-medica periods from seeming inexpressibly stupid. Dr. Alexander Monro, who gave the Anatomy instruction, was a University joke, no fool and yet foolishly content, out of indifferent laziness, to make shift with his grandfather's lectures, the very same that Charles's grandfather had listened to in Edinburgh in 1754, reading them word for word. Some of the students delighted to pelt him with dried peas as he drawled, "When I was in Leyden in 1719," but Charles was merely disgusted: "I dislike him and his lectures so much, that I cannot speak with decency about them."[1] Most students paid for more efficient extramural teaching, but Charles did not, thus missing that practice in dissection whose omission he always afterwards lamented. Instead, he and Erasmus spent more time than most in the

[1] *ML.*, i. 7.

University Library, and to his life's end, even after his friendship with Huxley, he was definite that "there are no advantages and many disadvantages in lectures compared with reading."

Disgust of another kind attended his two early visits to the Infirmary operating theatre; chloroform was then unknown, and he was so upset that he had to leave the room. After the second experience he refused to go again, and it was long before he could rid his mind of the memory. Perhaps it was no absolute bar to his prospective profession, but a setback it must have seemed. Some of the clinical lecture cases, too, made him queasy, and despite an attraction to others he could not feel that his interest was as compelling as it should have been. It was very disturbing. If he didn't especially want to be a doctor he *did* want to please his father, who was so eager that his sons should follow in his footsteps.

So work went on with dwindling enthusiasm and many thoughts of home and Maer; his and Erasmus's demands for news were so frequent and imperative that Caroline declared them quite troublesome. He wanted to know, in January 1826, his age next birthday, for if it was seventeen, as he rightly supposed, then he would have to spend a year studying abroad, since he could not take his degree at Edinburgh till he was twenty-one.

The end of the first year and the summer vacation came as joyous release. First he went tramping in North Wales with two friends, knapsack on back; they climbed Snowdon and felt themselves mountaineers. Then he accompanied a sister on a riding tour, still among the mountains. But when the end of August came, he must be off to Woodhouse and to Maer, not to miss a moment of the shooting, keeping eager record of every bird he brought down, and at Maer deserting the family day after day to tramp the heath with Wilcox. It was seventh heaven but for the niggling fear of his father's displeasure. "How I did enjoy shooting!" he wrote in his autobiography, "but I think that I must have been half-consciously ashamed of my zeal, for I tried to persuade myself that shooting was almost an intellectual employment."[1] Plainly he felt at once guilt and the need for self-justification.

With the autumn Erasmus went to Cambridge to take his medical degree and Charles returned to Edinburgh alone, thrown

[1] *LL.*, i. 43.

the more upon his own resources. His studies, which now included midwifery, practice of physic, and natural history, were no more interesting. He experienced a sudden jolt of attention on learning of the transportation of boulders by icebergs over great distances, for here at last was light upon the mystery of the Shrewsbury "bell-stone," but Robert Jameson's lectures on geology determined him to avoid the subject at all costs, and most of his life he remembered "that old brown dry stick" as the very type of the tedious lecturer. Jameson was the British leader of the Wernerian or Neptunist school of geologists which asserted the aqueous as against the volcanic origin of the basaltic rocks, in the great geological controversy of the day; and he would inveigh with all the acrimony of the partisan against the idiots who dared to think otherwise. The waters of the Lord covered the earth, and in the waters the foundations of the earth were laid; those who prated of fire as co-equal with the waters, let them be consumed in their own flames for the fools they were. The Vulcanists on their side were as blind and narrow, and Charles called a plague on both their parties. Water might drown fire, or fire burn water—it was all one to him, for both were identical in dullness.

Marine zoology, on the other hand, had at least its incidental pleasures—rambling along the coast in search of specimens, trawling with the friendly Newhaven fishermen, sailing out across the Firth of Forth to the islands and even to the far coast of Fifeshire. Once he was adventurously benighted on Inchkeith and had to seek shelter in the lighthouse. Among fellow zoologists he found some of his friends, mostly men older than himself. There was the lean, kindly, and emotional but restrained (Charles said prim) John Coldstream, working conscientiously for his degree though convinced he was but briefly for this world, yet "caring not if my peace is made with God. See to it, then, O my soul, and trifle not off the remainder of thy trust in vanities." [1] There were also Dr. Robert Edmund Grant, an experienced and skilful zoologist in his middle thirties, and very soon to become the first professor of zoology in University College, London; Ainsworth, whose enthusiasm and joviality offset his geological leanings; the botanist Hardie, a very promising young man; and William Macgillivray, the assistant-

[1] *Biography of Coldstream*, 12.

curator of the University Museum, who gave Charles rare shells for his collection and stimulated his awakening interest in birds.

Of them all Grant was, for Charles, the significant figure, for it was he who, walking one day in 1826 beside the tidal pools of the Forth, "burst forth in high admiration of Lamarck and his views on evolution."[1] Charles listened, quite astonished, but saying nothing. He did not know what to say, for though he had read his grandfather, and realized the similarity of Lamarck's views, he had not thought much about the matter. Twenty years before, when Carlyle was at Edinburgh, the *Zoonomia* was a common topic of controversy—the students would debate whether man derived from the oyster or the cabbage—and Dr. Duncan was not the only man in the University who still remembered its author and his works. But though the subject hovered in the air, stimulated by Lamarck's larger exposition and the tentative assertions of Goethe, Isidore Geoffroy Saint-Hilaire and others, it found little general favour, and only the rare bold enthusiast would sometimes appear to declare, as Grant now, his "belief that species are descended from other species, and that they become improved in the course of modification."[2] Charles had evidently thought such notions the especial property of clever but cranky grandfathers; he suddenly perceived that they might equally be held by sensible-seeming persons of one's own generation, individuals who already held one's liking and respect. He did not at the time think much more of the matter, but this was, one feels, the moment in which, as never before, evolution became for him a living and potentially credible doctrine.

Grant probably was also responsible for his joining, on November 28th, 1826, the Plinian Society, a student group holding small weekly meetings, to discuss natural history, in an underground room in the University. Grant was secretary, Coldstream one of the five presidents, and Charles soon became a member of the council. He attended regularly all that winter, speaking occasionally, once—but his remarks were not preserved—on specific characters and the principles of natural classification, and on March 27th, 1827, communicating two small but interesting discoveries of his own relating to the ova of the Flustra, or sea-mat, and of the skate-leech.

[1] *LL.*, i. 38.   [2] *Origin* (1872), xvii.

Finally, it was with Grant that he attended meetings of the Wernerian Society, at which too papers on natural history were read and discussed, the subjects ranging from mummified cats and the origin of meteoric stones (generated in the atmosphere, said some) to the ancient Jewish breeding of domestic poultry and the consistency of Cuvier's interpretation of the Geological Deluge with Mosaic testimony. He was there on the night — December 16th, 1826 — that Audubon's account of the American turkey-buzzard was read, the author himself appearing on the platform to demonstrate his method of fixing newly killed birds in desired attitudes, which must have interested Charles, then taking lessons in taxidermy. All that winter this strange man, who wore satin breeches and ruffled shirt on the American frontier but in Europe affected a picturesque backwoods costume, was the sensation of the city. Charles surely saw his great exhibition of drawings of birds, four hundred of them, a life-work, in the Royal Society building, though he missed the unfortunate lecture on the rattlesnake which gave rise to so many doubts of Audubon's honesty. But the mere appearance of the man, his wolf-skin coat, shoulder-length raven locks, handsome bird-like features, his suggestion of other lands and ways of life, his air of being Rousseau's natural man incarnated, together with the fascination of his brilliant drawings, may well have given new impulse to Charles's pleasure in natural history and his innate youthful desire to travel. He was also taken, by Leonard Horner, a solemn but amiable acquaintance of the Derby Darwins, to a meeting of the Royal Society of Edinburgh, where he was most impressed to see Sir Walter Scott.

Definitely by this time his desire to be a doctor, if ever more substantial than a reflection of his father's hope, had completely evaporated. He was doing less and less work. Medical subjects bored him. Once again, in May 1827, he escaped into holiday, the more eagerly perhaps that the time was spent in London rather than Shrewsbury. He dined at the Hollands' house with cousin Harry, and was shocked to hear Dr. Holland declare a whale's blood cold and to see him eat with his knife. Caroline was also in town, and there were gay visits to the theatre. But the great event was the expedition—apparently Charles's only cross-Channel trip—to the Continent with Uncle Jos, who was going to Geneva to bring home young Fanny and Emma from a stay with their aunt Jessie Sismondi. Caroline went too. The French weather was unkind, but improved by the time they reached Paris, where Charles had regretfully to turn back for his last term in Edinburgh.

He was now thoroughly at a loose end. It wasn't only that he didn't want to be a doctor, he didn't positively want to be anything in particular, nor did he see any especial necessity, for he had lately learned, possibly from Erasmus, that he would inherit at least enough money to support him in that moderate comfort which was all his unpossessive indolence demanded.

He had confided his general distaste to Caroline and Susan, and his father could not but realize his lack of eagerness. Robert must have been dismayed, angered, and distressed. Two years had been wasted. Well, if not doctor, what then? He could not stand by and see his son become no better than an idle sportsman, following vagabond at the heels of dogs, gossiping perpetually of shots and bags. The most obvious alternative occurred to him—the Church, where any man, wise or foolish, studious or sporting, could find a niche according to his talents. That summer the proposal was put to Charles—unexpectedly, for he asked time to think it over. There was much in the life of a *country* clergyman that appealed to him, but he had to determine also how far he could conscientiously affirm all the dogmas of the Established Church, for though he had no doubt of "the strict and literal truth of every word in the Bible," the broad Unitarian thinking of his family had inevitably affected him.

Throughout the vacation he pondered the matter over with the assistance of standard works on divinity. It was just at this time that Mackintosh said of him, "There is something in that young man that interests me," a remark Charles treasured with a pride revealing his need in this period of praise to correct his growing sense of inferiority.

Some such growth was inevitable in a youth of eighteen only too well aware of his lack of direction, whose only *passion* (for shooting) was despised by the elders in whose values he acquiesced. He did acquiesce because his mind was still superficial, neither intellectually nor imaginatively aroused, and when at length he agreed to his father's new suggestion, that was, like every action in his life so far, a negative acquiescence too. He had no definite objection to entering the Church; therefore he would make ready to do so forthwith. At the worst, Edinburgh was outworn, while Cambridge—where he was to proceed—would be new, unexplored.

One small obstacle had to be overcome, for since leaving school he had turned his back so resolutely on classics as to have forgotten even the Greek alphabet. All that autumn he crammed at The Mount with a tutor, and went to Cambridge accordingly only at the beginning of the Lent Term in the New Year of 1828.

III

All his life he was to remember Cambridge with a deep pleasure and affection. His years there were, he said in his autobiography, "the most joyful in my happy life; for I was then in excellent health, and almost always in high spirits." [1] He also asserted that, over the same period, "my time was wasted, as far as the academical studies were concerned, as completely as at Edinburgh and at school." [2] The two comments meet in a third, upon some of the gayer gatherings he attended: "I know that I ought to feel ashamed of days and evenings thus spent, but as some of my friends were very pleasant, and we were all in the highest spirits, I cannot help looking back to those times with much pleasure." [3]

That is the dominant note of all these years. Charles was sent to Cambridge to save him from becoming "an idle sporting man"—his life there was in fact the idle sporting man's last fling. Charles Darwin the scientist was not born either before or during the Cambridge period; at best the womb was preparing, and the vital semination occurred, if with a Cambridge colleague, only after he had left the place. So far as he had a purpose, it was an imposed one still, it meant nothing to his imagination. His freely chosen friends and pastimes, on the contrary, meant at least something, which is essentially why these, in themselves and in their consequences, seem in retrospect vastly more significant than any of his academic occupations.

Robert, one feels, might have reflected that if Edinburgh could not wean a sportsman from his idle ways, Cambridge was much less likely to do so. If there was a tolerably sharp division, more then than now, between the studious and unstudious, the

[1] *LL.*, i. 56.   [2] *LL.*, i. 46.   [3] *LL.*, i. 48.

reading and non-reading men, who should say which side of the fence any one individual would fall? Charles's natural inclination would almost inevitably be to the Sporting Men. He was not a Lounger or a Dandy, neither a Buck nor a Blood, the river did not draw him and he had no trace of the bitter viciousness of the Varmint Man. But among the Nimrods he could scarcely fail to find himself. Perhaps an avowed divinity student might hesitate to make himself over-conspicuous by jumping too suddenly into the long tight-fitting white trousers, the fancy waistcoats, and the low-crowned wide-brimmed hat which the set just then regarded as the very latest thing, but a hack for hunting, a gun, and even quite frequent visits to the Newmarket races were not likely to be frowned upon by anyone, least of all at Christ's, where his name had been entered, and where he took up residence after a preliminary two terms in rooms over Bacon's tobacco shop, then adjacent to Sidney Sussex College.

"More like a Christian than a dumb!" admired the old lady confronted by an intelligent dog, but Christ's in those days preferred a dumb to a Christian. It was a sporting college. Shaw, the senior tutor and a University proctor, was a keen judge of a horse, and the Newmarket course was as familiar to him as his college lawns; it gave him a homely feeling to see his charges also at the race-track. He turned an understanding eye on the general diversions of gentlemen, and was obligingly easy in his academic demands, few lectures being compulsory and few examinations stringent. Chapel services must be attended, but they combined speed and brevity with reverence in that tactful manner which is the exclusive possession of the Church of England. The organ was silenced; the Dean read alternate verses of the Psalms and prayers, daring any to respond; lengthy lessons were docked as short as a terrier's tail. Naturally the college drew like a magnet just those young men of comfortable means who wanted to enjoy their University years without working too hard.

It was, in short, the very place that Charles should have avoided.

He soon found himself at home, falling very quickly into a thorough sporting set which even included "some dissipated low-minded young men." He became a member of the Gourmet

—or Glutton—Club. There were long-drawn-out dinners of an evening. They sang. They played cards. They actually "sometimes drank too much"—once at least Charles discovered himself on the thither side of insobriety. There were other "very gay" gatherings. Francis Darwin, sixty years after, ventured the opinion that his father had "exaggerated the Bacchanalian nature of these parties," but that is in itself to magnify their character. Perhaps he did drink a little, bet a little, gamble a little. (That he sang is less probable, for though he liked music he had no ear for a tune.) But it was all essentially harmless. Whatever became of the low-minded young men, many members of the Gourmet Club were later to attain respectable eminence in Army, Church and civil life, and while two of Charles's more especial friends, Herbert and Whitley, both St. John's men, gave frequent large and lively parties, the respect he always expressed for "their excellent understandings and dispositions," and the general tone of his letters to Herbert, make it clear that they were far from being either rakes or idlers. Herbert, a Welshman, loved music, Whitley's enthusiasm was for art; Charles learned something from each. The point is that in a group of varied characters he found his friends among the "men of higher stamp." Had his character been a weaker one, it might have been otherwise, for he was still adrift, without orientation to either past or future, and yet a readily likable young man, amiable, sympathetic, full of fun and enthusiasm, his rather round but friendly features eager with interest in all that the passing changeful days might set before him.

His most important if not his most intimate friend, however, was his second cousin, William Darwin Fox, also of Christ's. He was the grandson of William Alvey Darwin, elder brother of Erasmus of Lichfield. He and Charles first met at Cambridge, but Robert knew the family from his visits to the old Doctor's widow Eliza, near Derby, where the Foxes had their home.

Erasmus had qualified for his M.B. and gone from Cambridge before Charles came; Fox was in his last year, leaving in the following June. So easily they might have missed each other, had Robert decided on another year at Edinburgh, or had work and pleasure kept them only a little longer apart. Then all might have been very different, for Fox was an essential link in the chain of his life. It was, Charles later said, *wholly* Fox's enthusiasm and handsome collection of butterflies which reconverted him to entomology, and on a new if scarcely less childish level of approach. It was no doubt Fox too who persuaded him to attend the Rev. John Steven Henslow's botany lectures, and Fox certainly who got him his first invitation to Henslow's Friday-evening gatherings.

Charles, it is true, might have come to "bug-hunting" anyway, for it was a fashion, almost a fever, among the Cambridge undergraduates just then, a kind of sporting contest to see who could boast the greatest number of varieties. That, he confessed, was all he cared for then; quantity and rarity were his criteria—scientific arrangement was quite beyond his care and ken. Purchase was reckoned legitimate, and prices were so good that not a few "lower class persons" made a summer living by collecting and disposing of their prizes about the colleges. Charles soon had his own private scout, later angrily dismissed for giving "first pick" to a rival. He was not willing for rare specimens to pass into the hands even of his closest friends. What glory it was to pounce upon *Panagæus crux-major* at the foot of a willow stump! What greater glory to discover, in an entomological text-book, a capture attributed to "C. Darwin, Esq."

Occupied thus, he did a minimum of work in these first terms. With Belles-Lettres—classics—he went no deeper than the compulsory lectures, "and the attendance was almost nominal." Christian Theology and Moral Philosophy he kept for a rainy day. Upon Mathematics he made a more determined attack, going to work at the subject in his first Long Vacation with a tutor at Barmouth, a favourite resort for Cambridge reading parties; but soon tiring and stopping short of the door of understanding to ramble, with a keen eye open for the bright flicker of a wing, across the close valleys and up the green rising slopes of the hills, stopping to turn over likely stones and popping worthy specimens into pill-box or alcohol-bottle.

It was on this holiday that he became especially intimate with Herbert, affectionately nicknamed Cherbury. They went everywhere together, walking daily in the hills, fly-fishing and sailing in the estuary and seaward to the tidal bank of Sarn Badrig. Charles loved this combined walking, talking and collecting. All the solemn and cheerful chatter of undergraduates in every age passed between them; the world might be destroyed and re-created in any hour. He damned the slave trade and the oppressors of the Poles, and nearer home Wellington's High Tory government as scarcely less a tyranny. There were the troubles in Ireland, arguments for and against Catholic Emancipation, war between Russia and Turkey, the new railways which were a topic of controversy everywhere. . . .

But neither new hobby nor new friendship could keep him from his older love, and he left Barmouth early to be at Woodhouse, Maer, and then Fox's home, Osmaston, for the September shooting. As ever, he enjoyed himself, placid content flooding his mind and drowning all thoughts of work left undone, of his future, of his father, stilling all doubts, as the parties walked hungrily, happily, homeward in the bright autumn evenings. Ah! if life could be all one autumn evening, a good day's shoot behind one, a good dinner ahead, the pleasantest, the dearest, companions about one, the dogs snuffling at one's heels, the comfortable weight of a gun beneath one's arm.

Nevertheless, the old could not submerge the new. At Woodhouse Sarah Owen chaffed him mercilessly for this new attention to smaller game, and from Osmaston he wrote joyfully back to Herbert, still at Barmouth, begging for further specimens of certain beetles which, he had just learned, were "some of the rarest of the British Insects, and their being found near Barmouth, is quite unknown to the Entomological world: I think I shall write and inform some of the crack entomologists."[1] So full was the letter of this "success," that only one brief paragraph noted his "contemptible" bag, a mere seventy-five head and a brace of black game in the first week—"but there are very few birds."

Back at Cambridge entomology was still primary, attendance at lectures desultory—but, Fox gone, Henslow was moving towards the foreground of the picture. Charles had been to some of his botany lectures in the spring, and found them interesting, but not sufficiently so to compel him to the subject. He preferred the incidentals of Henslow's course to its particulars: the occasional outdoor expeditions, which all might and many did join, sometimes afoot, sometimes by barge down the river, or best of all by four-horse coach to Gamlingay, bowling swiftly

---

[1] *LL.*, i. 172.

and joyfully along the level Fen-ward roads, with now and then a brief pause at some favoured country inn for a meal and to drink the toast of *Floreat Entomologie*; the soirées too at Henslow's home, more frequent and also more select, the favoured meeting-place for all in any way interested in natural history, professors, fellows and undergraduates coming together in an easy freedom to display and discuss any new specimens or views.

Henslow was a good host and a good friend, among the outstanding Cambridge figures and one of the most attractive, a handsome man of grave but tolerant expression, still in his early thirties though he had been Professor of Mineralogy for five years before his appointment in 1827 to the Botanical Professorship which he held till his death in 1861. His enthusiasm had made botany one of the most popular of academic subjects, and his treatment was reckoned so attractive, his personality so pleasing, that the same individuals would attend his courses year after year. Erasmus had often spoken of him with enthusiasm, as one who knew everything, and contact did not disillusion Charles. Henslow was nice to everyone, never presuming upon his position, or even his knowledge. Charles came to him one day bursting with news of an astounding botanical discovery. It was actually the commonest fact, and any other botanist, Charles said, would have laughed him to scorn. But Henslow "agreed how interesting the phenomenon was, and explained its meaning," making him "clearly understand how well it was known" without leaving in his mind any sense of mortification. Henslow acted better than he knew. A rebuff, the humiliation of mockery, might have turned Charles from botany and natural history, as well as from Henslow himself, for ever. As it was, their intimacy grew, though in this earlier period somewhat slowly, for it was not till his third year, when the older man was tutoring him for his final examination, that he occupied the place of favourite pupil, so frequently with the other that he came to be known about the University as "the man who walks with Henslow."

Meanwhile, entomology held the field. In February 1829 he was spending some days in London, informing "some of the crack entomologists" of his Barmouth trophies. He called on the Rev. F. W. Hope, first Professor of Zoology at Oxford and whom he afterwards called "my father in entomology," and had two engrossed days with his collection. He also met J. F. Stephens, whose cabinet proved even more magnificent, and who sent him away ecstatic with the present of no less than 160 new specimens. He spent the Sunday with Dr. Holland, rode in the park, and visited the Royal Institution, the Zoological Gardens, the Linnean Society, and "many other places where naturalists are gregarious."

While collecting still held first place, Cambridge had other pleasures too, and more sedate than earlier. Under Whitley's tuition he became a frequent visitor to the Fitzwilliam Gallery, the robust naked charms of Titian's Venus winning his especial admiration, and he bought what he hoped were choice engravings to decorate his new rooms at Christ's, one storey up on the south side of the first court, handsome panelled apartments looking out over the large green circle of the lawn. With Herbert, or alone, he would go often to listen to the anthem in King's College Chapel, under whose high and splendid roof Christ's counterpart sank to cellar-like insignificance, or to concerts to rejoice in the dominating symphonies of Beethoven and Mozart. A more notable luxury was to hire choristers to sing to him and his friends in his rooms, and all these things gave him great emotional delight, for all that he had so little ear for a tune that he could not recognize *God Save the King* when played out of time. He also liked to have Shakespeare's plays read aloud to him, and himself read frequently in the poems of Milton, Byron, Gray, and Shelley.

So far—all of 1828 and into 1829—there had been little mention of the purpose for which he had come to Cambridge, but now of a sudden he was expressing interest in the divinity books Fox was reading. What did Fox think of them? Evidently Charles was at least contemplating his projected vocation, and feeling some doubts, for he earnestly asked Herbert whether he (Herbert) could answer in the affirmative to the question in the ordination service whether he was "inwardly moved by the Holy Spirit." No, Herbert answered, he couldn't; and Charles agreed, "Neither can I, and therefore I cannot take orders."[1]

That might have seemed to settle the matter, but these doubts were voiced to no one else, and presently, if not stilled, were set aside. Probably Charles's hesitation was entirely negative; simply, he felt no inner impulse which he could accept as holy "to take upon you this office and ministration, to serve God

[1] *LL.*, i. 171.

for the promoting of His glory, and the edifying of His people." He cannot, it is clear, have moved far, if at all, upon the road to heterodoxy, and his real intellectual humility must have kept him from declaring himself against the piety of the respected Henslow, who was rumoured to hold odd religious opinions, but whom Charles found so whole-hearted in his orthodoxy as to "be grieved if a single word of the Thirty-nine Articles were altered"!

There were signs that he was finding Cambridge dull this second year, despite occasional parties and outings. He daily read a little, rode a little, walked a little. Somehow he wasn't too well, suffering intermittently a disagreeable eczema of the lips which could be so troublesome as to make him, that Long Vacation, turn back alone to The Mount from an entomologizing expedition through North Wales for which Mr. Hope from London had joined him at Shrewsbury. But there was Woodhouse to look forward to, and after that Maer. "For the rest of this summer I intend to lead a perfectly idle and wandering life."[1] There was also a Music Meeting at Birmingham, "the most glorious thing I ever experienced," he wrote to Fox, for Madame Malibran, lovely, gifted, youthful, sang there. "Nothing will do after Malibran, . . . a person's heart must have been made of stone not to have lost it to her." His sisters and cousins were there too, and they had a gay time. Too gay for him, for he reported in words which were in later years to become woefully familiar: "It knocked me up most dreadfully."[2]

That third winter he was reading for his Little-go. Mr. Shaw was no longer lord of Christ's, and Mr. Graham, his successor, was no Newmarketer. "Graham smiled and bowed so very civilly, when he told me that he was one of the six appointed to make the examination stricter, and that they were determined this would make it a very different thing from any previous examination, that from all this I am sure it will be the very devil to pay amongst all idle men and entomologists."[3] Nevertheless Charles put off most of his studies to the last few weeks, and the date of the examination found him tired and nervy, conscience-stricken for time wasted; but for all that there was little of repentance in his exultation when his success was declared. At once all his plans were for pleasure, for expeditions into the

[1] *LL.*, i. 179. [2] *LL.*, i. 180. [3] *LL.*, i. 179.

Fens, beetle-hunting, for a trip to London to call on Hope again, and Erasmus too, and to hear an opera. His new beetle-cabinet had arrived, he had two new and promising pupils in "the science"—if only Fox would pay his anticipated visit to Cambridge, what fun they would have together.

Among his newer and graver friends was Henslow's brother-in-law Leonard Jenyns, vicar of the Fen parish of Swaffham Burbeck, where Charles would visit him; he was a forbidding fellow at a glance, with his grim, sarcastic face, but acquaintance soon found him amiable, even jovial. He too had an entomological cabinet, and an ambition to make a complete collection of all the insects of Cambridgeshire; he was an ornithologist as well, with his own small "museum" of birds and smaller animals. Whewell—"Whuffler" to the irreverent—he met also at Henslow's, and would sometimes walk home with him at night when the gatherings were over, listening in respectful silence to the grave, vigorous dogmatic discourse of this Professor of Mineralogy of whom Sydney Smith said that science was his forte and omniscience his foible. There were others too, all older men, and on their private country expeditions Charles was often their only junior companion, a fact from which he afterwards inferred that "there must have been something in me a little superior to the common run of youths."[1] It was not apparent to him at the time; he enjoyed their company and solemn-jesting talk, but all their example did not make him see himself yet as embryo scientist.

Again summer, and again fly-fishing and beetle-hunting in North Wales. And again, of course, the shooting which gave him such unfailing pleasure. "Upon my soul, it is only about a fortnight to the 'First,' then if there is a bliss on earth that is it."[2] He had, certainly, occasional qualms, as when one morning at Woodhouse he came on a still-living bird which had lain in evident agony since the previous day. The thought of it troubled him all the rest of his holiday, and eventually, when back in Cambridge, he came to the resolve never to shoot again. It was a resolution which, as Herbert said, testified to his tender heart—but also to the shortness of his memory, for he had quite forgotten it before the following September. If he did not shoot then, it was for quite another reason.

In the next few months he was working for his B.A. examination,

[1] *LL.*, i. 55. [2] *LL.*, i. 167.

doing more hard studying than in the whole of his previous eight terms. He "brushed up" his Greek and Latin, and returned to algebra with detestation and to Euclid with some pleasure. He also read Paley—*Evidences of Christianity, Natural Theology, Moral Philosophy*—with close attention, and liked him, exactly as he liked Euclid, for his logical argument. Rarely, he said in 1859, had he ever admired a book more than he did the *Natural Theology*—he knew it practically by heart, and found its thesis both attractive and convincing. It was a work written, some have said, in direct reply to the *Zoonomia*, containing some of the earliest anti-evolutionary arguments (the theory of "appetencies," stimulating to change from within, was specifically dismissed as having no evidence to support it), but this aspect probably never entered Charles's head, for Erasmus's name and work were mentioned only incidentally, and though other commentators have thought that the *Natural Theology* first convinced Charles of the importance of adaptation in the organic world—presumably by its stress upon wide structural similarity with individual variation—clearly his mind in this phase was not seized by the evolutionary idea, for or against, while the degree to which he took Paley's premises on trust showed that *he had not at this time developed to the point of a fundamental questioning in any field.* He had as yet no intellectual individuality. That is the essential fact of all this period.

Too much of the other work wearied and disgusted him. How bored he was, how dispirited! He had not "stuck a beetle this term." Henslow's parties, even his hour of tutoring, were oases in this desert. Reading made him "quite desperate," preparing all his subjects simultaneously was "intolerable." "Really I have not spirits or time to do anything." [1]

Then came release. He sat for his examination in the New Year of 1831, and before the end of January knew that he had passed, not indeed with honours but without dishonour, for Christ's men, under Graham's goading, had done well, and he was tenth upon the list.

IV

But even if he had earned his degree, he had still to complete his qualification by two more terms of residence. So he remained

[1] *LL.*, i. 182.

"a Christian" till the summer, a welcome interval before embarking on his distinctively ecclesiastical preparations. It was, if this can be said of one link in a chain more than another, the greatest good fortune of his life. He had no set studies now, and Henslow, of whom he was seeing more than ever, took the opportunity to press upon him improving scientific books.

Two of these he declared in after years to have influenced his life as no other printed works, and though he was always prone to exaggeration, the claim may be allowed. The first—from which, he said, stemmed the whole subsequent course of his life—was Alexander von Humboldt's *Personal Narrative of Travels to the Equinoctial Regions of America During the Years 1799-1804*. Read to-day, it is apt to seem by modern standards, or even by those of Charles's own *Journal of Researches*, a rather prosy if capaciously informative account, but Charles fell instantly under its spell and never recovered; to the end of his life Humboldt remained for him "the greatest scientific traveller who ever lived." Almost the very opening took him back to schoolboy ambitions—"from my earliest youth I had felt an ardent desire to travel into distant regions, which Europeans had seldom visited"—and the second chapter, on the stay at Teneriffe and an ascent of the famous Piton or Sugar-Loaf peak, excited him beyond all bounds. He copied out the most thrilling passages, and on one of their most intimate country excursions read these aloud to Henslow and other friends with infectious enthusiasm. "We anchored after several soundings, for the mist was so thick that we could scarcely distinguish objects at a few cables' distance; but at the moment we began to salute the place, the fog was instantly dispelled. The peak appeared in a break above the clouds, and the first rays of the sun, which had not yet risen on us, illumined the summit of the volcano." So to come to it at dawn, its dazzling white cone projected upon "a sky of the purest blue, while dark, thick clouds enveloped the rest of the mountain," was to approach a paradise of wonders. "We easily conceive how the inhabitants, even of the beautiful climates of Greece and Italy, might fancy they recognized one of the Fortunate Isles in the western part of Teneriffe." To see it with their own eyes! Charles exclaimed, and some of the others, even Henslow, echoed his eagerness. Gaily they discussed the possibilities of such a trip. Later it occurred to Charles that they were less serious than they seemed, but he had no doubts then. He obtained an introduction to a London merchant who might have information of vessels sailing thither, all through April and May he was able to "talk, think, and dream" of little else, and even as late as July he was trying to learn Spanish—he found it "intensely stupid"—and hoping that Henslow's "Canary ardour" remained undimmed. "I read and re-read Humboldt; do you do the same? I am sure nothing will prevent us seeing the Great Dragon Tree."[1]

The other book, less striking in its immediate effect, was Sir John Herschel's *Preliminary Discourse on the Study of Natural Philosophy*, a cleanly planned and written systematic survey of the principles of scientific study, with some account of its static past and active present. The author stressed the limited nature of scientific inquiry—as concerned with the how, not the why, the action, not the origin, of things—and also its consistency with religious truth (truth was one, and great, and must prevail). He declared its interest, the possibility that "any well-informed person" might, more of will even than ability, contribute something "essential" to the common stock of knowledge, and also its practical benefits. He wrote too of its main divisions, of its inductive method, and of its high aim as distinctively "principles, not phenomena—laws, not insulated independent facts," but the earlier points meant, one suspects, the more to Charles at the moment. He was already aware of the interest at least of associated activities, he could even feel himself to have contributed something, if a mite only (*captured by* "*C. Darwin, Esq.*"); was there not, in this promise of benefit, a possible meeting-place for his pleasure *and* his father's hope of a useful life for him? Needs must turn clergyman if the devil drives—Henslow, Jenyns, were both in orders, but, in their degrees, "natural philosophers" also. If only one could have the sugar without the pill!

He was in some such mood when Henslow sought to turn him to geology (whose high importance among the sciences Herschel had particularly stressed) and, to seal these persuasions, introduced him to Adam Sedgwick, Cambridge professor of the subject, late president of the Geological Society and soon, in 1833, to be president of the British Association. One wonders

[1] *LL.*, i. 190.

a little that Charles had not met him before, for he was an outstanding University figure and an old colleague of Henslow, co-founder with him of the Cambridge Philosophical Society, a man of personality whose character appeared in his face—dark powerful features, grim but vigorous, dominated by almost Darwinian forehead and heavy brows. He is said to have been elected to his chair not for what he knew—he had at the time scarcely studied the subject at all—but for what he was, yet those who appointed him made no mistake; he was a good teacher and his field-work is part of geological history. He began his labours as ardent a Wernerian as Robert Jameson, but by 1831 had considerably modified his earlier dogmatic denial of the Vulcanists ("For a long while I was troubled with water on the brain, but light and heat have completely dissipated it." [1]), though he was by no means ready to admit the Uniformitarianism of Hutton and Lyell, the view that geology was to be explained not by sudden creation or extraordinary catastrophe but by the processes—subsidence, elevation, denudation, erosion, accumulation, volcanic or marine action, and the like—to be seen at work to-day. He deemed exact observation more important than any theorizing, and Lyell's *Principles of Geology*, the first volume of which had appeared early in the previous year, he thought altogether too speculative and extreme in its Huttonian fervour.

Charles was attracted both by Sedgwick and by geology. The latter, as Herschel had implied, was still in its first bright youth, and Charles seems to have felt at once now—what at Edinburgh he never glimpsed—its possibilities of adventurous discovery. He told a friend: "It strikes me that all our knowledge about the structure of our earth is very much like what an old hen would know of a hundred-acre field, in a corner of which she is scratching." [2] It was the simile of one who felt a world before him where to choose, and out of that sense of illimitation there came, perhaps, both his eagerness and that ever-living awareness of the *not-known* which characterized all his investigations for the rest of his life.

He received his degree on April 26th, but remained in Cambridge until June, and when he returned to Shrewsbury geology was more immediate than Teneriffe, of which he still

[1] *LL. Sedgwick*, i. 284–5. [2] *LL.*, ii. 348 n.

chattered till he wondered if his friends wished him there, though the trip had been postponed till next year. Throughout July he was amateurishly attempting a map of the strata about Shrewsbury—amusing himself too with even wilder speculation—in preparation for a working trip with Sedgwick through North Wales in August. Henslow had arranged it. At the last moment Sedgwick was asked to go instead with his friend Roderick Murchison. His biographers have thought it "unfortunate" that he refused to do so, for his later estrangement from Murchison might thereby have been prevented. In the larger scientific perspective, however, their regret seems ungrateful, for not only was the experience of the trip itself of importance in forwarding Charles's interest and knowledge, but one incident which preceded it was cardinal.

Sedgwick arrived at The Mount on August 3rd, and Charles lost no time in reporting to him (as to headquarters) the local discovery—by a workman—of a tropical shell in a gravel-pit. It was, if authentic, no less than a geological marvel, and as such Charles, collector of rare insects, rejoiced in it. Sedgwick took a different view, thinking it must have been left there by some passer-by wanting to be rid of it. Dismissing the matter, he added that had it really been embedded there, "it would be the greatest misfortune to geology, as it would overthrow all that we know about the superficial deposits of the Midland Counties."[1] Charles was astonished, flabbergasted; it seemed to him like denying a wonder just because a book said it couldn't be so. That is the view of the incident actually taken by one of Charles's American biographers.[2] Here on one hand was Charles, the true scientist, taking the fact as it came, on the other Sedgwick, the academic mind, rejecting the evidence because not strictly according to rule. It is a view which has a certain application to the minds of the two men, but which is a definitely misleading interpretation of the particular case. For Sedgwick, as it happened, was right, Charles quite wrong. More than that, it was a distinctive turning-point in Charles's mental development, and towards, not from, what Sedwick stood for. Quite suddenly, even as he blinked at the other's dictum, he perceived what Herschel, through his *Natural Philosophy*, had been driving at in his insistence upon "principles,

[1] *LL.*, i. 57.   [2] Henshaw Ward.

not phenomena," as the scientist's essential object of research. It was a decisive moment, for instantaneously all Charles's earlier attitudes to science were stood upon their heads. "Nothing before had ever made me thoroughly realize, though I had read various scientific books, that science consists in grouping facts so that general laws or conclusions may be drawn from them." [1] Hitherto he had been a collector only; now, on this August evening, Charles Darwin the scientist was truly born.

Sedgwick spent two days at Shrewsbury, talking so much of his bad health that Robert set him down a hypochondriac. On August 5th they set forth, Wales greeting them with a drenching thunderstorm. All that day along the vale of the Clwyd the hill-tops were hidden in cloud. Next day the weather was beautiful, and they reached Denbigh, subsequently making a traverse to St. Asaph, and so by gig to Conway. Tramping between or across the hills, Sedgwick would send his companion along a parallel route, directing him to make notes and collect specimens, an apparent mark of confidence which gave Charles great pride and pleasure, for he took his instructions at their face value, and only afterwards considered that possibly Sedgwick was more anxious to teach than to learn.

He was also in later years to recollect how the two of them passed many hours in Cwm Idwal, carefully examining the rocks for fossils, yet never once as much as noticing the marks of glacial action which a decade or so later he found everywhere telling their story as plainly as "a house burnt down by fire." The fact was that they were not looking for such marks, and only very rarely will the eye perceive what the mind does not anticipate.

Three weeks they spent together in this way, saying good-bye at last at Capel Curig, beyond Bangor. Charles set off alone to tramp across the mountains, by map and compass, to Barmouth, where some of his Cambridge friends expected him. They did not see him long, though, for September was at hand. The good resolutions of the previous autumn had evaporated and he would, he said in his autobiography, have thought himself mad to surrender "the first days of partridge-shooting" for any science. He arrived back in Shrewsbury, accordingly, on Monday, August 29th, on his way to Maer.

A letter was waiting for him. . . .

[1] *LL.*, i. 57.

CHAPTER V

OFFER OF ADVENTURE

I

IT was addressed in unfamiliar handwriting, and, tired with travelling, he took it without curiosity from his welcoming sisters. It had been waiting since Saturday, they said. Asking questions, answering them, he opened it, discovered that it was from Mr. Peacock, mathematical tutor of Trinity College, Cambridge, found an enclosure from Henslow, tried to take in the astounding, incredible contents. Henslow first: ". . . the offer which is likely to be made you of a trip to Tierra del Fuego, and home by the East Indies . . . a Naturalist as companion to Captain FitzRoy . . . the voyage is to last two years . . . I assure you I think you are the very man they are in search of. . . ." [1] Then Peacock: ". . . you may consider the situation as at your absolute disposal . . . the South Coast of Tierra del Fuego . . . the South Sea Islands . . . the Indian Archipelago . . . the ship sails about the end of September, and you must lose no time . . . the greatest anxiety that you should go. . . ." [2]

It was breath-taking. His mind had held no anticipation, save of Maer. Even Teneriffe had lain shadowy on the horizon of next year. Yet what was Teneriffe to this but a wayside inn on the road to all the world? In actual fact—the circumnavigation of the great globe itself! Better than Humboldt! Fifty, sixty hours these letters had lain here, and not a soul had dreamed! . . . Of course he must go. Did not Henslow say "the very man," Peacock "the greatest anxiety"? Of course, of *course* he must go!

He went through the letters again, read them aloud to his sisters, awaited their echoing agreement. But they did not agree. They displayed no enthusiasm at all. Caroline thought it a dreadful idea. What—go right away from them, from home,

[1] *LL.*, i. 192–3.   [2] *LL.*, i. 193–4.

from everything? Two years—it was eternity. Tierra del Fuego—that was the end of the earth!

Then "the tide came in," and the proposal must be laid before a soberer judge. Charles would have written that very night saying yes, yes, yes! But Robert delayed the decision till the next morning, then stated his objections quite clearly to his disappointed son. His point of view is evident. All he saw was a repetition of the Edinburgh business, a turning aside from a clerical as previously from a medical career. Charles's eagerness to go was once again the disinclination of "the idle sporting man" to settle to anything. What could this voyage do for him? After two years of wandering in wild places (Robert probably thought: amongst heaven knows what wild company) would he be more likely to settle to a steady life? Charles protested: it was not a turning aside, merely a brief and educative interlude; he had no intention to abandon the Church. The more reason, Robert retorted, for not going; the business could only be disreputable to him as a clergyman. Besides, it was a wild scheme on every ground. Clearly the place must have been offered to many better men before an unknown amateur naturalist who had done no more than collect a few beetles for his amusement. There must be some ponderable reason why the others had refused. The ship would be a small one, living conditions uncomfortable. Charles knew little—nothing—of the sea, and there would be no time to investigate or prepare. Probably Captain FitzRoy would find him quite unsuitable anyway. Why not abandon an utterly useless project, say No at once and have done with it? Robert did not command this, but it was his most urgent advice.[1]

Charles listened with a sinking heart. Perhaps it wasn't a command, but it was delivered with the vehemence and finality of one. Too often had he incurred his father's displeasure; he could not face it again. That very day he wrote to Henslow and Peacock, thanking them, saying how much he wished to go, but acquiescing in his father's objections—as he must, for "even if I was to go, my father disliking would take away all energy."

---

[1] It has been surmised on slender, some will prefer to say fantastic, evidence that Charles about this time was involved in sexual—masturbatory—difficulties of which Robert was aware, and that the latter feared the further complication of nautical homosexuality. Whatever the truth about Charles, the second idea may well have been in Robert's mind, for it might easily have seemed a very definite danger.

His disappointment breathed in every phrase, and Robert, perhaps thinking the matter effectively settled and not wishing his son to think him harsh, told him that if one man of common sense thought the project good, he would give his consent. They were expecting him at Maer the next day; let him see what Uncle Jos had to say.

He went to Maer . . . for the shooting, away from, not towards, FitzRoy and Tierra del Fuego. His back was turned on them, he believed, irrevocably.

But the Wedgwoods put quite a different face upon the matter. Not one of them but said he ought to go. Even Uncle Jos, when Robert's objections were set before him, had an answer for every one. He wrote to Robert that night, a letter interesting not least for its revelation of Jos's view of Charles at this time. Evidently he was distinctly dubious of the young man's clerical future. It was, he agreed, true enough that the undertaking would be quite useless though not, he thought, disreputable to him in his anticipated profession, and—these are the significant words—"if I saw Charles now absorbed in professional studies I should probably think it would not be advisable to interrupt them; but this is not, and, I think, will not be the case with him." His present leading interests were in fact much more naturalist than supernaturalist; that is, "in the same track as he would have to follow in the expedition." Considering the ways in which he was most likely to spend the projected period of the voyage at home, wasn't he just as likely—more likely, the implication was—to "acquire and strengthen habits of application" aboard the vessel? The same argument was applied to the point of the voyage unsettling and unsteadying him—was he to be deemed either settled or steady here and now? The other objections were soon disposed of. There was no real reason to assume the making of the offer to many others before Charles, neither would their refusals prove anything. Nor was there any need to presume undue discomforts. All in all, Jos's feeling evidently was that if the candidate for a curacy would benefit but negatively, the "man of enlarged curiosity," whom he also detected in Charles, might, afforded "such an opportunity of seeing men and things as happens to few," discover some more definite centre to his life than either Edinburgh or Cambridge had made plain.[1]

[1] *LL.*, i. 198–9.

Charles wrote too, reaffirming his submission to his father's decision but begging him to reconsider the matter in the light of Uncle Jos's comments, and the two appeals were despatched "by car" early next morning—September 1st—to Shrewsbury. Then the day's customary routine went forward, but Charles's heart was only half with his gun, and it was with relief that after a few hours, about ten o'clock, he was called back to the Hall. The more Uncle Jos thought about the matter, and the advantage to Charles, the more concerned he grew lest Robert should refuse to change. He decided therefore to follow his letter in person: would Charles accompany him? Would Charles! Anything was better than this waiting, waiting, hour after hour.

Soon they were away, and soon climbing the last steep slope out of Frankwell to The Mount, to find that the letter had not failed of its effect. Robert was a man of his word, and he himself had suggested consulting Uncle Jos. There was more talk. Charles was highly excited, tired from his Welsh trip and sleepless nights. He wanted to go, he could not bear the idea of not going, and yet he was oppressed by the need to make an absolute choice for or against. But the decision was made: Robert at last "most kindly gave his consent." Instantly Charles's sensitive conscience, stirred by gratitude, recalled his Cambridge extravagances; he would, he told his father "consolingly," be "deuced clever to spend more than my allowance whilst on board the *Beagle*." The conflict between Robert's doubts and Jos's reassurances was plain in the smiling answer: "But they tell me you are very clever."[1]

That same evening Charles despatched new letters to Peacock, Henslow, and Captain Beaufort at the Admiralty, and at three in the morning he was up and dressed, ready to make all speed to Cambridge before it was too late. The Wonder Coach took him to Brickhill in Buckinghamshire, where he hired a postchaise for the remaining fifty miles, reaching the Red Lion Inn too tired and late to do more than send a brief hasty note to Henslow by hand, announcing his arrival and asking when he might call next morning. "I trust the place is not given away."[2]

It was not, and the next day, Saturday, September 3rd, all seemed well. He and Henslow were busy in talk, and though he was warned against regarding the matter as settled until he

[1] *LL.*, i. 59. [2] *LL.*, i. 199.

had seen Beaufort and FitzRoy in London, his tone was confident when he wrote to Susan on the morning of the 4th. But uncertainty was not yet ended, for now FitzRoy, and rather mysteriously, began to waver. Wood, a Cambridge friend of Charles and a relative of FitzRoy, had written to the latter on the other's behalf, but something he said—perhaps the news that young Darwin was a Whig!—caused misgiving, for on Sunday Wood had a letter from FitzRoy "*most* straightforward and *gentlemanlike*, but so much against my going, that I immediately gave up the scheme." [1] There and then, again, the whole matter might have ended, for Charles almost did not go on to London at all. It may have been Henslow's indignation which impelled him to see FitzRoy face to face. He took the Monday morning coach Londonwards, and as he sped along the easy rise and fall of the shady roads tried to school his mind rather to thoughts of Maer and fox-hunting than of Teneriffe and South America.

He arrived early, and saw FitzRoy at once, before lunch. It was an odd interview. He liked the Captain instantly, with that intense admiration he was apt to feel for his friends. He afterwards wrote to Susan, with typical extravagance: "Captain FitzRoy is everything that is delightful. If I was to praise him half so much as I feel inclined, you would say it was absurd, only seeing him once." [2] The Captain at first looked at Charles more inimically. He didn't, in the familiar phrase, like the shape of his visitor's nose. A physiognomist (and ardent in that as in everything which interested him), he felt it to indicate insufficient energy and determination. He began, accordingly, to stress the disadvantages of the adventure—the poor or anyway plain living, the cramped quarters, the hardships. He talked of a personal friend, a Mr. Fletcher or Chester, whom he had promised to take, and whose presence, did he come, would leave little room for another. Then, as Charles's pleasanter parts and manner came to outweigh the drawback of his nose—not the handsomest feature of any of the Darwins—his tone changed a little. He turned to the happier aspects of the voyage. As it happened, he had heard only five minutes before Charles's arrival that Fletcher (or Chester) couldn't come after all, so that difficulty was disposed of. He would be very pleased if Charles would come, and would assuredly do all he could to make him comfortable, and to set

[1] *LL.*, i. 201. [2] *LL.*, i. 203.

books, instruments, guns and the like at his service. The storminess of the southern seas was, he modified, exaggerated. He was sure they would get on well, especially if Charles could "bear being told that I want the cabin to myself—when I want to be alone. If we treat each other this way, I hope we shall suit; if not, probably we should wish each other at the devil." [1] Charles, probably reading less into his remark than the modern Freudian psychiatrist, liked his frankness! Forthwith FitzRoy made formal application for Charles's appointment.

When Charles that afternoon wrote excitedly to Susan and to Henslow only one doubt remained. Once Teneriffe had seemed the world's desire; now it was a mere calling-point, and he was insisting on a world-tour or nothing. "Till that point is decided I will not be so." [2] He dined that evening with FitzRoy, however, and that second encounter settled the matter in his mind. He got on splendidly with his "*beau ideal* of a Captain," and though official confirmation did not come until Friday, from Tuesday forward he was busy "all day long at my lists, putting in and striking out articles," and, he said, really cheerful for the first time since the news of the voyage had come to him. He was writing to Shrewsbury for clothes to be prepared for him and books and instruments to be forwarded, shopping in London under the supervision of William Yarrell the newsagent-naturalist, consulting with FitzRoy and Beaufort, calling on Mackintosh and others, sending his news to Erasmus abroad, reporting with pride the Zoological Society's wish to make him a corresponding member. He wasn't too well, his hands troublesome with what seems to have been the same painful eczema as affected two years earlier his lips; he asked Robert's opinion of an arsenic treatment.

Thursday, the 8th, was necessarily a holiday, every shop being shut for the Coronation of William IV. Charles was childish enough to pay a guinea for a seat to see the procession, and was well pleased with his bargain, for the glittering show "was like only what one sees in picture-books of Eastern processions." He ventured on a Whig prophecy: "The King looked very well, and seemed popular, but there was very little enthusiasm; so little that I can hardly think there will be a coronation this time fifty years." [3] It thrilled him to watch the Life Guards driving

---

[1] *LL.*, i. 203.   [2] *LL.*, i. 202.   [3] *LL.*, i. 209.

the crowds off the roads, rearing their horses to frighten the people, but apparently hurting nobody. Later he viewed the illuminations, and thought them "much grander" than for the Reform Bill, though with little diversity, a few designs—"crowns, anchors, and W.R.s" being rather monotonously repeated. The streets were so crowded that even the carriages could only crawl like snails.

Most of the next day he was with FitzRoy. He also saw Captain King, who had been the other's superior officer on the earlier *Beagle* expedition and who volunteered the information that FitzRoy's temper was perfect. It was settled that Charles should go with FitzRoy by water to Plymouth to see the *Beagle*, return to London, and then pay his farewell visits to Cambridge and Shrewsbury. The sailing-date was October 10th. He felt very happy and very excited: "For about the first time in my life I find London very pleasant; hurry, bustle, and noise are all in unison with my feelings." [1]

II

Extraordinarily tenuous are the threads drawing a man to his destiny! Beyond question the voyage of the *Beagle* was the making of the Charles Darwin the world was to know, and yet link after link of the chain of connecting events held so barely that the wonder is it did not break, one might say, almost before it came into being.

Had Charles come to Cambridge but a few months later, he would never have encountered Fox, and so perhaps never entered Henslow's circle, or entered it quite differently and failed to attract attention. Had he come but a few months earlier, there would never have been those two extra terms in which he became still more intimate with Henslow, and under his influence read Humboldt and took up the study of geology. In that case there would have been no talk of travel abroad, no anticipation of visiting Teneriffe, or any of those other eager discussions which made it almost inevitable that Henslow's thoughts should turn to him when asked to suggest a candidate for the voyage. Again, had Charles not returned just when he did from Barmouth the letters making the offer might have come to him too late, while without Uncle Jos's backing and decision to go personally to

[1] *LL.*, i. 210.

The Mount, it seems likely that he would have accepted his father's verdict without effective appeal.

But if these links barely joined, no less precarious were those bringing the notion of the expedition into being, and only thereafter leading on to Charles. FitzRoy and His Majesty's Government were, of course, the ostensible effective forces, but three native Fuegians also played their parts.

Robert FitzRoy was a striking personality even as a young man. He was slight, darkly handsome, well-bred, of almost excessively aristocratic temper and bearing, inheriting both great gifts and fatal weaknesses from his grandfathers, the third Duke of Grafton and the first Marquess of Londonderry, father of Castlereagh. Born in 1805, he had entered the Navy in 1819 from the Royal Naval College, and—a lieutenant before he was twenty—saw early service in the Mediterranean and then as flag-lieutenant to Rear-Admiral Sir Robert Otway, commanding the South American Station. It was in 1826 that the *Adventure* and the *Beagle* began their coastal survey of these waters, from Rio de Janeiro south to Cape Horn and on the west side north to Chiloe and Chile. In 1828, Stokes, the captain of the *Beagle*, committed suicide in a fit of despondency, and FitzRoy took his place, quickly proving his unusual ability for the work. But his mind moved upon lines of broader schemes and larger curiosities. In the second month of his command he was already lamenting to himself his ship's lack of anyone possessing the necessary knowledge for geological and zoological studies, and "inwardly resolving, that if ever I left England again on a similar expedition, I would endeavour to carry out a person qualified to examine the land." [1]

These thoughts occurred to him in the Magellan Strait, and it was not far from there that four Fuegians—two men, a lad, and a girl—were later taken on board, some as hostages for a stolen whale-boat, another bought from his parents for a button. The intention was to return them ashore, but difficulties appeared, and characteristically FitzRoy was soon so seized by the idea of "the various advantages which might result to them and their countrymen, as well as to us, by taking them to England, educating them there as far as might be practicable, and then bringing them back to Tierra del Fuego," [2] that he resolved to do so on

---

[1] *Narrative*, FitzRoy, i. 385.   [2] *Ibid.*, i. 458-9.

his own responsibility and if necessary at his own expense. One of the men died of smallpox soon after arrival in England in 1830, but the others were cared for, received by the King and Queen, and given instruction in the simpler benefits of Christian civilization.

FitzRoy had presumed that the uncompleted survey was to continue after the briefest pause, and that the restoration of the natives would thus be easy; he was accordingly much perturbed to learn that the plan had been abandoned. His word was his bond, however, to whomsoever given, and he was actually in process of privately fitting out a 200-ton brig when "a kind uncle" spoke for him at the Admiralty with such effect that he was, after the briefest delay, reappointed to the *Beagle* for a second South American voyage.

The first intention seemed little more than the return of the natives, with a minimum of survey work to put a face upon the matter, but one thing led to another, and soon the programme included a two years' absence, and the carrying of a chronometric line right round the world. Again FitzRoy's enthusiasm was fired, and he determined to "spare neither expense nor trouble in making our little expedition as complete, with respect to material and preparation, as my means and exertions will allow." [1] From July 1831 the *Beagle* was under careful repair and reconstruction in Devonport dockyard, and it was as the work went forward that the Captain recalled his previous regrets, and applied to Beaufort, as the Admiralty Hydrographer, for permission to take with him some suitable "scientific person" both as a companion and to observe geologically and zoologically in "distant countries yet little known." Beaufort approved the idea and asked Peacock's advice. Peacock put the offer first to Henslow and then to Jenyns. Henslow wanted badly to accept, and, said Charles, "Mrs. Henslow most generously and without being asked, gave her consent; but she looked so miserable that Henslow at once settled the point." [2] Jenyns in his enthusiasm was actually packing when doubts of health and qualifications, together with thoughts of duty to his two livings, overcame his vision of far horizons and great opportunities. So the cup slipped from lip to lip! If neither, Peacock demanded of the two, then who? and together they declared for "Mr. Charles Darwin,

[1] *Narrative*, FitzRoy, ii. 17.　　　[2] *LL.*, i. 200.

grandson of Dr. Darwin the poet, as a young man of promising ability, extremely fond of geology, and indeed all branches of natural history."[1] A youth, Henslow particularized, who didn't know a great deal, perhaps, but would undoubtedly work. Accordingly Peacock and Henslow wrote away to Shrewsbury; Wood, meanwhile, forwarded his account, creating in FitzRoy his sudden, strange fear of irrevocably penning himself up for a period of years with a young man he might heartily dislike. On Friday, September 2nd, while Charles was speeding post-haste across England to Cambridge, FitzRoy was expressing all kinds of doubts to Beaufort at the Admiralty, wondering whether there really would be the scope he had imagined for such a person, even possibly inventing the mysterious Mr. Chester-Fletcher who might after all be able to go instead. It was then that he hurriedly wrote off to Wood "to throw cold water on the scheme," and it was not until Charles personally confronted him that, after his last anxious quibble about the shape of his visitor's nose, he withdrew his objections and the final link was joined.

III

The two took steamer together for Plymouth on the 11th, and had a quick and pleasant trip. The *Beagle* was still in the dockyard, mastless and in disorder, "more like a wreck than a vessel commissioned to go round the world,"[2] smaller too than Charles had expected, a bare hundred feet in length, a bare thirty wide. FitzRoy quieted his doubts, as to both lack of space and safety, and was able on the 15th to report his satisfaction. He was also introduced to the officers, and thought them "a pleasant set."

The return journey was made by coach, 250 miles in 24 hours—"wonderful quick travelling." The week-end in London, and he was off to Cambridge to bid his grateful good-byes to Henslow, who advised him to purchase the first volume of Charles Lyell's *Principles of Geology*—he would find it useful but must "on no account" accept its anti-cataclysmic advocacy—and at their parting gave him the very copy of Humboldt's *Personal Narrative* which had roused him to such enthusiasm earlier in the year. He also while in Cambridge settled up his undergraduate debts, and this together with the thought of all the new expense he was

[1] *Narrative*, FitzRoy, ii. 18-19. [2] *Beagle Diary*, 5.

incurring stirred him to an old dissatisfaction and to new resolutions of reform. He was away again on the 21st, reached St. Albans that night, and next morning caught the north-bound Wonder Coach to Shrewsbury for a last ten days at home, with a farewell visit to Maer to thank Uncle Jos and to say good-bye to his aunt, to his favourite cousin Charlotte, to Josiah, Harry, Francis, Hensleigh, Sarah, Frances whom he would never see again, and Emma. It was October 2nd, Shrewsbury Fair Day, when at last he shook his father's hand, kissed his sisters and his old nurse Nancy, patted his dog, and mounted the coach for London. The period of the voyage had now been extended, and he did not expect to see any of them again for more than three, perhaps even four, years, but his anticipation of adventure beat down any momentary sorrow he may have felt. There was so much still to be done, both in making purchases and in consulting with all kinds of "great guns in the scientific world." He had expected to sail for Plymouth by the steam packet on October 16th, but heard at the last moment that the *Beagle* would not set out until November 4th. "What a glorious day the 4th of November will be to me! My second life will then commence, and it shall be as a birthday for the rest of my life." [1]

He left London for Devonport on October 23rd, travelling by coach and so escaping the steam-packet's long week of hard struggle in the teeth of the south-west gales. When he arrived it appeared that the *Beagle* was not yet ready, was not likely to be ready even by the 4th. The birthday would have to be postponed. Actually it was only on the 12th that all aboard really began to look shipshape, and Charles to feel a pride in his new home, a three-masted brig of about 242 tons, broad in the beam with high bulwarks and raised upper deck, very seaworthy but inclined to roll badly. It had seemed at first sight an impossibly small boat to carry more than seventy persons, with all their needs, safely round the world and bring them back again, but its capacity was, he found with acquaintance, surprising. Admiralty instructions arrived on the 14th, but not till the 21st did he shift from his lodgings in the town to the poop cabin he was to share with not FitzRoy but Lieutenant Stokes, his clothes, guns, primitive instruments and books—these last two a matter mainly of some pocket magnifying-glasses, a microscope, a magnet,

[1] *LL.*, i. 214.

equipment for blowpipe analyses, an instrument for measuring angles, a few volumes on mineralogy, chemistry and geology, some on South America, Volume One of Lyell's *Principles of Geology*, Humboldt of course, and for recreation *Paradise Lost*. Little as it all amounted to, his first efforts to stow it compactly away brought back a gust of his "panic on the old subject, lack of room," but again FitzRoy reassured him. They were getting on very well together, looking forward to their long period of companionship.

The end of the month was now the sailing-date, and on the 23rd the *Beagle* crossed to Barnett Pool, a trip of not much more than a mile but Charles's first experience aboard a naval vessel under way, and he was duly delighted by the coxswain's piping and the fife to which the men pulling at the cables kept time, and impressed by the celerity and precision with which every order was obeyed. It was a smooth passage, and his stomach was not affected.

It was now a month since he had left London, and more than another month would go by before England's shore was left behind. The earlier period was not so bad. There were entertainments in the town, visits to Plymouth, explorations of the coast and countryside, boating about the harbour, a ride to Exmoor with Lord Barrington to geologize, breakfasts and dinners with the Captain and the other officers. He also read a good deal. His letters to Shrewsbury and to Maer were cheerful, gay even. But as the period drew on, with day after day of waiting, each morning a hope, each evening a defeat, his spirits dwindled, and always afterwards he recalled it as without exception the most miserable of his life. The weather, despite some sunny autumnal days, was mainly gloomy and often wet; the south-west gales, which held the westward shipping imprisoned week after week, blew almost continuously, whirling the brown wet leaves underfoot and filling the sky with great slow-moving banks of sullen grey cloud. Nothing pleased him. FitzRoy apart—and he was always desperately busy—the *Beagle's* officers seemed to him distant, stupid and formal, in their manners "like the freshest freshmen" and in general incredibly uninteresting for men who, if still young, had most of them seen somewhat of the world. He could almost weep when he thought of his family and friends, and how long it would be before he would see them again. Would he ever? he wondered sometimes, for he was "troubled with palpitation and pain about the heart, and like many a young ignorant man, especially one with a smattering of medical knowledge, was convinced that I had heart disease."[1]

Yet with it all he was determined upon going. He even refused, for all his private anxiety, to consult a doctor, lest he be told he was not well enough. With all his doubts and dubieties, both of the voyage itself and of ever setting foot in England again, go he would—"at all hazards." He could write that Jenyns was wise to have said no, and that he himself, had he been but a little older, "*never* could have endured it," but it did not enter his head to draw back. Not even his constant fearful expectation of seasickness could daunt him; he would welcome it could they but be on their way towards those Humboldtian vistas of his imagination. The anticipation of sailing always raised his spirits; it was delay that cast them down.

December 2nd brought one last unexpected visitor, Erasmus, just home from the Continent. They had a happy evening together, but next day Charles was busy from morning till night with final preparations for sailing on the 5th. He slept this night for the first time on board, having the usual landsman's trouble with his hammock, but being pleased to find that the rolling of the ship towards morning left his stomach undisturbed. Perhaps he would escape seasickness altogether! The night of the 4th was also spent on board, to be ready for an early setting-forth, but again the wind rose from the fatal south-west and this time so upset him that he quickly retreated to firm earth, dining that evening alone and quietly with Erasmus, and sleeping, "for the last time," on a firm, unshifting bed.

Daily now he awoke wondering what the wind would be. Daily he alternated between ship and shore, restlessly arranging and rearranging his belongings in the poop cabin, writing last letters to his friends, having repeated last suppers with Erasmus, finding each day more wearisome than the one before.

On December 10th the anchors were weighed, and at ten in the morning the *Beagle* sailed. Erasmus was on board, but went ashore as the vessel left the shelter of the breakwater. Charles knew all about the solemn emotions ascribed by Humboldt to such an occasion, but oddly—perhaps rather disappointingly—

[1] *LL.*, i. 64.

felt no more than had his destination been Calais. Wolf, it seemed, had been cried too often, and Wolf it was to be again. After a wild twenty-four hours at sea in a heavy gale—in which his sufferings from seasickness were worse than his worst imaginings, being given too a thoroughly nightmarish quality by the screaming wind, the shouts of the sailors, the cannonading anger of the sea—the *Beagle* dropped anchor once again in Barnett Pool.

Quickly Charles got on shore, ill and utterly wretched, and now with none to keep him company, for Erasmus had departed. It was in this state that all his father's doubts came full upon him. Was the venture really going to add to his happiness? "If I keep my health and return," he wrote in his diary, "and then have strength of mind quietly to settle down in life, my present and future share of vexation and want of comfort will be amply repaid."[1] *If—if* I keep my health, *if* I return, *if* I have strength—that is the stress of the whole sentence. Supposing the storms and sickness continued, supposing that not just for a few weeks, but perhaps for months, he remained unfit for work! There was so much to be done, and he so unprepared, so ignorant. Collecting and observation were but the beginning; he must read, read, read on all the subjects of his presumed interest. There were French and Spanish to be mastered, and mathematics, and he must read at least sufficient of his Greek Testament (on Sundays) not to forget, this time, *all* his classics. Some recreation he might allow himself, but above all he must make this new life one of discipline in thought and work. He knew how lazy, how lax, how remiss he had been in the past: now or never he must atone. "If I have not energy enough to make myself steadily industrious during the voyage, how great and uncommon an opportunity of improving myself I shall throw away. May this never for one moment escape my mind and then perhaps I may have the same opportunity of drilling my mind that I threw away whilst at Cambridge."[2]

With a few placid days his spirits recovered, but his determination remained. The weary waiting was resumed, but he had his happy moments, reading, dining, walking, or no more than strolling the deck, looking out over the sheltered Sound. On the 14th and 15th hopes were high, but the south-west wind was steady. The actual second start was made on the 21st, but

[1] *Beagle Diary*, 13. [2] *Ibid.*, 14.

it began badly by the *Beagle* running aground—though without damage—on a rock off Drake's Island, and another gale drove her back, with other vessels, to her anchorage. Charles succumbed again, but had a comfortable night at sea, and was even in a mood to find compensations in the delay, feeling himself to have been "broken in to sea habits" before having to face "the miseries of sickness," and having had the more time too to fit himself for his future labours.

Christmas Day was spent in harbour. After attending a church where an old Cambridge friend was preaching, Charles dined with the gun-room officers, but found their formality very unenlivening. The entire crew, except the officers, were drunk by nightfall, midshipman King having to stand sentry when the last staggering sentinel was put in chains for his insolence, and the next morning found the men and ship in an impossible state for sailing; all that could be done was to prepare for the following day.

The dawn of the 27th broke cloudy and dull, but dead calm, and the officers, scanning the sky, hoped for an easterly wind. To Charles the smoke from chimneys ashore spoke more plainly, first drifting easily upward, then taking, with the passing hours, more and more of a westward inclination. Catspaws rippled the water, the sails were set, and by noon the brig waited beyond the breakwater. Captain FitzRoy, Lieutenant Sulivan and Charles, having lunched ashore on mutton chops and champagne, came aboard about two o'clock. As the *Beagle* drove on before the fresh easterly breeze Charles again found himself regarding the retreating shore with a supreme indifference. By six the Eddystone lighthouse was lost below the horizon, and though the night was squally the morning was bright and the wind remained steady. Charles to his delight, was well all that first evening. But on the 28th he was sick all day. It was an unhappy time aboard, for FitzRoy was having the worst offenders of Christmas Day adequately flogged. He was no martinet, but he believed in discipline. If Mr. Darwin or other "inexperienced persons" thought that he exercised "unnecessary coercion," doubtless they would live to learn the efficacy of "harsh measures" applied at the right moment. Charles's real doubt seems to have been whether it was just to permit the men to get drunk and *then* punish them for it. He was, green in his hammock, in no mood to stomach either sight or sound of brutalities however warrantable.

CHAPTER VI

THE ADVENTURE BEGINS

I

NEVER would Charles forget those dreadful first days at sea. The officers talked consolingly of worse weather eluded, but their words meant nothing, for he could imagine no condition of greater suffering. The ceaseless rolling of the laden vessel in the heavy swell off the Bay of Biscay kept him almost continuously in his hammock, exhausted not by hours but by days of nausea and retching to a point of depression where he wished to heaven that Robert's first protests had prevailed and he stayed safe at home. He read when he felt able—Lyell, Basil Hall's *Voyages*, some Humboldt still—now and then struggled on deck for a few moments of fresh air, sea and sky spinning dizzily about him, or lay on the sofa in the Captain's cabin, FitzRoy tending him sympathetically or trying to distract him from his pangs with friendly conversation. But respite was short if sweet, for before he could start to gain strength the sickness would have him again, and his long body was ill sustained on the slight diet of raisins which was all his stomach could endure. He would not even get out of his hammock on January 4th to see Maderia in the distance.

He was by then, though, past the worst. Next day he was much better. The clouds were left behind, sun and sky were bright, the wind was easing, the swell was longer and more gradual. Spring was in the air. How superbly blue, dark yet brilliant, the water, how indescribably lovely the snowy foam of each wave's racing crest! He was able for the first time since sailing to bring up to date the journal he had started at Devonport—no great task, for he had little but his own distress to record.

The following day Teneriffe was sighted at dawn, and all that morning the *Beagle* bore him ever nearer to that very peak

ROBERT W. DARWIN

From *The Life, Letters and Labours of Francis Galton*. By the courtesy of the Cambridge University Press

THE ADVENTURE BEGINS

of his desire, whose actual view more than fulfilled his expectation, the dense clouds opening as they approached to show the snow-capped height in all its grandeur, twice as high, he exclaimed, as he had thought to look for it. At last, after all the Cambridge talks, plans, hopes, delays, he was here, where "rears huge Tenerif his azure crest," and in a few hours would stand on the island itself, ready to follow in Humboldt's climbing footsteps. He was so excited that though he sat down in his cabin at eleven o'clock to record the great event, he could stop only for a few lines before running on deck to gaze again upon "this long-wished-for object of my ambition." [1] Disappointment followed swift. They had been preceded by rumours of cholera in England, and the British vice-consul came alongside to announce a strict quarantine of twelve days. There followed "a death-like stillness" not only in Charles's heart but over the whole ship, till FitzRoy almost immediately announced his decision to set sail instantly for the Cape Verde Islands. Charles was utterly cast down. He might try to think that one volcanic island was very like another, but there was no aspect of Teneriffe, off which the ship was tantalizingly becalmed all the following day, which was not unbearably lovely, the cloud-girdled peak seeming of another world, the winds blowing about it the very airs of heaven! His Humboldtian hungers were not diminished but intensified by this first repulse.

A happy time followed. The weather continued good, the days were ever warmer; a general change to lighter clothing was made on the 10th. Charles quickly settled down to ship life, establishing his desired routine and getting to work without further delay. A small bag dropped astern caught for his study small sea-creatures whose exquisite forms and colours aroused an almost unhappy wonder at so much beauty for, it seemed, so little purpose.

He read a good deal, especially Lyell's *Principles of Geology*, which he regarded with an enthusiasm sweeping Henslow's warnings entirely overboard. Its great merit, he afterwards said, was that it re-created not only the knowledge but the vision of its reader. Science extends the perception of law ever deeper into the working of observed phenomena. Lyell's great achievement in geology was to establish unshakably the inductive method,

[1] *Beagle Diary*, 21.

the proper scientific procedure from known to unknown, the acceptance of law as constant and continuous. Huttonianly he argued from present to past, expounding past phenomena, practically entirely in terms of visibly existing forces. Previously, he declared, theological views diminishing the age of the earth had made such an idea inconceivable, and he described geology's long battle with such limitations, which in the seventeenth century rejected Hooke's suggestion of the extinction of species as a derogation from the wisdom and power of God, and in the eighteenth forced Buffon to renounce, as contrary to Moses, his conception of the adequate power of "secondary causes" to bring valleys and mountains into being, to destroy them and even the continents with them, and eventually to set others in their places. Asserting "the painful necessity of renouncing preconceived opinions," he especially stressed the "incalculable" duration of the geological periods, quoting with complete agreement Hutton's statement that "in the economy of the world I can find no traces of a beginning, no prospect of an end." [1]

That all this most deeply impressed Charles is plain from his frequent references to the immeasurable vistas of time evoked by much of the geological phenomena he was presently to see. It made hay, of course, of the Biblical chronology of Archbishop Ussher, and even if Charles had never taken that authority too seriously he doubtless, in this phase of unchanged parsonical intention, found Lyell the more acceptable for specifically dissociating geology from any "questions as to the origin of things," and for being so definitely deistic, always acknowledging—if at times rather like a gentleman nervously raising his hat to a strange lady he fears he ought to recognize—the dominant primary powers of "the Author of Nature." Seeking, like Hutton, to give "fixed principles" to geology, dispensing entirely with all "hypothetical causes," Lyell was anxious to preserve his views from any taint of evolutionary speculation, and decisively rejected, as without any foundation in observed fact, "the popular theory of the successive developments of the animal and vegetable world, from the simplest to the most perfect forms." [2] Characteristically, he swallowed heaven's signal intervention in admitting mankind's relatively recent creation (subsequent to the lower forms), then dubiously dodged the

[1] *Principles*, Lyell, i. 63.   [2] *Ibid.*, i. 153.

issue by declaring humanity to be so absolutely unique as not to constitute ground of argument against uniformity in the sub-human sphere.

Most of the book, however, was more particular, discussing the physical agencies causing geological change. It was a textbook, but one which reached out from recorded fact to instil a method and to suggest, though with hesitations, a scientific philosophy. All the matter was organized and expounded with a clarity amounting almost to genius, and it came to Charles's attention at a crucial moment when he was upon pins to prove what he uncomfortably felt to be his inadequacy, in training and knowledge, to unprecedented demands. He wanted a teacher—here was one. He learned the facts, he absorbed the method; if, remembering Henslow's warnings, he tried to dismiss the larger implications, his mind was not sufficiently set by previous study towards catastrophist dogmas to blind him to the beautiful simplicities of Lyell's conception.

By good fortune application followed fast to prove at least the facts and method. On January 16th he landed on St. Jago, one of the Cape Verde Islands, and there for three weeks had his first experience not only of "the glory of tropical vegetation" but also of unaided field work in a totally strange country. Each day he was ashore, walking, geologizing, collecting, observing, each evening returning to the ship bearing animal, vegetable and mineral specimens for systematic labelling, preserving, arranging. He worked hard and with unbounded enthusiasm, almost overcome by the profusion and beauty of material and his doubts whether he was really noting the right facts or whether anyone at home would ever have the patience to examine his prizes in detail. For all, he felt, *were* prizes, every goose a swan. Rowlett the purser and Bynoe the assistant surgeon rode with him about the island, gathering flowers and fresh-water shells. On briefer foot excursions he would collect in the morning and examine and record in the evening. The exquisite corals and strange marine creatures of the sea-shore most easily delighted him, but geology engrossed him more deeply. Carefully he essayed his first crude sketch-map, and one day, resting at the foot of a low lava cliff, the sun's glare about him, exotic tropical plants growing behind him, lovely corals visible in the tidal pools just beyond his feet, the thought came to him that observations of upheavals, subsidences, indeed the geology generally of all the countries he was to visit, would surely make a worthy subject for a book of authentic interest. Why not write it? The glow of prospective authorship, the sense of power bestowed by knowledge, swept over him and stamped the scene into his mind for ever. He also began to see useful work to be done in other directions. Already he was finding the printed descriptions of tropical creatures woefully inadequate. This omission too he might amend. . . .

The sailing from the island on February 8th thus found him full of aspiration, and though he was seasick on the 10th, a hasty note, written that day to his father and sent aboard the passing packet *Lyra*, was enthusiastic. He was unreservedly glad he had not missed the opportunity of a century. "I think, if I can so soon judge, I shall be able to do some original work in Natural History."[1] He stated many years later that he worked so hard aboard the *Beagle* "from the mere pleasure of investigation, and from my strong desire to add a few facts to the great mass of facts in Natural Science. But I was also ambitious to take a fair place among scientific men."[2] All three incentives were present from St. Jago forward, the second and third taking precedence as the pleasure of simple collecting dwindled with repletion.

Despite a recurrence of sickness, and consequent indolence, he was fairly constantly busy upon his specimens, his notes, and his diary—record of the events of each day, at this time set down each evening or the next morning, though later based on brief scribbled memoranda sometimes not written out for weeks or months, in one case the better part of a year.

He liked the imposed regularity of shipboard life, the impossibility of idleness, and he was more at home with the officers. They were all young men, Rowlett, in his mid-thirties, the eldest. Lieutenants Sulivan and Wickham, Stokes with whom he spent so much time in the chart-room, Rowlett, the midshipmen Arthur Mellersh and Philip King (son of Captain King), Augustus Earle the artist privately engaged by FitzRoy—for all of them he expressed the friendliest attachment in after years, and they in turn nicknamed him amiably "the Flycatcher" and "the dear old Philosopher" (aged twenty-three). Most of them at one time or another, and some frequently, accompanied him

[1] *Beagle Diary*, 434.   [2] *LL.*, i. 65.

ashore, walking or riding, and though Wickham would growl at the "damned beastly mess" made by the Flycatcher's specimens on his polished decks, his complaints were made and taken in good part. His liking for Matthews, the missionary accompanying the Fuegians, was qualified, but the only person for whom he expressed actual distaste was the surgeon Maccormick, who was returned home at the earliest opportunity because neither FitzRoy nor Wickham could get on with him.

The heat now began to be oppressive, and the weather uncomfortably squally, but Charles was well enough to go ashore at St. Paul's Rocks on February 16th, and on the 17th to receive the somewhat mitigated attentions of "Neptune's constables," who "came aboard" on the crossing of the Line, a ceremony FitzRoy disdainfully sanctioned as diverting the childish minds of the men.

There was but one more pause before South America, off the rugged island of Fernando de Noronha, where Charles wandered all day in the woods, thrilled by the miniature promise of the tropic forests of the mainland. Soon they were on their way again. The days were hot—but how pleasant, how peaceful, the cool nights as the vessel drove forward towards his Eldorado. The sea had been smooth practically all the way from Teneriffe; he could not but suppose his initiation over, his sea-legs found, seasickness a thing of the past.

So night by night he walked the short deck of the *Beagle* in a calm contentment with which there mingled still astonishment at being here at all. To have seen the sun northward at noon, and in a few days to stand in the New World, his utmost dreams come true before his eyes—and but February still! Six months ago he was holidaying in North Wales, with Teneriffe no more than a vision in the distance; and now the constellations in the gem-pointed skies, as he looked south-westward over the bows, where the waves and the wash whispered and rustled and splashed below in answer to the hushed singing of the wind in the sails and rigging, were wholly strange. Surely nothing hereafter should wholly surprise him!

Yet he sighed a little. Gladly he went forward, nevertheless he could not but think of the months, the years before he would set foot again in Shrewsbury, Maer, Cambridge. He wanted the whole adventure of the voyage, and yet oddly, at the same time, he wished it over. But a day or two ago he had written in his diary that save for the actual experience of tropical scenery his greatest pleasure was "in anticipating a future time when I shall be able to look back on past events; and the consciousness that this prospect is so distant never fails to be painful." [1] There were moments when the lonely ocean vistas, in which the sun seemed to sink immeasurably distant, made him acutely aware of the isolation of his situation. Yet if he ever thought for an instant of turning back, it was not as a real possibility. He was committed. He might write home acknowledging Robert's "wisdom in throwing cold water on the whole scheme," and admit that were his own advice asked in similar circumstances, he would be very cautious in encouragement, but now, as at Devonport, his way, irrevocably, was forward.

II

Always, too, the new day brought something to dispel the evening doubt. The coast of South America was sighted in the early morning of February 28th, a fine line of emerald vegetation along the horizon, and before noon the *Beagle* dropped anchor off Bahia, in the calm ship-studded waters of the bay. Charles thought it an almost incredibly beautiful vista: bright sunshine on the green woods rising up from the deep-blue sea, the white walls of the serried houses of Brazil's oldest city, the tall waving palms, the small sails of the canoes gliding about the harbour. FitzRoy preferred the view to the closer contact—the lower town at any rate, he said, stank—but if Charles could not deny the heat, the crowding, and the dirt, it was his mood to be enchanted, and he turned quickly to the wooded solitudes hemming the city round. Scarcely could he contain his joy. Here was all that Humboldt had promised him, and more. "I have been wandering by myself in a Brazilian forest . . . pleasure more acute than I ever may again experience . . . enough to make a florist go wild . . . nothing more nor less than a view in the Arabian Nights, with the advantage of reality." [2] He paused, as he walked, upon a thousand details, filled with a pulsing amazement almost touching terror; it was his deepest and most thrilling experience of "sublime" scenery, never forgotten,

[1] *Beagle Diary*, 38.   [2] *Ibid.*, 39-40.

never dimmed, and rivalled only by the exultation of Andean heights.

Magic days followed. He went riding with Rowlett, walking and shooting with King, botanizing, entomologizing, geologizing, and all with a freedom of conscience he had never known at Cambridge. It was "a new and pleasant thing for me to be conscious that naturalizing is doing my duty, and that if I neglected that duty I should at the same time neglect what has for some years given me so much pleasure."[1]

FitzRoy was making his first homeward reports, approving Charles as "a very sensible, hard-working man and a very pleasant messmate" who had taken to ship-life as a duck to water and was already exercising his "good sense, inquiring disposition, and regular habits"[2] to make the most of his opportunities. Charles's view of his Captain was quite as high. He liked him personally, as a man and for his gentle care during those worst days before Madeira, and he admired him too, more detachedly, for his mental and physical energy. "If he does not kill himself, he will during this voyage do a wonderful quantity of work."[3]

Yet here at Bahia they had a violent quarrel which almost sent Charles packing homeward there and then. Charles, before ever he set foot out of England, shared the family hatred of slavery and, though after some Brazilian experience he admitted that many slaves were probably happier than their English sympathizers supposed, his feeling never changed. There were too many, if exceptional, "atrocious acts"—even haunting screams heard "off" as he passed along town and country roads. His detestation found frequent expression in his diary and his published journal of the voyage, and years after his return he quite spontaneously "thanked God" that he would "never again visit a slave country."[4] FitzRoy's feeling was as different as his high Tory upbringing; in the name of property rights he would defend slavery, denying inhumanity, to the last ditch. Doubtless, in their close contact and inevitable discussion of South American conditions, their views had become known to each other, and when one day at Bahia the talkative Captain Paget of H.M.S. *Samarang* started reciting his own experiences of the ill-treatment

---

[1] *Beagle Diary*, 42.　　[2] *FitzRoy and Darwin*, F. Darwin, *Nature*.
[3] *LL.*, i. 229.　　[4] *Journal* (1845), 499.

of slaves, FitzRoy was intensely irritated by this innocent assault upon his position. When Paget had gone he returned to the subject, quoting the testimony of slaves given in the presence of their masters. Charles, bored by an enforced week aboard the *Beagle* with a poisoned knee, questioned the value of such evidence. FitzRoy exploded. Did Mr. Darwin doubt his word? Then, damn it, he would not sleep under the same deck with him! Charles would have gone straight ashore, but the gun-room officers begged him to mess with them. When FitzRoy had exhausted his anger upon the unfortunate Wickham he apologized sincerely and completely.

It was a characteristic incident, and not isolated. Charles had been involved in a similar outburst before sailing, when FitzRoy told a trivial lie, charged Charles with not believing him, got a flat answer—and after a brief silence admitted his fault. There was to be a later explosion as absurd and as furious. The officers aboard naturally suffered more directly and more frequently; the state of the Captain's temper was their first thought each morning.

These occasions were quite unimportant as quarrels, but deeply significant as manifestations of an irrational and uncontrollable, and in its momentary intensity almost insane, irritability. FitzRoy's case is clear. He was compact of nervous energy, of mingled sensitivity and passion; his aristocratic pride was directed as much against himself as anyone, forcing him always to give to his work a little more than he could, psychologically, afford, and thus keeping him always on the edge of overstrain and the bitter unmeasured anger of that state. He was never satisfied by his own labours, and seldom in consequence by those of others; the ideal ever outshone the real, driving him electrically onward, too often into trouble for himself and others. His family record, and his own history and distressing end—he was to cut his throat—suggest mental instability. The highest tributes were paid him by his colleagues, and Charles Darwin not least; all saw him as of the very highest character and talents, a man generous, affectionate, conscientious, a wise capable seaman, an admirable leader, a good friend, and in his normal state almost painfully just and honourable. Only when "out of temper" was he quite unreasonable, his "vanity and petulance" overcoming his customary candour and sincerity. During such attacks he had himself definite fears for his sanity. Yet he drew men to him, both officers and crew. None failed to recognize the fundamentally disinterested man whose thought—save in those paroxysms of demonic possession when suspicion, moroseness, the fury of shrilling nerves, beset and held him—was always for his ship, his men, and the spirit rather than the letter of his duty.

He was also something more: a keen, if limited, speculative intelligence, and this aspect, the impression of mental *force* rather than range, struck Charles as much as any. Writing home to his sisters soon after the quarrel at Bahia, he declared that he had "never before come across a man whom I could fancy being a Napoleon or a Nelson. *I should not call him clever*, yet I feel convinced nothing is too great or too high for him. *His ascendancy over everybody is quite curious. . . . Altogether he is the strongest marked character I ever fell in with.*"[1] The italicized phrases convey both FitzRoy's "ascendancy" over Charles himself and its limitation; Charles could admire FitzRoy's character and mental alertness; he was not, even in those early days, prepared to accept his convictions.

The question of FitzRoy's influence upon Charles in the five years of their intimate contact is an essential one. A recent writer on the subject deems it "indelible," and to have been exercised in two ways: first, by FitzRoy's "misreading of natural phenomena and fixity of religious ideas" acting "as an intellectual spur to Darwin's candid and inquiring mind," and second, by Charles's deep admiration causing "some retardation of the formulation of his evolutionary ideas for publication."[2] There is surely truth in both suggestions, if the latter influence was but one aspect of a larger caution. To appreciate the force of the former one need but glance briefly at FitzRoy's views as subsequently expressed by him in his account of the voyage, and compare them with Charles's developing ideas of the same period. By his own account FitzRoy in earlier years suffered no little disposition to doubt, reading for a while more among "Voltaire's school" and the geologists than in the Bible itself. But by 1830 he realized his error, and thereafter clung to the letter of the Word with all that earnestness which characterized his every action. Accordingly he appended to his more strictly nautical record of the voyage chapters, on "the early migrations

---

[1] *R. FitzRoy and C. Darwin*, Barlow, *Cornhill*. [2] *Ibid*.

of the human race" and "the Deluge," which demanded an absolutely literal acceptance and interpretation of every Biblical statement. To such geological liberalists as sought to see the "days" of the creation as "ages," he countered the problem: "Vegetation was produced on the third day, the sun on the fourth. If the third day was an age, how was the vegetable world nourished?"[1] Similarly he rejected the idea of successive conceptions at wide intervals of fishes, reptiles, mammals and mankind. Describing the Deluge he denied that any gradual elevation of land could account for certain phenomena (the alterations of aqueous and igneous strata) noted by Charles in the Andes: "These wonderful alternations of the consequences of fire and flood are, to me, indubitable proofs of that tremendous catastrophe which alone could have caused them"[2]—that is, the Flood plus consequent volcanic activity. Straining at this geological gnat he swallowed not only the camel but the entire Ark, dismissing "cavillers" who doubted whether the lion would consent to lie down with the lamb, with the assertions first that "He who made, could surely manage," and second, that even dismissing miracle ("though we should never forget that man is a miracle, that this world is a miracle, that the universe is a miracle"), the fright of the animals at the coming and actual Deluge would be sufficient to reduce them to submission and peace. And if one questions that, still "we are not told how many creatures died in the Ark. . . . Whether Job had himself seen, or only heard of, the leviathan and the behemoth, does not appear; but that those monsters were the megalosaurus and the iguanadon there seems to be little doubt."[3] He discussed also the migration of the animals from Ararat about the world, suggesting the possibility of subsequent geographical change, though in his chapter on human migration he presumed a geography substantially our own. All existing human races he declared to be descended from the sons of Noah, and thus the people like the beasts had all gone outward from Asia Minor to the uttermost parts of the earth, civilized when they started, and subsequently degenerating. In no country was he at a loss to discern "remains of Arkite observances," hinting that the word Shem or Chem, current among the Chilean Indians, had been

---

[1] *Narrative*, FitzRoy, ii. 661. [2] *Ibid.*, ii. 668.
[3] *Ibid.*, ii. 671 n.

"handed down from their ancestor of the Ark."[1] Sooner or later, he believed, every single Biblical statement would be proved factually exact.

A study of his earlier declarations would suggest that it was only towards the end of or after the voyage of the *Beagle* that his previously more open mind attained this condition of absolute rigidity, but he was throughout the period marching steadily to this point, the more urged on by what he felt to be the dangerous speculations of Charles, to whom in turn his obstructive dogmatism probably suggested a worser Professor Jameson. So they reacted each upon the other.

They certainly discussed points of scientific interest on a number of occasions, and with more dissent than agreement. On other topics too they were apt to be at odds, for FitzRoy was uncompromising in his conservatism and Charles stuck firm to his Whig principles. The first months of the voyage were the stormy period of the passing of the great Reform Bill against the opposition of the Tories and their allies the Lords; Charles might "hurrah for the honest Whigs" in the privacy of letters to England, but discretion as well as the pressure of new interests may have helped him to thrust his earlier keen political interest towards the background of actual discussion. It was safer, or anyway easier, to talk about the phrenology of the cuckoo.[2]

Their last evening at Bahia, March 17th, Charles took a "farewell stroll" ashore with young Philip King. It was a clear, quiet close of day: the lovely prospect lay in a windless hush, a picture long to shine clear in his memory. His hopes burned undiminished, as he paid what he deemed a "lasting farewell" to his first New World port of call.

[1] *Narrative*, FitzRoy, ii. 400.
[2] Dr. Edward J. Kempf, in his analysis of this period of Charles Darwin's life, suggests a deeper source of mutual strain, positing Charles as psychoneurotic and sexually unadjusted, and FitzRoy's irritability as due to sexual suppression. " Homosexuality is a serious problem among seamen." The suggestion may be held in mind, if it be added that there is no evidence at all of even unconscious homosexual tendency or strain beyond the plain fact of FitzRoy's tension, the possibility (if Kempf's preceding arguments are accepted) of Charles's lack of adjustment, and the obvious—to sailors familiar—situation. Charles's statement, solemnly quoted by Kempf, that the Captain was difficult to live with " on the intimate terms which necessarily followed from our messing by ourselves in the same cabin " really carries one nowhere.

CHAPTER VII

SHIP AND SHORE

I

SOUTHWARD the *Beagle* sailed in fair weather. Not once was Charles seasick, and only once "rather uncomfortable." He had quite found his place on board, was on the best of terms with all, and enjoyed the respite from the continuous excitement of tropic shore and forest. He liked the daily routine—the hurried breakfast at eight o'clock, the morning work on his specimens at the chart table opposite Stokes or King, dinner, rather more formal, with FitzRoy at one, then perhaps more work or, were the sea heavy, lying in his hammock to read until tea at five, whereafter all the long evening waited for other labours, or more reading, or talk with the others, or walking the short confined deck, and seeing the ever-enchanting because strange southern constellations above. Objective events were few. He saw a water-spout. He caught a shark. At the Abrolhos islets he collected every plant in flower. Once Sulivan shouted below the thrilling news of a Grampus Bear to port. Eagerly Charles rushed on deck, to be greeted by roars of laughter. It was April 1st!

The 4th, and they anchored in the superb harbour of Rio de Janeiro beside the flagship of the Admiral commanding the South American Station, to whom FitzRoy must report. Now Charles received with "ecstasy" his first letters from home. Next day he and Earle explored the town, crowded with well-dressed merchants, ragged negro fruit-sellers and porters, fat priests under high cone-shaped hats, sailors chaffing the black women, the more respectable ladies aloof in their safe carriages. This day Mrs. Eliza Darwin, widow of Dr. Erasmus Darwin, "expired calmly," Dr. Robert Darwin at her bedside. Charles dined "in great spirits" with several young Englishmen enlisted under the Brazilian flag.

He and Earle forthwith found lodging in the near-by village of Botafogo, the smooth curve of the lovely bay stretching before them, the pointed peak of Corcovado rising steeply behind, and with this a new phase of the South American experience began. Of the twenty remaining months of the *Beagle*'s stay on the East Coast, he was to spend practically a year ashore, away from the ship often for many weeks at a time. Essentially it was an educative period. He was trying his wings and finding his feet, his mind receptive rather than formulating. He was busiest simply at gathering material. But all the while there was a deep mental strengthening, a constant increase in intellectual capacity. It was his time of greatest, of least qualified, enthusiasm. Despite occasional qualms he was looking forward; not, as later more and more, just homeward. The initial Humboldtian impulse was still fresh and keen and free. He was ever busy with eager labours, observations of sea, sky, earth, animals, birds, insects, people, plants, rocks, clouds, climate, winds—all the vast ceaseless inexhaustible phenomena of the living world which presented themselves to eye and ear and clamoured to the mind, evoking scientific, social, aesthetic, even political, speculations, deliberately tentative and happy to be so. All was grist that came to his mill, even though the miller was inclined to, capable of, no more than occasional trial workings.

Setting off almost immediately on a fortnight's riding trip inland to the north of Rio de Janeiro, he found a prospect where only man was vile, and as at Bahia turned the more to the tropical forest, whose "sublime grandeur" constantly excited him to wonder. He sought to analyse this feeling. What was there here, he asked, "that can bring the delightful ideas of rural quiet and retirement, what that can call back the recollection of childhood and times past, where all that was unpleasant is forgotten," and concluded that "until ideas, in their effects similar to them, are raised, in vain may we look amidst the glories of this almost new world for quiet contemplation." Lacking any such element, the observing mind must be "wrought to a high pitch, and then no delight can be greater; otherwise your reason tells you it is beautiful but the feelings do not correspond."[1] It was an amateur and incomplete but also genuinely perceptive consideration of an immediate sensation.

[1] *Beagle Diary*, 60.

He was settled in Botafogo again, with all his things brought ashore (the *Beagle* returned to Bahia), before the end of April, ready to spend May and June making a complete collection of the local insects. But larger issues were impressing him. He was deeply struck amid this tropical luxuriance by both the fertility of nature and the relentless cruelty of its fate. He was impressed too by life's variety and persistence; there seemed scarcely any condition under which some living form could not thrive.

He walked a great deal along the shore and among the near-by hills and mountains, making geological notes and rough sketches. He thought how much he owed to Sedgwick, but Lyell was his acknowledged master now.

Usually he collected one day, and worked on his specimens and his notes the next. But there was social pleasure too—dinners with the Admiral, with Mr. Aston the English minister, with the merchant Mr. Cairnes and his delightful household. There were naval occasions—the Admiral's inspection of H.M.S. *Warspite*, divine service on the battleship, a regatta in which the *Beagle*'s boats failed to shine. When the *Warspite* cleared for action "the pomp and circumstance of war" so filled Charles's soul that he "almost wished for an enemy"; even during service he felt more patriotic than Christian, and softer emotions woke only when the band after dinner played airs from Mozart.

The gusto of his letters to Fox, Henslow and Herbert show how pleasantly the weeks passed at Botafogo, yet he was pleased when the *Beagle* was in port again and he could go aboard. It was like coming home to hear the men singing forward, and he was glad when they sailed on July 5th for the Rio Plata, headquarters for the rest of their stay on the East Coast. The band on the *Warspite* played them out of the harbour with "To glory you steer"—more appropriately than usual. Yet had the candidate been picked, who could have hesitated? Charles himself, leaning unobtrusively on the high bulwark, waving to the cheering sailors of the battleship, would have been the first to name FitzRoy, marked for distinction by birth, capacity, and even by achievement. (*A bathroom, a razor, a cut throat before breakfast.*)

Southward again they sailed in high spirits, growing—Charles among them—their manly beards in preparation for the virgin lands they would explore. But a series of gales sent Charles to his hammock, and he was relieved when the twenty days' passage was over. The idleness annoyed him almost as much as the nausea.

The extent of his sickness on the *Beagle* is sometimes minimized. Once the pangs of the first few weeks were over, his diary contained only occasional references to illness, and it might be supposed that for months at a time he was continuously well and active. This was his own recollection; he denied that seasickness could have caused his later weaknesses, and appeared to assert that he had never been really ill with it after Teneriffe, though sometimes "uncomfortable" in heavy weather. But he himself wrote to Henslow in July 1834 that "at sea to this day, I am invariably sick, excepting on the finest days,"[1] and in 1836 he declared that he suffered more from seasickness then than three years earlier. Several of his shipmates spoke in unequivocal terms of his "constant suffering," and in the last year of the voyage FitzRoy considered his health so poor from "confinement and seasickness when under way" as to be on that account alone "most anxious to hasten as much as possible"[2] the *Beagle*'s home-coming! There is no doubt but that anything less than the finest weather served to send Charles to his hammock, where reading and reflecting were his utmost activities. Over the entire period it was a serious restraint upon his labours, yet not without its compensation; for, despite the unfavourable conditions, it gave him, when not too bad, the more time for contemplation not of particular facts but facts in their mutual relations, for, indeed, speculation leading towards synthesizing hypotheses.

II

The three weeks in the Rio Plata, "a wide expanse of muddy water," were spent unsatisfactorily in preparation for the first short southerly cruise. Political troubles ashore kept everyone mostly on board, and Monte Video, when they did land, proved as dull as dirty. They crossed once to Buenos Ayres, failed to heed a warning gun, and were fired on by a guardship. FitzRoy in a fury—shared for once by Charles—returned the message that had he known he was entering an uncivilized port he would have "had his broadside ready for answering." Skirts aflutter,

[1] *ML.*, i. 14. [2] *R. FitzRoy and C. Darwin*, F. Darwin, *Nature*.

they returned to Monte Video, thence to send a British frigate to demand an apology or sink the offending vessel. Charles hoped for the latter, but his martial ardour found some abatement in going ashore with part of the crew to protect the people of Monte Video from rebelling black troops. The excitement—there was no battle—sent him back on board with a headache.

He wrote many letters and prepared his first consignment of specimens for dispatch by packet boat to Henslow. He apologized that it was not larger, but protested he hadn't been idle. There were several hundred items in all, and who thought the geological pieces small was welcome to "try carrying rocks under a tropical sun." If the plants were poor, there were nevertheless almost complete collections from the Abrolhos and St. Jago, not to add the many spiders and beetles from Rio de Janeiro; there would have been more but for the care he was taking with his notes.

He really had been very industrious. In addition to his more personal diary, portions of which he sent home to his family at intervals when opportunity offered, there were his records in his little red-covered notebooks of every specimen, these alone filling 600 small quarto pages in the first thirty months, and 2000 before the voyage was done. His aim was precision, as in one typical note on a small lizard: "Centre of the back yellowish-brown, sometimes with a strong tinge of dark green; sides clouded with blackish-brown; in very great numbers under stones; makes a grating noise when taken hold of; after death loses its darker colours."[1] The journal and the geological notes were to prove of the utmost future value; the biological matter was less successful owing to his lack of skill in dissection and drawing, and a good deal of this was really waste material. Then there were the specimens themselves to be sorted, ticketed, catalogued, and packed for dispatch in every manner of container from pill-box to cask. It was indeed no idler's harvest.

III

The cruise was brief, preliminary only to the real adventure. August 19th the last case and letters went aboard the packet, and that afternoon the *Beagle* sailed; October 26th she was back at the same anchorage. But the two months' interval saw important

[1] *Zoology*, iv. 27.

developments. Gales kept Charles much to his hammock, and he landed with relief at Bahia Blanca, a military post consisting of a few huts and a wall. He reaped his usual harvest along the coast, but also found a whole hoard of fossil shells and bones embedded in soft rock at Punta Alta. Some of these remains were evidently of large animals—a skull rather like that of a rhinoceros; a jawbone, with tooth, of "the great ante-diluvial animal the Megatherium"; several bony plates very similar, save in size, to those of the living South American armadillo. The tooth of a horse particularly puzzled him, for he had supposed that creature to be a modern importation, and could not understand what might have led to its previous extinction—a point to which his thoughts were often to recur.

He was so eager that when he and some others were one night marooned ashore by bad weather without food or shelter, FitzRoy thought the episode enough to damp *even* Mr. Darwin's love of adventure.

To forward the survey work FitzRoy hired two small sealing schooners, putting Stokes and Mellersh on one, Wickham and King on the other, and when these were ready to carry on, the *Beagle* turned north again.

A month at Monte Video and Buenos Ayres saw some frivolities—the last for a long while to come: visiting English and Spanish homes, admiring the handsome Spanish women; attending the Opera and a grand Presidential Ball. But he was distinctly moody. Often he longed for home, wondered how he could, whether he could, endure the months and years before he would see them again. On the *Beagle*, he wished himself ashore; ashore, he would be at sea again. "Hurrah for Cape Horn and the Land of Storms," he wrote to Henslow on their setting out in late November, hoping to see no more of Monte Video for months to come.

There was good weather at first. They met the schooners, gave them orders for the next three months, continued south into cold, gloomy, gusty days. Charles took to his hammock until the excitement of reaching the coast of Tierra del Fuego revived him. He thrilled to think that here was country never before traversed by Europeans, and it was from such imaginative impulse—*the sense of the unknown awaiting exploration*—that the thought came to him, in Good Success Bay, that he could employ his life no better "than in adding a little to Natural Science."[1] He scarcely realized the implications of his thought. It wasn't, after all, inconsistent with his " distant prospect of a very quiet parsonage," seen " steadily " still, " even through a grove of palms "[2]; some of the scientific men he had admired most were clergymen themselves.

Little less exciting was the first close meeting, on December 18th, with the wild human inhabitants of that barren countryside. They seemed like "the troubled spirits of another world," and "without exception the most curious and interesting spectacle" he had ever seen. He couldn't, he declared, have conceived without actual contact "how wide was the difference, between savage and civilized men; it is greater than between a wild and domesticated animal, inasmuch as in man there is a greater power of improvement."[3] If FitzRoy said then, as later, how disgusting it was "to consider ourselves even remotely descended from human beings in such a state," when "even the mental contemplation of a savage" was "disagreeable, indeed painful," Charles must have agreed, convinced as he was that "no lower grade of man could be found," even to the point of arguing, against Sulivan's more sympathetic view, that the Fuegians were *specifically* different from Englishmen and lacking the same high potentialities.

FitzRoy was anxious to land his own three Fuegians, with Matthews, on their native shores. The *Beagle*, therefore, left Good Success Bay on the 21st, and doubled the Horn on the calm bright evening of the 23rd before a fine east wind. But a westerly gale arose, and they saw the Horn again—this time "in its proper form, veiled in a mist, and its dim outline surrounded by a storm of wind and water," rain and hail beating about it, and great black clouds ominously filling the sky—before shelter was found in a protected cove for Christmas Day. It was the last day of the year when they ventured out, only to spend a fortnight vainly striving for westing. Charles was desperately ill. On January 11th they were within a mile or two of the desired anchorage at Waterman Island, but again were driven back at the last moment. The gale intensified, the worst FitzRoy had ever experienced. For two days the vessel rode it perilously out. On the morning of the 13th the ship rolled its lee bulwark right under the water,

[1] *LL.*, i. 90. [2] *Beagle Diary*, x. [3] *Journal*, 228.

but recovered with the loss of one boat and some damage to Charles's specimens.

From that moment the storm abated, and at midnight anchor was dropped under False Cape Horn, without twenty miles of the true Horn passed nearly four weeks earlier. It was decided to make no further effort to reach York Minster's country, he agreeing to go with the others to Jemmy's home, which could be reached by boat.

The expedition set out from Goree Sound on January 19th, FitzRoy, Charles, Matthews and twenty-five others in three whale-boats and the heavily laden yawl. They made their placid way along the Beagle Channel, discovered by FitzRoy on the earlier voyage, and Jemmy was at last returned, with his companions, to his people. Animals, everyone thought, might have been more demonstrative in greeting. Matthews too was set ashore with his goods, three huts built, and adjacent ground dug and planted. Then one whale-boat and the yawl returned to the ship, while the other boat, with Charles and FitzRoy, continued exploratively westward. The crews slept on shore at night, guarding against native attack, and Charles, taking his midnight turn, was filled with a solemn sense of the party's utter remoteness and cosmic insignificance amid these wild hills and seas. On January 29th, opposite the later-named Mount Darwin, disaster threatened them when a mass of ice fell into the water while all were on shore and at dinner, creating a great wave which, dashing up the beach, would have swept both boats irretrievably away, had not Charles and a few others rushed to hold them fast.

They began their return on February 3rd, and approaching Matthews' camp were alarmed to see the natives ashore decked in tartan and other familiar articles. Matthews was alive and well, but he had been robbed and threatened with violence, and was persuaded to go back to the *Beagle*, the three Fuegians remaining a little sadly behind. Goree Sound was reached again on the 17th, but a fortnight of storms and surveying delayed their departure for the Falkland Islands, where they anchored on March 1st at Port Louis in the East Island for a five weeks' stay.

It was a dismal time. The Islands had been formally a British possession about a fortnight; they were barren, treeless, with only a few miserable inhabitants. Whaling vessels were practically the sole visitors. A few days after arrival Hellyer, the

Captain's clerk, was drowned while shooting for FitzRoy's collection, and the funeral depressed everyone and perhaps crystallized their leader's growing impatience. At their present rate, he felt, they would never be done. Too much time was wasted in returning to ports for refitting; a consort was needed to act as supply ship and assistant, and he therefore bought at once, on his own responsibility but anticipating Admiralty approval, a third and larger schooner, naming her the *Adventure* for old time's sake, and sending her forward to meet Stokes and Wickham.

Charles did some shooting ashore, but there was little game, and he had lost his old love of slaughter for its own sake, and turned the more eagerly to geology. It fascinated him, posing everywhere he went problems for his reason to solve, giving him his first extended experience of pure mental effort and satisfaction. He continued to apply Lyell's methods, and found them, as before, far superior to any others. Nothing on the island interested him more than the famous "streams of stones," relics of some remote cataclysm. In the Andes he was to see evidences of the overthrow of whole mountains, mountain-ranges even, like tumbling ninepins, "but never did any scene, like those 'streams of stones,' so forcibly convey to my mind the idea of a convulsion, of which in any historical records we might in vain seek for any counterpart." [1]

By this time he had read, probably more than once, the slimmer second volume of the *Principles of Geology*, which had reached him during the last visit to Monte Video together with letters not much more than five months old. It interested him, even though it seemed to stand rather outside the scope of his immediate activity, concerned as it was less with practical geology than with theoretical problems of species, evolution, human origins and the relations of organic and inorganic life.

The argument on these points centred principally about the Lamarckian view of transmutation, but outlined it only to reject it with what was, considering Lyell's privately expressed open-mindedness on the subject, a surprising decisiveness. In contemporaneous letters to intimates he allowed the widest scope to mutability; in print he declared all his studies to support immutability, showing each species to be endowed at its creation

[1] *Journal*, 256.

with its permanent attributes and organization. Some slight accommodation to new conditions there might be, but it was minor and limited, with extermination rather than change waiting always as "part of the regular and constant order of Nature." [1] From the geological evidence he sought to solve the problem of the simultaneous or successive appearance of species, favouring the latter and suggesting limited distribution as showing later creation. For humanity especially he declared a relatively recent origin, urging great caution in accepting the longer estimates of the antiquity of human remains, but seeming to refrain deliberately from any definite statement of a personal view.

It was in his second chapter that he summarized the train of thought which might lead "the student" to a contrary view. To begin with, he declared, the generic groups seem as eternally distinct as species themselves, but the lines of demarcation lose their clarity when intermediate forms are examined, and as likely as not the student becomes doubtful, a state of mind soon turning from genus to species. Seeing how some individuals vary, and how peculiarities are inherited, the question arises how far such variations might not extend in the course of ages and under influence of changing environment. Incertitude grows, but is checked by the fact of sterility between hybrids. Sooner or later the case of a prolific hybrid comes to his notice, and at once all nature is seen as in flux, without stability. Then, in this mood, "he encounters the Geologist, who relates to him how there have been endless vicissitudes in the shape and structure of organic beings in former ages—how the approach to the present system has been gradual—that there has been a progressive development of organization subservient to the purposes of life, from the most simple to the most complex state—that the appearance of man is the last phenomenon in a long succession of events—and, finally, that a series of physical revolutions can be traced in the inorganic world, coeval and coextensive with those of organic nature." These views seem to confirm his doubts, and speculation is set free. "He gives the rein to conjecture, and fancies that the outward form, internal structure, instinctive faculties, nay, that reason itself, may have been gradually developed from some of the simplest states of existence,—that all animals, that man himself, and the irrational beings, may have had one common origin;

[1] *Principles*, Lyell, ii. 141.

that all may be parts of one continuous and progressive scheme of development from the most imperfect to the most complex; in fine, he renounces his belief in the high genealogy of his species, and looks forward, as if in compensation, to the future perfectability of man in his physical, intellectual, and moral attributes." [1]

Lyell for his own part here rejected such a line of thought because he would not allow the possibility of deviation beyond the specific boundary, or a more than temporary hybrid fertility, and Charles, not having given much thought to the matter, did not find it difficult to agree with him and to be as true an immutabilist as any. He had on the East Falkland Island some discussion with FitzRoy on the specific identity of the only indigenous quadruped larger than a mouse—a wolf-like fox resembling but also differing from the Patagonian fox. Charles was sure that it was a distinct Falkland species. FitzRoy was doubtful, citing known cases of the modifying influences of climate, food, and habit. Both were talking well within limits of sure faith in the abiding reality of species, but of the two it was Charles who put least faith in the possibilities of even slight change by adaptation.

The path he was to tread had been shown to him, but marked No Thoroughfare, and he had, dutifully, not thought to set foot upon it. Yet he could not but wonder. He often puzzled over that inescapable similarity between some of the fossil remains found at Punta Alta and still living species. It was, to say the least, very odd.

IV

Before the end of April the *Beagle* had joined the *Adventure* in the Rio Plata estuary, and while these set about local surveying and refitting Charles was set ashore for a two months' stay at the pleasant village of Maldonado, east of Monte Video. He had a twelve days' riding trip inland, but most of the period he was collecting locally—birds, animals, insects, snakes, every variety of flowering plant. He seemed happy, for once indifferent even to delay, and he felt "a complete landsman" when he went on board again on June 29th. Landsman he was to be most of the latter half of the year. It was July 24th when the vessel sailed

[1] *Principles*, Lyell, ii. 20-21.

south again; landing at the mouth of the Rio Negro river, he bade good-bye to the *Beagle*, practically till December.

The four months' interval saw three overland excursions, of which the first, from Patagones on the Rio Negro to Buenos Ayres, broken at Bahia Blanca, was the longest and also the most dangerous, for he was passing through country actively threatened by marauding Indians, and he and his guide Harris, a local trader, welcomed the protection of gaucho companionship. Charles enjoyed the trip through wild, barren scenery, the pauses for a drink and a cigarette, the camping at night under the strange stars amid the death-like stillness of the endless plains. They spent three days at the camp of the barbarically impressive General Rosas, future dictator, then warring cruelly but effectively against the Indians. Plainly Charles doubted whether this slaughter of unfortunate natives was worth while merely to put "white gaucho savages" in their place. The "friendly" Indians about the camp seemed to him clearly of the same race as the Fuegians, though without the degeneration caused by the latter's less favourable environment. Whatever he thought of the Falklands fox, he did not deny variation of species mysteriously connected with environment; in fact, it seems from this time to have been increasingly clear to him.

At Bahia Blanca he made contact with the *Beagle* again, excavating at Punta Alta—"a perfect catacomb for monsters of extinct races"—and for recreation read a Spanish translation of the trial of Queen Caroline.

Only that August a military outpost between Bahia Blanca and Buenos Ayres had been wiped out by Indians, and he and his guide moved cautiously from one soldiers' camp to another across the dreary plain, inhabited by partridges, small foxes and the inimical carrion vulture, and only occasionally broken by the sudden humps of barren mountains. At Buenos Ayres, he had a week's rest, then started northwards, still riding, for Santa Fé. He arrived there on October 2nd, feverish from the heat, gave up his idea of a longer eastward tour, and had instead "a fine sail of 300 miles down that princely river the Parana." Rosas' soldiers were now giving the Indians a rest, and besieging Buenos Ayres instead. Charles got into the town, but soon retreated to Monte Video. In the latter part of November he made his last journey inland, up the Uruguay estuary, then east and south again,

Work went on continuously. He could not stay to hear all the sermons in stones, nor read to the end the books in every brook, but he caught their texts as he hurried by and set them down for further consideration. Between shooting, skinning, preserving, entomologizing, zoologizing, geologizing, fishing, netting, questioning, noting, walking, talking, riding and boating, he had little time to spare. More than one consignment of cases went off to Henslow, a new instalment of diary for the pleasure of his family at home.

So busy did he find himself that he engaged a servant, a lesser member of the crew released by FitzRoy. Charles thought Covington odd, but he could shoot and skin birds, and was willing. Before committing himself to this expense, however, Charles wrote to Robert to point out his need for assistance, his economy and the trivial nature of the proposed addition. He even tried to diminish it to vanishing point, by arguing that it might prove impracticable anyway. The letter was an essay in self-exculpation, and this note appears in almost all his communications with his father throughout the voyage. It was as though he never could forget that he had come without Robert's full approval, and felt constantly the need to display himself in the best, excusing light. He wrote to one of his sisters, acknowledging a note from Robert: "Give my love to my father. I almost cried with pleasure at receiving it; it was very kind thinking of writing to me." [1] And the following passage, with its visibly self-conscious declaration of a high impersonal aim, an elaboration of his thoughts at Good Success Bay, though also addressed to a sister, was surely penned with one eye on the tide that came in about dark—"I trust and believe that the time spent in this voyage, if thrown away for all other respects, will produce its full worth in Natural History; and it appears to me the doing what *little* we can to increase the general stock of knowledge is as respectable an object of life as one can in any likelihood pursue. It is more the result of such reflections than much immediate pleasure which now makes me continue the voyage, together with the glorious prospect of the future, when passing the Straits of Magellan, we have in truth the world before us. Think of the Andes, the luxuriant forest of Guayaquil, the islands of the South Sea, and New South Wales. How many magnificent and char-

[1] *LL.*, i. 245.

acteristic views, how many and curious tribes of men we shall see! What fine opportunities for geology and for studying the infinite host of living beings! Is not this a prospect to keep up the most flagging spirit? If I was to throw it away, I don't think I should ever rest quiet in my grave." [1] In other words, cease, dear father, to judge this venture by that which *I* may or may not draw from it (*if thrown away for all other respects*), and consider that so much vaguer, that so much less precisely estimable thing, what, what *little*, may be added to the general stock of knowledge (as respectable an object of life as even a prospective, an accomplished, clergyman, can in any likelihood pursue)! Always his messages to Robert were couched in these terms: "I pray forgiveness from my father."

Now again, in December, the *Beagle* was about to sail, to return to the Rio Plata no more. Charles was glad. He had had enough of the gaucho democracy of these parts. He was tired of the uninteresting plains of the Pampas. Not Teneriffe, but Cape Horn, was the gate to marvels now. "In truth, the world before us."

V

A fair passage south was delayed by the lagging of the *Adventure*, and a stop to adjust her rigging was made at Port Desire. There Christmas Day was celebrated with sports ashore. The Patagonian coast was sterile, indeed a desert, but Charles was deeply interested to find on the high plain inland shells of a kind still common in the ocean, proof that "the plains of gravel have been elevated within a recent epoch above the level of the sea." [2] The mental excitement of such speculations brought him increasing pleasure; thinking on these matters he could walk the wildest desolation in a glow of happiness.

While the *Adventure* work went on, the *Beagle* surveyed southward to Port St. Julian, and there Charles and FitzRoy went ashore, with others, to find fresh water. It was a long fatiguing fruitless day, and Charles, who walked farther than any, was badly exhausted. But three days later he was on shore again busily collecting shells.

After surveying eastern Tierra del Fuego for another month, while the *Adventure* charted the Falklands, FitzRoy took his

[1] *LL.*, i. 245–6. [2] *Journal*, 193.

ship into the Beagle Channel, anchoring on March 5th at Woollya, where Jemmy Button told a sad story of the flight of York Minster and Fuegia with all their common possessions. Gone his gloves, his shoes, even his clothes! But though much thinner, he was well, had a pleasing wife, displayed no wish to return to England. It beat Charles what life offered these dirty, ill-fed savages, more like wild beasts than human beings, living by instinct rather than reason. "Although essentially the same creature, how little must the mind of one of these beings resemble that of an educated man. What a scale of improvement is comprehended between the faculties of a Fuegian savage and a Sir Isaac Newton!" He questioned: "Whence have these people come? Have they remained in the same state since the creation of the world? What could have tempted a tribe of men leaving the fine regions of the North to travel down the Cordilleras, the backbone of America, to invent and build canoes, and then to enter upon one of the most inhospitable countries in the world?" He did not know the answers, but clearly, since the number of the Fuegians was apparently not decreasing, they must find sufficient satisfaction in life, of whatever kind, to make living seem worth while to them. Nature, he supposed, "by making habit omnipotent, has fitted the Fuegian to the climate and productions of his country."[1] Thus, the adaptability and persistence of even human life was made plain to him.

He was as well now as he ever could be at sea in sometimes stormy weather, yet about this time, perhaps because his more technical notebooks demanded the closer attention, his diary entries began to grow perfunctory, and less regular. In the first year and more they had been actually daily; now, frequently, days were omitted completely, or covered only by the most general statement.

Woollya was left behind on March 7th, Jemmy's good-bye beacon fading into the distance. A last visit was paid to the East Falkland Island. Fitzroy was depressed, but Charles did not share his mood, even though "for what I can see this may last till we return a fine set of white-headed old gentlemen."[2]

Contact was made with the *Adventure*, then again the *Beagle* sailed for the mainland, anchoring on April 13th off the Santa Cruz river for the last adventure of the East Coast.

[1] *Beagle Diary*, 213.   [2] *LL.*, i. 250.

This was a boat expedition up the Santa Cruz. On the 18th the party set out—FitzRoy, Charles, and twenty-three officers and men with the three whale-boats. Their purpose was to trace the river's course back to the Andes. The lower country was sterile, dreary. The nights were cold, with hard frosts. The boats had to be towed by the men against the swift current, and progress was slow. There were high awkward banks and dangerous rapids. Armed guards had to be posted against possible Indian attack. On the 26th Charles's geological interest was wakened almost for the first time since the start by the lava capping the cliffs—poured forth, it was clear to him, from Andean volcanoes when the whole plain lay deep under water. He shot a condor, eight feet across, and guanacos for fresh meat. The river-banks drew closer and the current became stronger. On May 4th, food nearly gone and the river unexpectedly bending south, FitzRoy gave up his plan and took a small party on foot westwards, where at noon observations were made from rising ground. The Andes were visible all along the western horizon, lovely snow-capped peaks seeming in the clear air no more than a few miles distant. There was a general feeling of much time and labour for little gain, but Charles thought it all very satisfactory, from affording "a most interesting section of the great tertiary formation of Patagonia."[1] The next morning the return began, the boats in the first day covering a week's hard upward pulling. By mid-day on the 8th they were aboard the *Beagle* again.

A last farewell to "the sterile plains of the Eastern side of S. America"—and they were off. Near Cape Virgins the *Adventure* joined them, bringing letters no more than seven months old, and the third volume of Lyell's *Principles of Geology*, published a year before. Mid-June they sailed together for the West Coast, through the Magdalen and Cockburn Channels.

[1] *Journal*, 226.

CHAPTER VIII

THE WORLD BEFORE US

I

It was amid squalls and gales that the two ships came eventually under full canvas into the Pacific. Charles had day after day of prostrating illness. From the 10th to the 28th of June he made no entry in his diary. On the 27th the popular Rowlett died, and the funeral service was the next day. "It is an aweful and solemn sound," wrote Charles, "that splash of the waters over the body of an old shipmate." [1]

That midnight the *Beagle* dropped anchor in the port of San Carlos, Chiloe, the *Adventure* joining her two days later. They lay there till July 14th, regulating chronometers and receiving provisions, Charles walking ashore in renewed delight at a tropical abundance recalling Brazil. But the weather was indifferent good, and he was not sorry to be away to Valparaiso.

Once again he had been making use of long hammock hours to study Lyell's work, and with the more attention at finding in this third volume support for his own speculations on the Santa Cruz formations. Nothing there had impressed him like the great pebble bed, two hundred miles across and fifty feet thick, and all made up of innumerable fragments "derived from the slow falling masses of rock on the old coast-lines and banks of rivers," dashed and rolled and rubbed into smooth round pieces, carried afar and at last all brought together in this vast consistent layer. The mind was stupefied "in thinking over the long, absolutely necessary, làpse of years," and even *that* was but the most recent of many similar eras, and even *these* had been all, as their fossils showed, within the Tertiary Period of mammals, birds, flowering plants and trees. Only to think of it so was to feel the wonder and terror of the infinitudes of geological time.

[1] *Beagle Diary*, 231.

Lyell's new volume touched on this very matter, if obliquely, the author regretting to be charged, by a friendly critic, with asserting that "the existing causes of change have operated with absolute uniformity from all eternity." In his view to set the earth's beginning far back was not to deny that it had one. Millions of ages were but the infinitesimal part of eternity! True, it had been urged that as he admitted man's creation he might allow the earth's also, but he demurred that this "cannot warrant us in presuming that we shall be permitted to behold the signs of the earth's origin, or the evidences of the first introduction into it of organic beings." In vain do we seek to assign limits to the universe; wherever we look we see proofs of Creative Intelligence, foresight, wisdom, power. "To assume that the evidence of the beginning or end of so vast a scheme lies within the reach of our philosophical inquiries, or even of our speculations, appears to us inconsistent with a just estimate of the relations which subsist between the finite powers of man and the attributes of an Infinite and Eternal Being."[1] Despite a certain equivocal eagerness to have his Flood as well as drink it, Lyell opened the gates of a past of immeasurable duration.

Whatever he may have felt about the equivocations, Charles seized with avidity on the main view, the release into endless time. A more limited chronology might well have forced him, with the pebble beds in mind, to have rejected anti-catastrophic gradualism as practical impossibility. But given sufficient time, was not anything possible?

It is likely that what absorbed him most, however, was the more purely geological account of formations and classifications. He was very much the practical geologist still, asking no more from "philosophical" discussion than that it leave him free to observe and note and theorize—theorize the more boldly in his imperfect knowledge of accepted geological doctrine.

II

The arrival at Valparaiso, a lovely vista of low white houses, set off by green trees and a vivid red soil, backed by steep hills, and all bright in morning sunshine under a clear blue sky, seemed a return to civilization. The continued fine weather, after so

[1] *Principles*, Lyell, iii. 383-5.

many weeks of gale and rain, was almost incredible. Ashore there was not only welcome but, and among the English residents, educated society. Charles was delighted to find, for the first time in South America, informed students of the sciences, even of geology, to be asked, indeed, his opinion of Lyell's newest work. And there in the distance, promise of delights to come, lay the white bastion of the magic Andes. There were letters too, and one especially which filled him with delight: Henslow's acknowledgment, only eighteen months old, of the first specimens sent from Monte Video. He had already received a later letter from Henslow, and the absence of any reference to the consignment had filled him with fears that it had been so bad that his friend thought silence the better part. Now all was well again. Henslow's clear approval jumped his spirits to the highest. Replying, he tossed into a happy letter all the busy interests of his mind: geology of Patagonia, of the Andes, doubts on existing classification of corallines, the abundance of work awaiting him.

Some months were to be spent at Valparaiso, and accordingly winter quarters were established ashore where the officers might settle to their survey records. So much waited to be done that FitzRoy sent Wickham in his place to make official calls. Clearly he was again in a state of extreme nervous tension, and in his restless unreason his plans grew ever more grandiose, embracing a return to Tierra del Fuego, a detailed charting of the coasts of western Patagonia, Chile, and Peru, and finally New Zealand and most of the Pacific. God knew when England would see them again, for even 1837, three years away, was to find them only a little ahead! Whatever the others thought, they dared not dissent. Charles found it his "most comfortable reflection" that "a ship being made of wood and iron, cannot last for ever, and therefore this voyage must have an end."[1] White-headed indeed—he had better have said bald-headed!

From August 2nd he was living in a suburb of the town with an old Shrewsbury schoolfellow, Richard Corfield, but soon set off northward along the coast, then inland to the lower slopes of—at last—"the very Andes." And now on every hand he discerned evidence of the relatively recent "residence and retreat of the ocean" before a continental elevation, and again and again he was lost in awe at the thought of the amazing forces thrusting

[1] *ML.*, i. 18.

upward these vast mountains and mighty plains. He speculated as to the marked lack of animals; it might, he thought, "be owing to none having been created since this country was raised from the sea."[1] The tour was leisurely but energetic, he and his guides circling round to the far south.

Nearing Valparaiso again he began to feel unwell and reached Corfield's house only with difficulty on September 27th. He was several weeks in bed. His illness has never been defined; Robert, when the symptoms were described to him, could make nothing of them. Charles in later years was inclined to attribute all his subsequent bad health to it (not always: his view wavered), but then and there he only recorded: "It was a grievous loss of time, as I had hoped to have collected many animals."[2]

When he visited the *Beagle* again in early November, he found many changes and FitzRoy in far from happy state. Lately news had come of the Admiralty's brusque refusal to sanction the purchase or even hire of the *Adventure*, or to shoulder any of its expense. FitzRoy was ordered to discharge her as soon as possible. The blow was a heavy one, to his pride and projects even more than to his pocket. He fell into a mood of such total depression that he wondered whether he was going mad, and refusing to listen to his surgeon's plea that he was no worse than overworked, he "invalided," Wickham succeeding him as captain and all thought of the Pacific trip being abandoned.

Charles considered returning home at once. Then he felt that he must, while on the spot, see more of the geology of Chile and Peru.

Then the situation changed again, to Charles's relief. FitzRoy was induced to withdraw his resignation. But his mood was permanently changed. "I could myself no longer bear the thoughts of such a prolonged separation from my country as I had encouraged others to think lightly of."[3] They would make the Pacific crossing after all, but would cut short their South American work.

It is not surprising that FitzRoy now broke out again. Charles afterwards set the incident at Concepcion, but it could not have occurred there and almost certainly took place at Valparaiso towards the end of this first visit. The Captain and other officers had been much entertained by families ashore, and

[1] *Beagle Diary*, 236.   [2] *Ibid.*, 249.
[3] *Narrative*, FitzRoy, ii. 362.

FitzRoy despondently complained of having to give "a great party to all the inhabitants of the place." Charles palliatingly questioned the need. FitzRoy exploded. Yes, he shrilled, Mr. Darwin was exactly the sort of man who *would* take favours and feel no obligation. Without a word Charles got up and returned to his lodgings ashore. When he went on board some days later FitzRoy received him "as cordially as ever."

The *Adventure* discharged, the *Beagle* returned south alone to survey the Chonos Archipelago. Charles went with Sulivan by whale-boat along the sheltered eastern coast of Chiloe, the active volcanoes on the mainland standing in splendid view. The Chonos Islands, where a stormy month was passed, were as barren and dreary as Tierra del Fuego, but the sense of wonder and novelty sustained him. "There is an indefinite expectation of seeing something very strange, which however often it may be baulked, never failed with me to recur on each successive attempt. Everyone must know the feeling of triumph and pride which a grand view from a height communicates to the mind. In these little frequented countries there is also joined to it some vanity, that you perhaps are the first man who ever stood on this pinnacle or admired this view." [1] The granitic formations of this area stirred him to wonder, too. Granite was, he said, the fundamental rock, "the deepest layer in the crust of this globe, to which man has been able to penetrate." He added: "The limit of man's knowledge in any subject possesses a high interest, which is perhaps increased by its close neighbourhood to the realms of imagination." [2]

Christmas was spent in dismal surroundings, a dreary occasion; their principal New Year emotion was anticipation of "the Pacific Ocean, where a blue sky tells one there is a heaven—a something beyond the clouds above our heads." [3] They came north, slowly, to San Carlos again. Among Charles's finds, a minute barnacle especially excited his curiosity by its unique character; he set it aside, with many other matters, for future study. At Valdivia the forests clustered so close that Charles began to yearn for the wide freedom of the plains of Patagonia! His mind played with zoological problems and theories, but often days went by without diary-entry of any kind.

But a new excitement was in store.

[1] *Journal*, 343.   [2] *Ibid.*, 345.   [3] *Ibid.*

III

On January 20th, when they were still at San Carlos, Mount Osorno, seventy-three miles away on the mainland, had been seen in fiery eruption, flame and dark masses flung high into the night, torrential lava tracing red angry lines all down its flanks. Later they heard that other volcanoes as much as three thousand miles to the north had been in simultaneous activity. On February 20th, about half-past eleven in the morning, Charles was on shore with Covington, resting in the forest. He was lying down when suddenly the earth seemed to rock beneath him—the sensation, he thought, was like that of skating over very thin ice—and the leaves above and about him whispered in strange secret agitation. Leaping to his feet, he had no difficulty in standing, but the shuddering motion made him giddy. The tremors lasted two minutes or more; the feeling of insecurity passed more slowly. He hurried back to Valdivia. The wooden houses of the town had suffered little. During the day smaller shocks were felt, and sea and river rose and fell rapidly many times. The earthquake was the severest that anyone in Valdivia could remember. The excitement drove him direct to his diary, and he but echoed the common thought when he wrote: "I am afraid we shall hear of damage done at Concepcion." [1]

So it proved when they arrived there on March 5th. The whole coast had been rocked from Chiloe to Copiapo, and Concepcion, and its port Talcahuano, suffered worst of all. Both towns, Charles and FitzRoy saw as they rode inland, lay in utter ruin, the houses of the port razed by the great wave which had swept over them from the sea, battering, smashing all before it, those of Concepcion tumbled to heaps of brick and stone and splintered beams. It was with a kind of terrified excitement that Charles heard the survivors' stories of the swift instinctive rush for safety to the open spaces, with buildings crashing down at the refugees' heels, the sky darkened by clouds of dust, fires breaking out on every side, robbers stealing with one hand and crossing themselves with the other while new shocks shuddered the ground they stood on, the terror, at

[1] *Beagle Diary*, 278.

Talcahuano, of the dreadful wave. He was deeply stirred. "Earthquakes alone," he wrote, "are sufficient to destroy the prosperity of any country. If, for instance, beneath England the now inert subterranean forces should exert those powers which most assuredly in former geological ages they have exerted, how completely would the entire condition of the country be changed! What would become of the lofty houses, thickly-packed cities, great manufactories, the beautiful public and private edifices? If the new period of disturbance were first to commence by some great earthquake in the dead of the night, how terrific would be the carnage! England would at once be bankrupt; all papers, records, and accounts would from that moment be lost."[1] Violence and rapine would take the place of order and authority; famine, pestilence, and death would follow. "Who can say," he questioned, "how soon such will happen?"[2]

The emotional and imaginative quickening producing these speculations was equally apparent in the more strictly geological sphere, for he found all about him unmistakable evidence of new changes in land and sea levels—the process he was convinced of in theory now demonstrated in fact before his eyes. About the Bay of Concepcion land had risen two and three feet; at an island thirty miles away rocks recently under water stood now ten feet above it. It is always one thing to theorize and another to experience. Charles was now, really for the first time, seeing geology in *act*. So to one sensation followed another: "It is a bitter and humiliating thing to see works, which have cost man so much time and labour, overthrown in one minute; yet compassion for the inhabitants is almost instantly forgotten, from the interest excited in finding that state of things produced in a moment of time which one is accustomed to attribute to a succession of ages. In my opinion, we have scarcely beheld since leaving England, any sight so deeply interesting."[3] His mind was still full of it long after Concepcion—they stayed only three days—had been left behind.

Apart from the earthquake it was natural that geology should fill his thoughts, for from Valparaiso, which had felt the shock but suffered no damage, he was quickly away to Santiago, and thence to cross the Andes by the high Portillo Pass. It was

[1] *Journal*, 373.   [2] *Beagle Diary*, 282.   [3] *Journal*, 376–7.

mid-March, winter was at hand; extra food had to be taken and mules as carriers. All along their way they met troops of cattle being driven down from the higher valleys. They pressed on too quickly in fact to give time for all the observations he desired to make, and he could only note the outstanding features: the flat valleys floored with, he believed, ocean-laid gravel, the vivid stratification of the wall-steep hills, the piles of detritus declining from the mountain sides into the valleys. And here, once again, the vastness of geological time came home to him as he listened to the mountain torrents carrying down their countless pebbles, and thought how "whole races of animals have passed away from the surface of the globe during the period throughout which, night and day, these stones have gone rattling onwards in their course." [1] He was expressing just the same feeling, though from another point of view, when he afterwards wrote, in relation to this same trip, that "daily it is forced home on the mind of the geologist that nothing, not even the wind that blows, is so unstable as the level of the crust of this earth." [2]

Breathing became difficult on the 21st, but not so much so that he could not forget it in his delight at finding fossils on a high ridge. The views of the shining snow-capped mountains under the intensely blue sky were superb, breath-taking in their silent magnificence. Save for the high shifting specks of the few wheeling condors, it was, utterly, undistractedly, a geologist's world. "I felt glad that I was alone: it was like watching a thunderstorm, or hearing a chorus of the *Messiah* in full orchestra." [3]

The plain of Mendoza beyond the pass was dull and he was glad to return over the easier Uspallata Pass. He crossed the higher ranges so easily as to make him wonder at travellers' tales of danger, and came back more convinced than ever of the recent elevation of the Andes and his ability to prove it. He wrote to his sister that it was the most satisfactory if also (father, forgive me!) the most expensive of his excursions. The harvesting in full swing upon the fertile Chilean plain brought many thoughts of Shropshire, though he "missed that pensive stillness which makes the autumn in England indeed the evening of the year." He was happy to think the voyage's term really fixed, to write

[1] *Journal*, 386.   [2] *Ibid.*, 321.   [3] *Ibid.*, 394.

to Henslow that early September would see the last of South America, and that a year thence they would be home.

He was in Valparaiso, with Corfield, by April 17th, and six days later carried on board the *Beagle* the "good news" of its commander's promotion to post rank. FitzRoy, still depressed, only grumbled at the lack of recognition for Stokes and Wickham. The ship sailed that night for Coquimbo, Charles returning ashore to go by land.

It was a disappointing journey. Barren, sterile, uninteresting are the words which recur in his account of it, but he noted some cases of human and vegetable adaptation to environment, and along the seventy-five miles of coast-road he kept a keen eye open for signs of elevation. At Coquimbo he and FitzRoy took some flea-infested rooms together, and further specimens were sent to Henslow before he continued north to Guasco and Copiapo, where the *Beagle*, after a last return to Valparaiso, was to take him aboard for Peru, the Pacific, home. He sighed for fresh green turf and spring flowers as relief from day after day of desert plain. Beside this, Patagonia had been a land of plenty! His expeditions from "miserable" Copiapo were made without excitement; even the mountains in this part were "very tame." When he went aboard the *Beagle* on July 5th Wickham was in temporary command, FitzRoy having gone to the rescue of H.M.S. *Challenger*, ashore south of Concepcion. It was a characteristic episode, FitzRoy threatening a dilatory superior officer with court-martial and taking the matter entirely into his own hands; he rejoined the *Beagle* at Callao, the port of Lima.

These last two months on the West Coast brought little diversion or relief from tedium. The passage from Iquique to Callao was uncomfortable: the ship rolled, the weather was gloomy, every landscape dull. In the Atlantic he had longed for the Pacific. Now the Atlantic had all the virtues—brilliant days, cool evenings, glorious skies, bright stars. Callao was filthy, its inhabitants drunken and depraved. He liked Lima better— the dark-eyed women, their tight gowns showing their firm full figures, seemed to please him most. "Certainly they are better worth looking at than all the churches and buildings in Lima."[1] But he spent most of his time on board writing up his diary and geological notes for the past six months.

[1] *Beagle Diary*, 332.

He wanted to get on, to see the real, the mid, Pacific, and, in particular, the coral islands, for already, without having as much as sighted one, he had conceived a theory of their origin. Lyell at the end of his second volume of the *Principles* had summed up all that was known of the subject, strongly supporting the view of their growth upon submerged volcanic craters. It had occurred to Charles, however, that if the South American coast, the entire continent, could be slowly elevated, why should not the bed of the Pacific Ocean be slowly subsiding, the reefs rising gradually upward from the sinking mountain-tops?

England now was often in his mind. Shrewsbury, Maer, Cambridge; his sisters, his cousins, his friends—nothing could compensate for their distance. Nevertheless some aspects of his return disturbed him. "This voyage is terribly long," he wrote to Fox from Lima. "I do so earnestly desire to return, yet I dare hardly look forward to the future, for I do not know what will become of me." He referred to Fox's personal position—in orders, married, settled in his own home—as "above envy," but: "I do not venture even to frame such happy visions. To a person fit to take the office, the life of a clergyman is a type of all that is respectable and happy."[1] Clearly some change was taking place in him, obscuring that "distant prospect of a very quiet parsonage" he had seen so plainly three years before. Thus far his future had been very much left to take care of itself. The presumption on his sailing had been that he *would* return to enter the Church. Robert expected it. He expected it. The nature of his present doubts remains totally obscure; probably even he himself could scarcely have defined them. But there they were.

FitzRoy's spirits, on the other hand, had been improved by his *Challenger* exploit, and almost his last act in Peru was the purchase of yet another small schooner to carry on the survey of the Chile coast. But he was keeping to his time-table. On September 7th the *Beagle* sailed north-west for the Galapagos Islands. The South American coast-line sank swiftly out of sight.

IV

The Islands were nearly twelve hundred miles from the mainland, but the passage was a quick one, and the first peaks of

[1] *LL.*, i. 262.

Melville's blighted Encantadas were sighted on September 15th. Five weeks were spent amid the group, a period as important to Charles as any of the voyage. He had looked forward to the visit mainly in the vain hope of finding an active volcano for close inspection!

They landed first on Chatham Island, on a shore which FitzRoy thought "fit for Pandemonium" and someone else more precisely compared to the cultivated parts of Hell. Stunted trees apparently dead, meagre ill-smelling brown brushwood, and unwholesome weed-like flowers barely subsisted on the black broken heaps of crumbling lava, over which innumerable crabs, darkly hideous yard-long lizards, and huge swaying tortoises scuttled, scampered and crawled at one's approach. The birds were extraordinarily, even stupidly, tame.

The coast region of Charles Island was similar, but it had a fertile inland area supporting more than two hundred persons. Lawson, the Vice-Governor of this settlement, talked to Charles about the more famous inhabitants of the group, the giant tortoises with their old men's faces, fixed unwinking gaze, and almost hypnotically rhythmic walk, saying that these creatures differed so from island to island that he could tell at a glance from which any one of them came. It shows the dormant quality of Charles's speculations that he "did not for some time pay sufficient attention to this statement," collecting "blindly" and mixing the specimens from this island and that indiscriminately together, never conceiving that places so closely adjacent, "formed of precisely the same rocks, placed under a quite similar climate, rising to a nearly equal height," could really be "differently tenanted." [1] Regarding the group as a unit, he noted in his diary the interest of discovering "from future comparison to what district or 'centre of creation' the organized beings of this archipelago must be attached." [2] When the significance of Lawson's remark first came home to him remains obscure. The awakening incident was his own observation of specific differences in the mocking-thrushes shot on various islands. One of his accounts would suggest that this took place actually on Charles or perhaps Albemarle Island, and that he thereafter paid especial attention to their collection, but this is contradicted by other statements and above all by his failure to

---

[1] *Journal*, 394. [2] *Beagle Diary*, 337.

notice similar variations among the ground-finches, which he mingled so carelessly that FitzRoy's private collection had to be utilized for tentative local identification some while after the return to England. "It was not *intentionally* (his italics) that I brought the different species from different islands." [1]

Albemarle Island, "the mainland of the archipelago" but sterile and uninviting as Chatham, was visited next, and last of all James Island, its interior fertile like that of Charles Island.

By October 19th the surveying parties were drawn together again, and the next day they were off. West-south-west they pressed for nearly twenty days, making their regular 150 miles between noon and noon, with never a sight of land or vessel. Pleasant days they were of bright sunshine under a cloudless sky, white sails set full above, blue water all about. There was little to do and less to say, however; in a fortnight Charles wrote only twice in his diary.

November 15th they anchored off Tahiti, even then the most noted of the South Sea Islands. The first view, under the heavy clouds of early morning, was disappointing, but as the sun rose the sky cleared, and the high sharp peaks of the imposing mountains, the massed green, the graceful palms, the vivid blue sea breaking white upon the reef, appeared before them dazzling and delightful. The men who came out in welcoming canoes were fine athletic figures, and handsome too, but Point Venus, where they landed, belied its name, for the shambling, ill-dressed, flat-nosed women they met there pricked the bubble of their expectation. All thought with FitzRoy: "Either they are not as handsome as they were said to be, or my ideas are fastidious." [2] But the friendliness of natives and English missionaries alike made their stay a pleasant one. Charles pursued his usual activities, also inquiring into the moral state of the islanders. FitzRoy considered the missionaries over-severe; Charles thought they probably knew their business best. They visited the Queen on official business, and she came aboard the *Beagle* for a firework display and singing by the seamen. FitzRoy regretted he could not fire a royal salute for fear of disturbing his chronometers.

Now, November 26th, for New Zealand. Again for more than three weeks they sailed uninterruptedly south-westward, and Charles could not but exclaim at the sheer immensity of this

[1] *Journal*, 629. [2] *Narrative*, FitzRoy, ii. 512.

vast stretch of ocean wherein the islands were but the merest infinitesimal specks. And now "the meridian of the Antipodes" was passed, and "every league, thanks to our good fortune, which we travel onwards, is one league nearer to England. These Antipodes call to mind old recollections of childish doubt and wonder. Only the other day, I looked forward to this airy barrier as a definite point in our voyage homewards; but now I find it and all such resting-places for the imagination are like shadows which a man moving onwards cannot catch. A gale of wind, which lasted for some days, has lately given us time and inclination to measure the future stages in our long homeward voyage, and to wish most earnestly for its termination." [1]

That wish dominated him in all these last months. He could at times work freshly enough, but constantly he looked to his home-coming with almost unbearable eagerness. This had been the part of the voyage which he had most ardently anticipated: it had been delayed too long. He was tired. The physical exhaustion of still frequent seasickness was bound to tell on him. He had had a bellyful of new sights; the remaining course of his life shows that his was not a nature which sought external change or movement, and the labour and strain of these busy years, without any sustained rest, were also having their effect. It was something of an effort to be enthusiastic about, or even to like, new places, and sometimes it was more than he could make. He was always glad to bid anywhere farewell.

In New Zealand he certainly found little remarkable beneath his visiting observation. The country round about the Bay of Islands, their one part of call, was bad for walking, indigenous animals were almost non-existent, birds were few, the English residents—apart from the friendly missionaries—were disreputable, the Maoris were dirty and treacherous, and government did not exist. He was only happy when stray glimpses of the missionary settlements reminded him of England. They sailed on December 30th in expectation of a gale, but feeling that any gale was better than another hour ashore!

Sydney, the red revolving beacon of whose lighthouse was sighted near midnight of January 11th, 1836, was another matter. All had heard of the progress of this "Capital of Australian civilization," and Charles was filled with patriotic pride to see

[1] *Journal*, 496.

how much had been done here in a few years; how little, comparatively, in South America in centuries. "My first feeling was to congratulate myself that I was born an Englishman."

A further view brought doubts. He visited Bathurst, 120 miles inland, and though the roads were well made and the views often impressive, the broad impression he received was again of sterility rather than richness. He wouldn't choose to live here, anyway. The convict associations were lowering, there was no culture; were, in fact, all the social drawbacks with none of the economic possibilities of the United States.

The trip had nevertheless its interest. He saw something of the aborigines, and thought them superior to the Fuegians. Considering their rapid decline in numbers, he noted the instantaneous effect of lessened food supply in checking native population. Spending a few days on a sheep-farm, he saw "the famous Platypus" at play, and also recorded an incident demanding quotation as one of the most interesting exhibitions of the workings of his mind at this time on primary problems of geographical distribution: "I had been lying on a sunny bank and was reflecting on the strange character of the animals of this country as compared with the rest of the World. An unbeliever in every thing beyond his own reason might exclaim, 'Two distinct Creators must have been at work; their object, however, has been the same and certainly the end in each case is complete.' While thus thinking, I observed the hollow conical pitfall of the lion-ant: first a fly fell down the treacherous slope and immediately disappeared; then came a large but unwary ant; its struggles to escape being very violent, those curious jets of sand described by Kirby as being flirted by the insect's tail, were promptly directed against the expected victim. . . . There can be no doubt that this predacious larva belongs to the same genus with the European kind though to a different species. Now what would the sceptic say to this? Would any two workmen ever have hit on so beautiful, so simple, and yet so artificial a contrivance? It cannot be thought so: 'One Hand has surely worked throughout the universe.'"[1] He added: "A Geologist perhaps would suggest that the periods of Creation have been distinct and remote the one from the other; that the Creator rested in his labour."[2] Plainly he

[1] *Journal*, 526. [2] *Beagle Diary*, 383.

himself, when he wrote the words, was far from clear what his own view was.

He returned to the *Beagle* tired, and it was now that FitzRoy expressed his anxiety to get him soon home, where he would be bound to rest. They had some pleasant evenings at Hobart, in the house of the Surveyor-General, and the Tasmanian interior impressed them more favourably than almost anything Australian. Their last glimpse of that continent, a week's desperately tedious anchorage in King George's Sound, enlivened only by a native corroboree for which Charles paid the piper with unlimited boiled rice and sugar, made them glad to see the last of it. "Farewell, Australia!" Charles apostrophized on March 14th, "you are a rising infant and doubtless some day will reign a great princess in the South; but you are too great and ambitious for affection, yet not great enough for respect; I leave your shores without sorrow or regret." [1]

v

Now indeed they might feel that they were homeward bound. The gaps in Charles's diary become still more evident. At sea there was nothing to say. He had written to Henslow from Sydney that he was tired of seeing new places unless they had some special scientific interest, and he didn't, he said, expect to find many such between Australia and England. After a fortnight's run in poor weather, the white beaches and emerald lagoons of the Keeling Islands rested and refreshed his spirit; he felt their tropic spell, and, more than that, as he pressed his inquiry, making observations all about the reefs, he seemed to find confirmation for his theory.

More sickness, and then, at the end of April, Mauritius, with music and, for the first time since Valparaiso, a measure of European culture: the benefits of civilization amid entrancing tropic scenery. He rode on an elephant and enjoyed what he called an "idle and dissipated" time. Three weeks again of sea, silence, and sickness ("I positively suffer more from seasickness now than three years ago" [2]) reached South Africa, a sorry contrast: "I never saw a much less interesting country." [3] The one bright spot was a social week in Cape Town. He

[1] *Journal*, 538.   [2] *LL.*, i. 224.   [3] *Journal*, 579.

dined out, accompanied Dr. Andrew Smith on geological rambles, met Maclear the Astronomer-Royal, and thoroughly enjoyed discussing Lyell and languages at the pleasant home of Sir John Herschel, whose good-nature made up for his lack of manners.

In Cape Town he and FitzRoy found interest aroused by their reports of missionary work in the Pacific, and on the way to St. Helena they set down their views—Charles contributing extracts from his diary—in a *Letter containing Remarks on the Moral State of Tahiti, New Zealand, etc.*, which urged all support of missionary work and was signed by both of them. Returned to the Cape by another vessel, it appeared in the *South African Christian Recorder* in September 1836. They had, too, a larger collaboration in view. Both were busily writing, Charles at his geology, FitzRoy at his account of the voyage. The latter had had, at his request, some portions of Charles's diary read aloud to him, and had suggested that he make use of it to fill out his own account, no doubt with due acknowledgment. Charles modestly agreed, if his friend thought "the chit-chat details of my journal anyways worth publishing." [1] No more was done in the matter—perhaps because the bulk of the diary was already at Shrewsbury, for there was clearly no idea then in Charles's head that he might make a book of it himself. It may even have been FitzRoy's suggestion that set the notion there.

At St. Helena he turned away from the Napoleonic associations, which depressed him, to exalting geology, which once again so engrossed him that he felt almost sorry when, on July 14th, the *Beagle* went her way. Nearing home, he seemed to gain a second wind, and at Ascension his spirits were raised yet higher by the report, in a letter from his sisters, of Sedgwick's prophecy to Robert that he "should take a place among the leading scientific men." [2]

His pleasure softened the blow of FitzRoy's sudden decision to complete his chronometrical data by another visit to Bahia. Yet if at one moment he thought it would be pleasant again to see Brazil and walk in its abundant forests, at others he felt that it was almost more than he could bear: "I loathe, I abhor the sea and all ships which sail on it." [3] He wanted to be at home,

---

[1] *R. FitzRoy and C. Darwin*, Barlow, *Cornhill*.
[2] *LL.*, i. 66.   [3] *LL.*, i. 265.

with "not one single novel object near me!" He wanted to get to work. He trusted that in future, he wrote to Susan, "I shall act as I now think—as a man who dares to waste one hour of time has not discovered the value of life."[1] The phrases are odd and indeed cryptic. It is not at all clear whether the latter part of the sentence presents the substance of his thought ("I shall act in the belief that a man who dares etc.") or only its motive-force ("I shall forthwith act according to the beliefs I now hold, because I believe also that a man who dares etc."). It may be merely a pious aspiration thrown in for his father's delectation. It may on the other hand, if the latter of the two interpretations is accepted, conceal much deeper meaning—that he has definitely arrived at new conclusions, whether scientific or religious, and that he means, in the strength of the confidence given him by Sedgwick's good opinion, to base his life on them henceforward.

At Bahia, in early August, he found his pleasure in tropical scenery as keen as ever, taking a lingering farewell draught of its strange intoxicating beauty in the endeavour to "fix in my mind for ever an impression which at the time I knew sooner or later must fail." He consoled himself that though the details would be lost, yet something would remain: "like a tale heard in childhood, a picture full of indistinct, but most beautiful figures."[2]

A few last days were spent at Pernambuco, sheltering from ill weather, but he liked neither the place nor the sullen inhabitants, and he was glad to be away, from it and from them and from Brazil, glad even to be seasick, sobeit he was on his homeward way. They crossed the Equator on August 21st not far from their first southward transit, and soon were once again at Porto Praya. But now they went by way of the Azores, leaving Teneriffe to the eastward, and it was here, on the island of Terceira, that he was able at last, relatively almost on his own doorstep, to fulfil his ambition of visiting an active if minute volcano. Now for the last time he recorded the "good news" that the *Beagle* was ready to sail, and after one fruitless final call for letters at St. Michaels, "we steered, thanks to God, a direct course for England."[3]

It was a quick passage of seven days, but through continuously

[1] *LL.*, i. 266. [2] *Journal*, 591. [3] *Beagle Diary*, 425.

bad weather, and he ended the voyage almost as painfully as he began it. Sunday, October 2nd, 1836, there was storm up the Channel, drenching rain driving before buffeting south-west winds. It was dusk at least when the brig dropped anchor in sheltering Falmouth Harbour, dark by the time he came ashore, beardless as he sailed, asking about coaches for the north, for Shrewsbury. He was weak, even ill, pale under his tan from the days of severe sickness, it was late, the storm continued unabated—he caught, nevertheless, the night mail-coach eastward. Through five years he had longed for England. Now it meant little more than firm ground under his feet; Shrewsbury was home. Next day, as the coach dragged its way over the steep West Country hills, he felt terribly tired, but by the 4th he was recovering, the weather was improved, the way easier, and his spirits rose to see again the green fields and autumnal woods and orchards of Gloucester, Worcester and, at last, Shropshire. He wondered that his fellow-travellers could sit so stolidly disregardless of the superbly, movingly beautiful panoramas that opened before them ceaselessly. The wide-stretching vistas of grass softer and fresher than any he had seen these five years, the little villages with their stone churches and thatched low cottages, market-towns dreaming in the sunshine, steeply rising hills wooded or turfed to the top, rivers or streams sparkling and whispering in every valley, here sheep, here cattle, there the ploughmen with their horses turning over the good red earth for winter sowing—the variety was endless, the repetition ever fresh. "The wide world," he wanted to cry out, "does not contain so happy a prospect as the rich cultivated land of England." [1]

He was in Shrewsbury that midnight, but would not disturb the family, so slept at an inn, and only next morning entered The Mount unannounced just before breakfast, to greet his father, his sisters, his nurse Nancy, with excited happiness. Robert, shaking his hands, cried in surprise: "Why, the shape of his head is quite altered!" [2] but the girls found him only a little thinner. He had been absent from them just five years and three days. He went out to speak to his dog, wondering whether it would have forgotten him altogether, but it gave him precisely the same placid recognition, no more and no less,

[1] *LL.*, i. 270. [2] *LL.*, i. 64.

it had been wont to give him any morning he came to call it for a walk. He might, he felt, never have been away!

The others did not think so. The labourers about the stable and garden felt a celebration to be in order, and before night were roaringly drunk. Charles was content just to be home again, to find everyone so well and as he had left them, to talk over the thousand things that never got said in letters, to hear the news of Erasmus, now settled in London, and of Maer, to read the welcoming notes awaiting him from cousins Charlotte and Elizabeth. He dashed off at once an excited few lines to Uncle Jos as "my first Lord of the Admiralty," expressing his joy to be back, hoping to see all at Maer quite soon, sending his good wishes to everyone. He could manage no more that day. "I am so very happy I hardly know what I am writing." [1] Even a word to Henslow must wait till to-morrow.

VI

Soon after his home-coming, probably in his first quiet days at Shrewsbury, he brought his diary to conclusion with a judicial summing-up of the profits and losses of his adventure. He was quite definite that no one should undertake such a business save for a purpose: its merely momentary compensations were inadequate to balance the lack of comfort, the loss of rest and of social companionship and cultural amenities. Seasickness could be a very serious matter, he knew too well, and even when one was fit the long weeks at sea grew wearisome. The scenery of the various new countries held its unique delights, no doubt, but omitting what was special to the tropics might not equal grandeurs be encountered within European limits? And yet, when he looked back on it all, how many memorable sights there had been: the sublime primeval forests of Brazil, the wastes of Tierra del Fuego, the sterile and yet in their vastness impressive plains of Patagonia. Wide views from high mountains, primitive man in his savage haunts ("one's mind hurries back over past centuries, and then asks, Could our progenitors have been men such as these?" [2]), the exotic stars of the South, water-spouts, great glaciers, volcanoes in fiery action, solid rock

[1] *ML.*, i. 28. [2] *Journal*, 60.

shuddering in earthquake, the strangeness of coral islands—these he especially recalled, though they must have been but few among many more: the white peak of Teneriffe, the sedan chairs in the streets of Bahia, the *Beagle* phosphorescent in storm off the Rio Plata, mile-long clouds of butterflies filling the air far out to sea, the cries of unhappy slaves, the Buenos Ayres guard-ship firing (good God!) on the British flag, the fossils at Punta Alta, Cape Horn in calm sunlight, Jemmy beside his beacon waving them good-bye, nights on the pampas with the gauchos, the strata of the eastern plains of South America telling their story in ever-plainer language, the white albatross flying serenely up the wind in the severest gale of all, the sunlit peaks of the Andes seen in distance, their snowy tops all about him as he climbed to the pass, the ruins of Concepcion, the noble athletes of Tahiti, the Australian platypus and the lion-ant, the lagoon of Keeling Island by moonlight. . . . They had no end!

He *was* unequivocally glad that he had gone. Nothing, he thought, could be more improving to a young naturalist, for it must stimulate him to active thought: "as a number of isolated facts soon become uninteresting, the habit of comparison leads to generalization."[1] Admittedly one stayed so short a while in any place that impressions were bound to be cursory, and "hence arises, as I have found to my cost, a constant tendency to fill up the wide gaps of knowledge by inaccurate and superficial hypotheses," nevertheless he had "too deeply enjoyed the voyage not to recommend any naturalist . . . to take all chances, and to start on travels by land if possible, if otherwise on a long voyage."[2]

Sedgwick's words to Robert had cheered him at Ascension, and he cannot have been in Shrewsbury long before learning that Henslow had valued his letters so high as not only to communicate them to the Cambridge Philosophical and London Geological Societies but also to print some of them as a pamphlet for private circulation, and that Sedgwick had written also, and most handsomely, to Dr. Butler of how he was "doing admirable work in South America, and has already sent home a collection above all price. It was the best thing in the world for him that he went out on the voyage of discovery. There was some risk of his turning out an idle man, but his character will now be

[1] *Journal*, 608.   [2] *Ibid.*

fixed, and if God spares his life he will have a great name among the naturalists of Europe." [1]

It was fine praise, and complete vindication. When Charles now took for granted his foremost need to set his collection in order and make available the fruits of his labour, no one, not even Robert, could say him nay.

[1] *LL. Butler*, ii. 116.

CHARLES DARWIN IN 1840
(From a water-colour by George Richmond, R.A.)
From *The Life, Letters and Labours of Francis Galton*. By the courtesy of the Cambridge University Press

CHAPTER IX

LONDON: THE FIRST FRUITS

I

ONE would like to know exactly when it first became clear, to Charles and to his father, that he had neither wish nor intention to enter the Church. Aboard the *Beagle* his piety had been impeccable, the far from impious officers chaffing him for his solemn citation of the Bible as a standard authority on current morality. From South America he was still seeing his future in terms of that "very quiet parsonage," and despite the doubts assailing him at Lima there is no indication that these might not have been overcome as completely as his doubts at Cambridge. His Christian faith at the time of his home-coming was, he long afterwards testified, "as firm as that of Dr. Pusey," and he did not abandon it till he was forty. Actually the thought of a Church career seems to have been still in his mind some time after his return, and the probability is that it was never explicitly rejected. Simply, there supervened the more immediate task of dealing with his harvest of specimens and notes, and the other matter receded, eventually to vanishing-point, as that work extended from year to year.

He hoped, optimistically, it would not delay him long, and set busily about it, meaning to waste no time. His first letter to Henslow, dated October 6th, said in effect: I am home; who will help me order my geological specimens?—and after a week at The Mount, without leisure for even the most hurried visit to Maer, he was off to Cambridge to talk the matter over before meeting the *Beagle* at Greenwich on the 28th to see the last of his treasures safely ashore. There were the friendliest of reunions, with Henslow at Cambridge, with Erasmus in London. He stayed with the latter in his rooms at 43 Great Marlborough Street, just to the south of Oxford Street, and there was much gadding-about and excited talk. Hensleigh Wedgwood and his wife, Frances

Mackintosh, married in 1832, were close at hand and eager to see him. Nor did he fail to renew old scientific contacts and make new ones, finding himself greeted on every hand with ready interest. He visited Brown and Yarrell, dined at the Linnean and Geological Society Clubs, saw Richard Owen, Broderip, Lonsdale the assistant secretary to the Geological Society, Professor Bell the zoologist, and many others. He was, to his delight, quickly elected a Fellow of both Geological and Zoological Societies.

Most thrilling of all was his meeting with Lyell. He wanted to meet no man more, and no man more splendidly fulfilled his expectation. Their very first encounter, to which both had looked forward for months, was compact of pleasure and excitement. Lyell at this time was nearly forty, a handsome, frank-featured man with engaging quizzical glance and friendly smile, the quick humour and easy manners of the naturally sociable being. Of independent means, he was a model of personal disinterestedness, anxious only for science's advancement, the clarification of knowledge and ideas, and he united interest and caution in sound judgment. Charles, as disciple, came to worship and remained to talk. How flattering to find one's very master listening so eagerly, taking every suggestion so seriously, pausing patiently to understand its every implication! How electrifying his reception of one's tentatively proffered explanation of coral-reef formations as due to ocean-bed subsidence! He put first, in his low inexpressive voice, his customary cautious questions, anxious to get the matter completely clear in his mind. Then, as he realized its entrancing simplicity, the conviction of its truth possessed him, and he leapt to his feet and actually pranced about the room, hooting and chuckling with delight. Away, his own volcanic craters—Truth must be served! Eagerly they discussed it. Charles till now had scarcely realized the importance of his theory, or how, if it could be established, the distribution of coral reefs might serve as index to areas of subsidence and elevation. Lyell urged him to make it matter for a paper for the Geological Society, and as soon as might be. Few of the members of that divided body would jump to it as he had done, but that was the lot of pioneers: Truth *would* be served!

Naturally Charles was delighted by such approval; it did him, he said, a world of good, keeping him encouraged at a moment when he needed some such aid. For he soon found the prospects of getting his specimens attended to unpropitious. The zoologists were particularly unhelpful, Mr. Owen and Dr. Grant (the Edinburgh friendship renewed) alone showing interest. The British Museum had totally neglected the collection presented to it by Captain King six years before. The only thing, it seemed, was to assemble all the trophies at Cambridge and then see what could be done with them. That was what Henslow urged, and though Lyell, in the enthusiasm of new-found friendship, was "all for London," he too eventually agreed. Everything was despatched there by November 6th, and there Charles, a few weeks later, was to set to work, but first he must visit Shrewsbury again, and then, at last, with Caroline, Maer. There had been changes. Aunt Elizabeth had been an invalid since an epileptic fit and fall in 1833. Fanny, youngest of all but Emma, had died in 1832; and dear Charlotte had married that same year and had her own home now. But Uncle Jos was there, and Aunt Elizabeth too, and the younger Josiah, and Harry, Frank, Elizabeth and Emma, and if the old Tag-Rag days were gone for ever, they could be happy still, full of high spirits, talking, questioning, laughing, walking together, greeting the cousins and neighbours who crowded in to meet the returned wanderer.

December 10th he was back in Cambridge, spending a few days with the Henslows before settling in lodgings in Fitzwilliam Street. He felt at first a melancholy, missing the old faces, finding Christ's wholly inhabited by strangers. But dining there he was soon at home again, entering as of old into the after-dinner jests and wagers. He was really very happy all these first few months simply to be once more in England. He had already (October 19th) received his degree as Master of Arts, a purely formal business.

Meanwhile he was at work, ordering his collections and still seeking to place them in safe hands. Sometimes he wondered whether all his trouble, notably with the fossils, had not been for nothing; he would have given the lot to the Paris museums but for a sense of national obligation since he had got them aboard "a King's ship." But Owen reassured him, and prospects improved; he was eventually able to dispose of most of the material to good effect.

The problem of the publication of his diary also bothered him.

FitzRoy's suggestion still stood, but Charles's sisters, Catherine particularly, urged its separate, complete printing. Dr. Holland was consulted, and depressingly thought it unworthy of appearance in any form. Emma thought him no judge at all; she was with Catherine, whose plan was quickly adopted. Charles accordingly, after his day's work on specimens, would settle in the evenings—if he was not dining out, at Christ's, or with Henslow or Sedgwick—to copying, recasting, expanding, his original entries. He made good progress. The original diary consisted of some 189,000 words, of which about 126,000 were to be retained, another 100,000 being added to make up the volume presently published as the *Journal and Remarks*. The omissions were mainly of personal matter: records of his own feelings, comments on shipmates or hosts and companions ashore, details of sea-life and minor movements on land. The period of waiting at Devonport and the first weeks at sea were cut completely, and the partial rearrangement by place rather than time, most marked in the South American East Coast chapters, disposed of a number of passages. The additions consisted almost entirely of "popular" scientific treatment of natural phenomena recorded in his various more technical notebooks—inserting in the first four chapters, for instance, paragraphs and pages on the zoology and geology of St. Jago, Brazilian geology, sea-discoloration, flatworms, climate, phosphorescent insects, butterflies, beetles, ants, spiders, the lazo and bolas of the Argentine gauchos, the climate and zoology (especially birds) of Brazil, and the salinas of the Rio Negro. These insertions comprised about half the 40,000 words so far set down.

Other matters interrupted, notably a paper read to the Geological Society in London in early January, on *Proofs of recent elevation on the coast of Chile*. He went up to town two days before, on the 2nd, to discuss parts of it with Lyell, dining there and meeting Mrs. Lyell's sister, mother, and father, that Leonard Horner whom he had known in Edinburgh. The reading of the paper was the occasion of his formal admission as a Fellow of the Society, and in the following month he was elected to its Council.

In early March his work at Cambridge seemed done; he must settle now, at least for a while, in "dirty, odious London." Though he loathed the very sight of the smoke-grimed walls opposite, he took rooms close to Erasmus at 36 Great Marlborough

Street, and moved in on March 13th. The next day he read to the Zoological Society a paper on the South American ostrich.

Work now pressed upon him. It was quite settled by this time that he was to have the third volume to follow King's and FitzRoy's accounts of the two *Beagle* voyages, and completion of his material seems to have been regarded as urgent. He concentrated on it, working many hours a day, permitting himself only the professional leisure of visits to the Geological Society—where he took active part in discussions and sometimes proved himself a match for bores and fools—but becoming more and more aware, as he turned his notebooks over, of the work still awaiting him. A couple of years at least, he estimated pessimistically. Walking the "hated" London streets, he lamented his exile from both Shrewsbury and Cambridge. He took what consolation he could from the hospitality of Lyell, Owen, Babbage, the more intimate companionship of Hensleigh, Erasmus, his visiting sisters. Erasmus had already begun his long friendship with the Carlyles, and Charles would occasionally encounter them at No. 43, but with all his customary generosity he found Thomas a boor and Jane hardly a lady.

II

He was, as the *Journal and Remarks* went ahead, finding himself wading in deep waters. He had sailed upon the voyage a convinced creationist, to whom the immutability of species was axiomatic, and in my opinion he never decisively renounced that view while still abroad. "Vague doubts occasionally flitted across"[1] his mind—as he turned his trophies over (the fossils especially), as he read Lyell, as he wakened to the oddities of Galapagos distribution—but they were isolated, unorganized, never uniting in any single substantial point of view.[2]

It was only now that, as he wrote, or pondered over his notebooks, these doubts grew less vague, became concrete, all but

[1] *ML*., i. 367.
[2] Mrs. Nora Barlow (*Nature*, September 7th, 1935) quotes a passage from one of his *Beagle* notebooks seeming to indicate that, either at James Island or while still in the Pacific, he realized the significance of the Galapagos phenomena as "undermining the stability of species"—his own immediate phrase. Though its date is uncertain, the note shows how far his questioning could occasionally go without, as I believe, taking a stand upon its intuitions.

irresistible. During this time his brain was busily speculative on many matters—the conditions of insect luminosity, the activity of lightning, presence and absence of trees, origin of salt lakes, relation of earthquakes to weather: a hundred such phenomena—and naturally his questioning extended also to these problems of species persistently presenting themselves: causes of specific distribution and extinction, specific tendency to slight variation, specific interdependence, the hereditary nature of acquired characters. His contemplation and consideration of the fossil bones found at Punta Alta, La Plata, Southern Patagonia, pressed upon him "confirmation of the law that existing animals have a close relation in form with extinct species,"[1] and the geographical as well as chronological succession of types, as also adaptation to environment, was illustrated "by the manner in which closely allied animals replaced one another in proceeding southwards over the continent."[2] But it was all very tentative; he was much more concerned with the disappearance than appearance of species, and even the late chapter on the Galapagos Islands, though almost all newly written, did not carry him much further. He described the geographical and geological conditions, the flora and the strange and evidently aboriginal reptilian fauna. The variation, in both birds and reptiles, of species from island to island was remarked, and he regretted his failure to realize this fact in time to keep his collections separate. He noted the generally American—though also distinctively individual—character of the islands, and in writing of the unusual tameness of the birds set down their commoner wildness elsewhere to acquired habit become hereditary. In general, however, he recorded the facts with no more than a minimum of comment.

Clearly he was at this point picking at the problem rather than facing it, held back partly no doubt by his characteristic caution, but also by the sheer haste of writing which prevented him from fully grasping the hints and pointers with which the book bristled—shoots of undeveloped ideas needing time for growth. It was, this *Journal*, the work of a man who had observed widely and recorded clearly, who had been deeply impressed by the variety, fertility and persistence of life, and by the endless cycles of inorganic and organic change everywhere in progress, as unceasing in the centuries as in the instants of time; and whose mind was

[1] *Journal*, 209.   [2] *LL.*, i. 82.

restlessly playing over the whole field of his observation, attentive yet still so dazzled by the wealth and range of his material that he found it hard to settle to particular considerations. The book was in fact the process, not the fruit, of clarification. This undoubtedly is the explanation why—though the matter was moving towards the front of his mind all these spring and early summer months—it was not until July 1837, when the manuscript of the *Journal* was completed and out of his hands, that he started his first notebook of facts and comments bearing on transmutation of species.

It was a subject, as has been indicated, of old and honourable ancestry. Notions of evolution, whether by survival of the fitter in the struggle for existence, or by the action of some inherent perfecting principle, had haunted the minds of men intermittently since the days of Greek speculation, kept alive through many centuries by the devoted study of the classics. In more modern times Bacon, Descartes, Leibnitz, Kant, Lessing, Herder, and Schelling had all looked over a philosophic fence at it. Diderot, Robinet, and later Oken and others, guessed at its mechanism out of a void of factual ignorance, each seeming to anticipate elements of the final Darwinian solution. Buffon, the first great naturalist among modern evolutionists, glimpsed the power of the struggle for existence and the extinction of the less fit, but left it to Erasmus Darwin and then Lamarck, keen observers both and each in his own way knowing something of the evidence, to make the first coherent statements of their resemblant schemes of one great organic evolution of all species branching from a single beginning by individual variation, influence of habit, inheritance of acquired characters, and selection by struggle and survival or extinction. Buffon had laid stress on the direct effect of environment in producing change; Erasmus Darwin and Lamarck both saw that influence as indirect, changed environment changing habit, and changed habit shaping the creature anew by its own "perpetual endeavour" to satisfy its basic needs. Each stated a case rather than substantiated it. Goethe, Treviranus, Naudin and St. Hilaire were all advocates of evolution, and the last-named in 1830 brought the subject into public controversy with Cuvier before the scientists of Paris. In England Dr. W. C. Wells, who knew the *Zoonomia* and was probably influenced by it, read before the Royal Society, as early as 1813, a paper on "a White Female, part of whose Skin resembles that of a Negro," which definitely stated, though with narrow application, the working of Natural Selection, and that same principle was again perceived, and much more definitely applied to the general evolution of species, by a Mr. Patrick Matthew in an appendix to his volume on *Naval Timber and Arboriculture*, published in 1831.

Evolution was in the air. While Charles was far away on the other side of the earth, Browning was writing *Paracelsus* and Tennyson embarking upon his *In Memoriam*, each of which embodied or implied the essential evolutionary idea of growth and unfolding, not only "from life's minute beginnings, up at last to man," but also in mankind itself, so that "in completed man begins anew a tendency to God." Emerson too, in America, was expressing even more definite views, derived from Lamarck.

Yet if Charles only partook of a growing mental movement of his day, his debt to most of his precursors was either non-existent or relatively small. In particular he knew nothing of Wells or Matthew, and the Greeks as ever, and for that matter most of the Continental writers, were Greek to him. His early reading of his grandfather's prose and verse had given him his first introductions to the evolutionary hypothesis, and probably left him more of a residue than he was ever willing to acknowledge. Lamarck he knew, one gathers, at least to a late date (well after 1837), only by hearsay, perhaps enthusiastically from Dr. Grant at Edinburgh, more critically by way of Lyell. Grant's own belief in the descent of species from other species, with improvement by modification, must have impressed him ten years before, and now probably he found Lyell, in personal talk, less opposed to evolutionary ideas than parts of the *Principles* had led him to imagine. That train of thought set forth in Volume Two as leading the uncritical student towards error may have taken on a new cogency in his reflections. How nicely, after all, it did account for many puzzling things. Despite the deadweight of general zoological opinion against it, it would at least bear looking into, once this blessed *Journal* was out of the way. . . .

III

He could not give all his time to it. Other matters intervened continually. Two papers had to be read to the Geological Society in May, one on the fossil deposits of Punta Alta, the other—product of Lyell's enthusiasm—*On certain areas of elevation and subsidence in the Pacific and Indian Oceans, as deduced from the study of coral formations.* He was able, though, to incorporate the latter in the *Journal* as the bulk of Chapter 22. There was also the business, now arising, of making public the zoological fruits of the voyage. To do the job properly, under the supervision of competent zoologists, the slow troublesome business of obtaining a government grant was absolutely essential, for Charles, though he looked for no profit, was determined to spend no more than his time upon it.

There were further preoccupations. Emma Wedgwood, in May, seemed to be hinting to Hensleigh's wife at rumours of his possible engagement to a London girl whom Susan and Catherine had met and liked. It may have been no more than chaff, but marriage was in Charles's mind, if the ascription of some undated notes to this period is accepted. In these he set forth the advantages and disadvantages of the married state. On the one hand: "Children (if it please God)—constant companion (and friend in old age)—charms of music and female chit-chat." On the other: "*Terrible loss of time*, if many children forced to gain one's bread; fighting about no society." But: "What is the use of working without sympathy from near and dear friends? Who are near and dear friends to the old, except relatives?" Then the conclusion: "My God, it is intolerable to think of spending one's whole life like a neuter bee, working, working, and nothing after all. No, no won't do.—Imagine living all one's days solitarily in smoky, dirty London house—Only picture to yourself a nice soft wife on a sofa, with good fire and books and music perhaps—compare this vision with the dingy reality of Gt. Marlboro' St.

"Marry, marry, marry.

"Q.E.D."[1]

The time and place and the loved one, however, were not yet brought together, and Miss —— passes from the picture, no more than a glancing reference in a single letter.

By June the *Journal* was mostly done and an easier period began. He visited Shrewsbury, perhaps to celebrate Caroline's long-awaited engagement to cousin Josiah (they were married, in the quick fashion of those days, before the summer was out),

[1] *Emma Darwin*, i. 277.

but was soon back, staring out at "the same odious house" on the other side of the street. William IV died and young Victoria became Queen, but Charles, also making history, was more concerned with his notebooks. A Government grant of £1000 for the zoology volumes was promised him after a pleasant interview with the Chancellor of the Exchequer. In August the first *Journal* proof-sheets were coming to hand. All seemed going well.

He was tired, though, weary of London still, and bothered by the same palpitations which had troubled him six years before at Devonport. On medical advice he set all work aside and went holidaying at Shrewsbury and Maer. At the latter place the family was enjoying *Pickwick Papers*, reading it aloud. Emma thought it "not at all too low," and Charles acclaimed Samivel the very prince of heroes. Uncle Jos set him to reflecting on the part played by earthworms in the formation of mould, with the result of a paper read to the Geological Society in November and a volume on the subject published in the year before his death. He was worrying while at Maer whether to accept the proffered post of honorary secretary to the Geological Society, and doing all he could to evade it, fearing the burden of the labour for so slow a writer as himself; on the urging of Whewell, Sedgwick and Henslow he finally withdrew his objections, and occupied the position from February 1838 until February 1841.

From Maer he went to stay with Fox in the Isle of Wight, and was back in London, and much better, soon after the middle of October, preparing the prospectus of the *Zoology of the Voyage of H.M.S. Beagle*, which was to appear first intermittently in thin sewed parts, then bound up in five large quarto volumes. Owen was to do the Fossil Mammalia; George R. Waterhouse (who is said, quite incredibly, to have been invited to accompany Charles on the *Beagle*) the Recent Mammalia; John Gould the Birds; Jenyns (at Charles's insistence, since no one else offered) the Fish; and Thomas Bell the Reptiles. Charles would edit and superintend, contributing introductory and other notes. His principal hope was to get the work right off his hands in order to settle down to his geology, but a year later he was finding it still "a millstone round my neck."

The *Journal* was now in final page-type, and as he leafed it over he was divided between irritation at the stupid errors of the "goose" employed by Colburn the publisher as proof-reader, and astonishment at realizing himself an author. Between now and publication in 1839 bound or unbound copies passed from hand to hand among his friends. Owen read it. Horner praised its style. Lyell's father enjoyed every page of it. Quite unknown to the author, young Joseph Hooker was sleeping night by night with it under his pillow, that he might take it up the moment he awoke.[1]

IV

Zoological and geological labours went on through the winter and spring. So did the species notes. By February 1838 the first notebook was full, another instantly started. Already he had gone far, conviction sunk deep; he had grasped the essence if not the machinery of the matter. For example: "If we choose to let conjecture run wild, then animals, our fellow brethren in pain, disease, death, suffering and famine—our slaves in the most laborious works, our companions in our amusements—they may partake of our origin in one common ancestor—we may all be melted together."[2] "We"—not the beasts only, but men also, he saw, even so early, "must come under the same law."[3] Why not? After all: "The different intellects of man and animals not so great as between living things without thought (plants), and living things with thought (animals)."[4] He knew the objections which would be raised, the demand for proof of intermediate forms—the "missing links" of later fame—and considering this he coined a striking metaphor which deserves to be better remembered: "The tree of life should perhaps be called the coral of life, base of branches dead; so that passages cannot be seen."[5] He revealed how geographical distribution had impressed him: "Countries longest separated greatest differences—if separated from immersage, possibly two distinct types, but each having its representatives—as in Australia."[6] He reflected on extinction as the result of specific failure to change in a changing world;

---

[1] He and Charles met soon after this in Trafalgar Square, being introduced by one of the *Beagle*'s officers, and again, in January 1838, at the College of Surgeons, when Hooker made known a visiting American, the youthful Asa Gray.
[2] *LL.*, ii. 6. [3] *LL.*, i. 93.
[4] *LL.*, ii. 6. [5] *Ibid.* [6] *Ibid.*

he really had, at this time, could he but have seen it, the whole key in his hand: "With respect to extinction, we can easily see that variety of ostrich (Petise), may not be well adapted, and thus perish out; or, on the other hand, like Orpheus, being favourable, many might be produced. This requires principle that the permanent variations produced by confined breeding and changing circumstances are continued and produce according to the adaptation of such circumstances, and therefore that death of species is a consequence . . . of non-adaptation of circumstances."[1]

Adaptation was the outstanding fact, implying the capacity to change—variability. "If *species* generate other *species*, their race is not utterly cut off."[2] Variation among domesticated creatures was apparent enough, and so too was the man-controlled means of extending and preserving it: selection for breeding. Nothing could be plainer. The problem was to find the natural equivalent of such domestic selection—*a natural selection*. A hundred times he confronted the problem and retreated from it baffled.

He did not, though, fail to discern the importance and imaginative power of his conception of a single branching genealogy. To show that "all mammalia were born from one stock, and since distributed by such means as we can recognize,"[3] seemed to him a matter comparable only to Newton's exposition of gravity, and containing no less of grandeur. There was here, on a vastly greater scale, that *simplicity* of concept which Lyell had hailed in the coral-reef theory. There was, he felt, no branch of biology, psychology, and even metaphysics which it would not invigorate.

The moment of his first complete realization we do not know—whether it came to him suddenly in a flash of insight subsequently verified, or he to it slowly through weeks and months of growing perception. But there must either way have been times when he saw clearly and trembled at what he saw, not in fear but in exultation, the exultation of the moment—perhaps the highest known to human experience—when chaos falls into order, and opposed patterns into a single harmonious scheme, when, to take a phrase from a very different context, "the avalanche is no longer meaningless, since in nature everything has a meaning."[4] Of such occasions Kropotkin has written unforgettably: "There are not many joys in human life equal to the joy of the sudden

---

[1] *LL.*, ii. 8. [2] *LL.*, ii. 7. [3] *LL.*, ii. 9. [4] Tchehov.

birth of a generalization, illuminating the mind after a long period of patient research. What has seemed for years so chaotic, so contradictory and so problematic, takes at once its proper position within an harmonious whole. Out of the wild confusion of facts and from behind the fog of guesses—contradicted almost as soon as they are born—a stately picture makes its appearance, like an Alpine chain suddenly emerging in all its grandeur from the mists which concealed it the moment before, glittering under the rays of the sun in all its simplicity and variety, in all its mightiness and beauty. And when the generalization is put to a test, by applying it to hundreds of separate facts which seemed to be hopelessly contradictory the moment before, each of them assumes its due position, increasing the impressiveness of the picture, accentuating some characteristic outline, or adding an unsuspected detail full of meaning. The generalization gains in strength and extent; its foundations grow in width and solidity; while in the distance, through the far-off mist on the horizon, the eye detects the outlines of new and still wider generalizations. He who has once in his life experienced this joy of scientific creation will never forget it; he will be longing to renew it." [1] The passage as a whole may have stricter application to subsequent periods of Charles's thinking, but it is worth quoting here, at once, in relation to this first wide (however incomplete) perception, for it makes clear what was unquestionably all his life a strong impelling factor. Men do not pursue abstract knowledge in entire disinterestedness, but for their own ever-renewed and ever-deepening satisfaction. *He who has once in his life experienced this joy of scientific creation will never forget it; he will be longing to renew it.*

There were times too, no doubt, when he saw clearly and trembled at what he saw, not in exultation but in fear, his heart sinking as he surveyed the inimical faces and forces of the surrounding, the overwhelming, world of orthodoxy, ready, he must have realized, to cry out upon such blasphemy against God the Father Almighty, Creator of Heaven and Earth and all species living upon the earth or in the waters under the earth. There wasn't, seemingly, one friend to whom he could talk freely. Not even Lyell. So, for years, he spoke no word. For twenty years he printed no word. The most he did was to let it be known that the subject of species interested him, that it was his "prime

[1] *Memoirs of a Revolutionist*, Kropotkin, ii. 5-6.

hobby" and that he hoped "some day" to "do something" on the matter. In fact he began very early collecting facts not only on species generally but definitely on Man and Reproduction. If lower species were mutable, Man must be so too, and Reproduction, surely, was the key to the mystery. Notebooks soon proved inadequate, and folders, each clearly marked with its special subject, became the receptacle of copious notes tossed in at random and left to await further consideration and digestion.

Work alternated with pleasure. On March 7th he read to the Geological Society possibly the most important of his papers of this period: *On the Connexion of certain Volcanic Phenomena and on the formation of mountain-chains and the effects of continental elevations.* It was faithfully Lyellesque. In May he was unwell, but a three days' trip to Cambridge did him good. He stayed with the Henslows, spent a morning with Jenyns, dined with Sedgwick, and attended Sunday evening service at Trinity Chapel. Early in June Catherine Darwin and Emma were in London, just next door, on their way home from Paris, and he saw them every day, aware, it appears, of a new and not wholly cousinly attraction to Emma. Later, Francis Galton, grandson of old Erasmus by his second marriage, was in town for the Coronation of the Queen; he stayed with Charles, and they went out together to see the crimson-decorated streets. Young Galton paid thirty shillings for a seat in Pall Mall, but Charles saw the procession from a friend's house. The crowds seemed much more enthusiastic for young Victoria than ever for Sailor William. That over, he was off to Scotland by steam-packet, a thoroughly enjoyable trip with no shadow of sickness. It was his first visit to Edinburgh since his student days. He went by gig and jolting cart to Loch Leven and then Glen Roy, where he spent a week in superb weather and scenery examining the strange geological phenomenon of the "parallel roads," in explanation of which he invoked the action of the ocean against the orthodox lake theory. He clung to his view for twenty-three years before acknowledging his error.

He returned by sea from Glasgow to Liverpool, and was soon at Shrewsbury, where he began writing his Glen Roy paper and observing, desultorily, the fertilization of flowers by insects in pursuance of previous speculations on the part played by cross-fertilization in keeping species constant. He also started a note-book on metaphysics, but it did not get very far, for, as he said in later life, he "had not a metaphysical head." He did, however, some unspecified reading on the subject, and religious matters seem to have been exercising him, so that it is to this period that we must most probably ascribe his realization of the impossibility of his becoming a clergyman. There was, though, no sudden change. His disbelief first in the Old Testament, then in the miracles of the New Testament, and finally in Christianity as a divine revelation came to him so slowly as never even to distress him particularly. He did not want to give up his faith, and would day-dream of sudden discovery of incontrovertible proof of its authenticity; but gradually towards 1850 these idle fancies ceased, and he accepted his agnosticism without pain or protest. Basically, one feels, he, like Gallio, cared for none of these things; he had other, nearer preoccupations, and religion remained just one among other subjects to which he would have given more thought had he had more time.

One holiday event was a brief call at Maer. He meant now to ask Emma to marry him, and would have done so there and then, but feared her refusal.

Back in London, health still unsure, he was trying a new routine of working two hours, taking exercise, then returning for another two hours. He would dine at the Geological Club, or the staider Athenaeum of which he was now a member, and generally find a fellow geologist or naturalist to talk to. Lyell, writing from Scotland in September, asked when the book on coral reefs and volcanoes was likely to be out. Charles optimistically hazarded "at least four or five months; though, mind, the greater part is written." He must have been referring to his notes, for the formal writing actually began only in October.

Personally he wished all this geological stuff out of the way to give his time to species, for now in this same month his views on the subject took a sudden step forward. He had been reading fairly widely, from metaphysics to poetry. He even read *The Excursion* twice through, and among other books took up, "for amusement," Malthus's famous *Essay on Population*, in essence an assertion of the tendency of population to increase many times faster than subsistence, being kept in check only by famine, sickness, and war. There was, in effect, a constant "struggle for existence" in which the weaker perpetually perished. Instantly Charles perceived that here was that natural selection he had been seeking: "under these circumstances favourable variations would tend to be preserved, and unfavourable ones to be destroyed. The result of this would be the formation of new species."[1] He had now a definite theory about which to group his collected facts, yet still he refrained from any immediate attempt to formulate it. The anxiety to avoid prejudice, he called it; that, and the pressure of other affairs.

[1] *LL.*, i. 83.

CHAPTER X

LONDON: MACAW COTTAGE

I

HE could no longer, he suddenly felt, wait to know Emma's mind. He went unexpectedly to Shrewsbury, and thence, with Catherine, on Thursday, November 8th, to Maer. He told no one of his intention, but early on Sunday asked Emma to marry him. She did not hesitate in her consent.

She had hoped for this as long as he, or longer, had often wondered about his feelings. He seemed sometimes so fond, but then he was so affectionate by nature, and so demonstrative in his affections, that perhaps it meant nothing. That week in London in June it had seemed to her he did not care at all. That August day had been different, though; they had been so happy together, and she had felt that if only they could see more of each other he might really come to love her. His proposal was a shock, leaving her almost too bewildered to feel much happiness. She went as usual to Sunday School, found herself beginning to talk nonsense, and made an excuse to come out. There was the usual houseful of guests, but they told only her father, Elizabeth, and Catherine. The others suspected, but Charles and Emma both looked so miserable, he with a bad nervous headache, that all thought she must have refused him. But the news came out at bedtime, and there was a celebration "talking it over till very late" in Hensleigh's and Fanny's bedroom. They laughed and chattered and were happy, feasting on thickly buttered bread foraged from the pantry. The absent Charles's qualities were reckoned over—his straightforwardness, affection, kindliness, essential temperance. Of course they would live in London where he was so important and busy. It would be a wrench to go so far from Maer, but thank heaven for the London-Birmingham railway, opened this very year. "Charles is so fond of Maer that I am sure he will always be ready to steam down whenever he can." [1]

Early next day Charles and Catherine were off to Shrewsbury with the news, and Robert was as pleased as Uncle Jos, who was soon writing of marriage settlements, promising a bond for £5000 and an allowance of £400 a year "as long as my income will supply it, which I have no reason for thinking will not be as long as I live." [2] Aunt Elizabeth's failing mind ran rather to wedding-cakes. Charles, back in London, was soon receiving the congratulations of the Lyells and Carlyles, paying his first visit to Cheyne Row. (Jessie Sismondi, who "knew" that Emma was to be a Darwin "by her hands," was especially grateful to him for saving her from Erasmus, whom Aunt Jessie unreasonably disliked.) On all hands he heard Uncle Jos's words repeated—that he had, in Emma, "drawn a prize" indeed.

It was, for once, true. There is no period of Emma's long life when she does not appear as attractive in the extreme. Born on May 2nd, 1808, nine months before Charles, she was the youngest of the Wedgwood children, and she and her sister Fanny, two years older, were the inseparable pets of the family, the "dovelies" or, in more impish mood, Pepper and Mustard. She was the positive one of the pair, a high-spirited, active, pretty child, impulsive, honest, a little self-willed. Tidy she never was, and never became; her mother called her unavailingly Miss Slip-Slop. She was educated mostly at home, by her sisters Elizabeth and Charlotte, but there and on visits abroad she learnt a little French, German, and Italian. She read, she rode, she played the piano with a notably fine touch, she mastered needlework and archery, drawing and dancing with an equal ease. Often she was away on visits to Cresselly (a pleasanter place now that old John Allen had gone the way of all flesh) or Parkfields and, later, London. She first went abroad in 1818, spending spring and summer in Paris, some of the time with father, mother and sisters, some of it at a French school. When barely fourteen, she went with Fanny for a year to a young ladies' establishment at Paddington Green, a semi-rural village, and, returning, became teacher in the Maer Sunday School which, held in the laundry, was the sole schooling of the sixty poor children attending it. She made up little stories in simple

[1] *Emma Darwin*, ii. 7.   [2] *Ibid.*, ii. 3.

EMMA DARWIN IN 1840
(From a portrait by George Richmond, R.A.)

From *Emma Darwin : A Century of Family Letters*. By the courtesy of C. G. Darwin and John Murray

words for them, and these were printed in large type; later she was to use them to teach her own children to read. She was confirmed in 1824, as a matter of custom; religious "enthusiasm" was never encouraged in the Wedgwood home and she betrayed no inclination towards it. Most of 1825 and the winter of 1826–27 she was abroad, at first with the family travelling in France, Switzerland, and Italy, then with her Aunt Jessie Sismondi at Geneva. The latter visit was very gay, with dancing and music, fine dresses and laughter, and back at Maer her happy life went on, with balls, bazaars, and archery meetings. The problems of life were how to eat slippery, buttered pikelets between dances, or to contrive "by a little juggling" that someone else should win the archery prize for once. She was never vain, but she had grown, despite her large Wedgwood build, into a very handsome girl, and in her middle twenties confessed to finding her unsought suitors tiresome. Men liked her frankness and lack of affectation.

This happy carefree childhood ended in 1832. That year Fanny died, her first real sorrow, and in the next year her mother, afflicted since 1822 by recurring epileptic fits, became a permanent and increasingly pathetic invalid. Emma helped Elizabeth with the nursing. Most of the others had married and gone, while father was now Member of Parliament for Stoke-on-Trent in the first "reformed" House of Commons, and a good deal away.

Yet gaiety went on, in visits to London, to The Mount (but the Doctor really talked *too* much, "especially the two whole hours before dinner"), to the Manchester and Worcester music festivals, where she heard Malibran a few days before her death, and young Clara Novello, already famous at eighteen. She dearly loved music, and had lessons from both Moschelles and Chopin; careless of practice, she yet played a little each day for her own pleasure to her life's end.

She was twenty-eight when Charles returned from his long absence, and as warmly attractive in this period as any, with her good features well set in her rather rounded face, her pleasing expression, her amiable mouth, her handsome figure and pretty hands. Her hair, parted in the middle and drawn smoothly down into bunches of curls upon the ears, was a rich glossy brown; her eyes were grey. She had a pleasing manner, self-possessed, direct in address, yet sympathetic and easy-going in smaller matters, taking life much as it came and missing none of the pleasure it might bring. "A prize," indeed. Charles thought so too; he was never, at any time, to think otherwise.

He was impatient for a brief engagement and quick wedding, but Emma was reluctant to leave Elizabeth to tend their parents alone, for Uncle Jos was now ill with a palsy, and the mother's mind clouded as her body weakened. Charles left the decision in Emma's hands, but hoped for the best. Their marriage would be, he said in a phrase not altogether new, the commencement of another life; he was eager to begin it. Soon the date was fixed for the New Year. There was great discussion of where they should live. Emma preferred the country, but Charles thought London necessary anyway for the present; let them live there for a while and make the most of it, then when his affairs were more settled look about for a permanent home elsewhere. Work, actually, was pressing him. Owen's mother had died and that seemed bound to delay Part One of the *Zoology*; he must hurry on the other sections. There were also notes to be added as extra pages to the printed *Journal*. Every morning he was at his desk; every afternoon he and Erasmus would walk the muddy, foggy London streets looking at houses to let. Emma came up in December to stay with Hensleigh, buy clothes, and inspect properties. Charles proposed a visit to the theatre—but their choice fell on *The Tempest* and they all found it tiresome.

One small and ugly but convenient house—No. 12 Upper Gower Street—they rather liked despite its garish decorations: Macaw Cottage they called it, and this, practically at Christmas, they settled upon. Charles moved in on January 1st, 1839, with Covington to look after him. He wanted to finish his Glen Roy paper before leaving London again.

The wedding day drew near, fixed first for the 24th, then the 29th, and he suffered the inevitable apprehensions. His tailor thought a blue coat and white trousers most correct, but he held out for strictly travelling clothes. He paid one quick visit to Emma, and returned full of doubts. He thought of Maer as he had always known it, the continuous bustle of happy family life, relatives coming and going—and on the other hand of how his thoughts were set in solitude and quietness. His years on the *Beagle*, he felt, had accustomed him to loneliness and made him a little in love with it. How could Emma, coming from such a home, be happy with a man absorbed in the dry abstractions of the scientist? He hoped she could, indeed he did, but he doubted, or anyway wondered. The last two days and nights in London he was prostrated by a dreadful headache, but the "railroad" journey on the 25th marvellously dispatched it, and he wrote in much better spirits from Shrewsbury the next day.

They were married at Maer Church by the vicar, another Wedgwood cousin. It was all very quiet, and they left at once for London and Gower Street, arriving that evening to find the house "blazing with fires" and looking very comfortable.

II

Settling down to married life was a slow if happy process. Shopping, visiting, entertaining, going to the play—all took time. They "slopped through the snow" to buy at Broadwood's the grand pianoforte which was Uncle Jos's wedding-present. They held their first at-home—and had two callers. They gave their first dinner-party with Erasmus and Hensleigh and his Fanny as guests; Covington wore his best livery and was declared quite an Adonis. They called on Dr. Holland, saw Macready in the part of Richelieu, attended dinners which neither of them especially enjoyed. In late March Sedgwick called "and was very pleasant," and on April 1st, which was Easter Monday, the Henslows arrived for a brief stay. That evening the Lyells, Mrs. Lyell's sister, and Robert Brown and Fitton, acclaimed respectively the greatest botanist and greatest geologist in Europe, came to dinner. But greatness did not shine. Fitton kept them waiting, Lyell talked so low no one could hear what he said, Brown hardly talked at all. Mrs. Lyell and Mrs. Henslow played up bravely, but Charles took a full day to recover.

One evening that month he took tea with FitzRoy, now more than three years married, and some time since a proud father. Charles liked the baby, and was charmed by Mrs. FitzRoy, but of the Captain found himself more than ever critical as "a man who has the most consummate skill in looking at everything and everybody in a perverted manner."[1] He was shown

[1] *R. FitzRoy and C. Darwin*, Barlow, *Cornhill*.

some of King's journal and thought it confoundedly dull; FitzRoy was still working on his volume, but hoped to finish in time for complete publication in June. Charles's subsequent contacts with him were to be few, which was as well in view of their widely diverging outlook, obvious even in their accounts of the voyage. Remembering some of the tirades on the *Beagle*, Charles probably both sighed and smiled to hear of his quarrel with and challenge to a fellow-Tory after the General Election of 1841, an absurd affair ending, but for a flurry of pamphlets, in a Gilbertian scene outside a London club, the other refusing to strike FitzRoy but begging him to consider himself horse-whipped, and the belligerent naval officer responding by knocking this gentle aggressor to the ground with an umbrella. But he probably had Charles's entire sympathy in his failure in 1845 as Governor of New Zealand, for the outcry against him largely resulted from his taking the part of the natives against unscrupulous white settlers, if also in some degree from his intense reserve and excessive piety, both growing with the years.

Visitors to No. 12 were mostly scientific, and the Lyells were always especially welcome, for the wives could gossip while the husbands discussed. Brown, Sir Roderick Murchison, Sir John Herschel (home from the Cape) all came calling. Sydney Smith was met at Dean Milman's house, and found inevitably amusing. Carlyle called occasionally, but more often on Erasmus; Charles thought him wrongheaded on most topics ("his views about slavery were revolting") and fantastic on science.

Emma that spring was far from well in the early stages of her first pregnancy, and Charles too suffered repeated bouts of ill-health all the year. When they took three weeks off in May at Maer and Shrewsbury he consulted Robert rather anxiously about his health. The Coral Reef work went very slowly ahead, and the account of the Volcanic Islands, though announced in April by the *Gentleman's Magazine* as in active preparation, had been set aside indefinitely. Endless charts had to be consulted, every published book on the Pacific islands carefully scanned if not completely read. There was also, always, his Geological Society work, which he could never take lightly.

By midsummer the sewed parts of some of the *Zoology* studies were beginning to appear, and the four volumes (King's, FitzRoy's,

Charles's, and a mainly statistical appendix) of the *Narrative of the Surveying Voyages of His Majesty's Ships Adventure and Beagle between the Years* 1826 *and* 1836 were out at last. Volume Three, Charles's *Journal and Remarks*, found an instant public —Dr. Holland even, seeing it in print, handsomely recanted— and before the year's end Colburn had brought out a second issue of it alone, with new half-title and title-pages tipped in. Yet a third issue followed in 1840, and by 1842 some 1337 copies had been sold in all. But Colburn was a publisher as enterprising as successful (several of the best-known authors of the day were on his list), and Charles's only financial transaction in connection with these editions was his resentful payment in 1842 of £21, 10s. for presentation copies, a small number of the first issue having been specially bound for this purpose.

Late August and all September he spent first "idling" at Shrewsbury and Maer, then an interested onlooker at the Birmingham meeting of the youthful British Association, which at all times he supported steadfastly against the scorn ("drollery," "humbug," "the British Ass") of the older scientists and the superior newspapers. He returned almost looking forward to the winter, and to sitting quietly of a foggy evening with Emma by his own warm fireside, hearing only distantly the dulled murmur of the traffic in adjacent thoroughfares. Uncle Jos visited them in early December; then Elizabeth came to be with Emma at the birth, on December 27th, of her first child, a boy, William Erasmus, otherwise Mr. Hoddy Doddy.

Emma was soon up, she and the child well. Charles was the invalid, sliding into a prolonged illness which was to prevent sustained work all through 1840 and into the late summer of 1841. He took his enforced idleness amiably enough on the whole, amusing himself by studying young William's "expression of the emotions," a subject he had taken up even before his marriage, convinced of the "gradual and natural origin" of every form of human and animal expression. The confusion of his Coral Reef manuscript fretted him though, as it did to accomplish barely two days' work on it in nine weeks. He nagged desperately at his notes, then in June set them entirely aside for thirteen months, and though sometimes he would turn to his species notes, he could not force his mind to their constructive arrangement. Through most of 1840 he was absent from the Geological Society meetings, and in February 1841 resigned the secretaryship, though keeping a seat on the Council and becoming in 1843 a Vice-President.

He had a bad second winter. Emma nursed him attentively (she once said that nothing married one so completely as sickness), but she was again pregnant, with Anne Elizabeth, born on March 2nd, 1841. That spring and summer he improved, but when he visited Shrewsbury in June and July his father had little hope of a full return of his strength for years to come, and the question of moving from London to the country revived with some urgency. It seemed to him to savour of a final retirement; he wrote to Lyell: "It has been a bitter mortification for me to digest the conclusion that the 'race is for the strong,' and that I shall probably do little more, but be content to admire the strides others make in science."[1] But from this point the decision was taken, the intention announced to their friends, and house-hunting went on persistently until both Charles and Emma were weary of it.

The difficulty of finding a suitable property was increased by a return of Charles's diffidence towards his father in money matters. He wanted to buy a house outright. Robert urged living there first, even for some years. He was talked out of that, but Charles went about the business fearing all the while a paternal charge of extravagance, though it appears without foundation, for a few years later Robert scoffed at the idea. Kempf has analysed Charles Darwin's life largely in terms of a sustained father-son conflict, even to setting down his convivial Cambridge life as a protest against Robert's abstinence, and more importantly ascribing his long reluctance to make his species views public to his fear of further distressing and incurring the wrath of that father whom he had twice disappointed in failing first to pursue his medical studies and then to enter the Church. Kempf's presentation of his case is ingenious but airy (and sometimes fantastic), resting on too little substantial evidence, but there does seem from beginning to end an exaggerated subservience in Charles's attitude to his father, as though he never quite grew out of his early awe for the godlike being who knew and saw all, even to the workings of one's secret mind. A disproportionate humility was, however, the

[1] *LL.*, i. 272.

note of his relation to many others; he always tended to be excessively self-depreciatory.

Much better as he felt towards the autumn it was a strain for him to entertain even Erasmus or his cousins, and he could work only a few hours daily at his desk. Nevertheless the Coral Reef manuscript was done at last, and sent to the printer early in January 1842. In February he was delighted to meet his old idol Humboldt at a breakfast specially arranged for them at Murchison's house. The seventy-three-year-old traveller was in high spirits, and talked to Charles rapidly and continuously, with many vociferous compliments, for a full three hours. The younger disciple left quite exhausted but still venerating. In March he was at The Mount again, delighting in the fresh spring weather, the first crocuses bursting in bright blossom along the garden borders. Then he was back in London, and the proofs of *The Structure and Distribution of Coral Reefs: Being the First Part of the Geology of the Voyage of the " Beagle "* were corrected by May, when he wrote in his diary that it had cost him twenty months of labour over three and a half years, a period he thought the least productive of his scientific life.

It was, in comparison with the time spent on it, a small book, its main text containing some 60,000 words, and the appendix, describing briefly "every existing coral reef, except some on the coast of Brazil," half that length. Its principal biographical interest, like that of the two other geological volumes following it, was in its early exhibition of Charles's powers of generalization, his ability to establish a common pattern linking a multitude of miscellaneous facts, and also as an example of his method of apparently facing and disposing of every objection while leading the reader steadily forward to final complete acceptance. Facts were his firm stepping-stones, ordered with care and clarity; and if this lay in the very nature of scientific procedure, one may still wonder whether his special attention to it did not spring in part from the self-doubt inherent in his extreme modesty, an unwillingness to claim what he could not, he felt, absolutely *show*.

All the book's theory was repeated directly from the Geological Society paper which had been printed in the *Journal*; only the arrangement was better and there was much more supporting evidence. The three types of coral reefs—atoll, barrier, and fringing—were defined and shown to differ fundamentally only in situation; the varied conditions favourable to their respective growths were stated; other theories and their difficulties were declared; his own view of the slow subsidence of the ocean bed giving time for upward growth was presented, together with proofs of local subsidence and arguments for extensive subsidence; it was then shown how this view accounted for all peculiarities of structure; and a definition of the area of subsidence was suggested. Only at the very end, in the recapitulation, did the imaginative force which had impelled the whole break forth to evoke "a magnificent and harmonious picture of the movements, which the crust of the earth has within a late period undergone. . . . We there see vast areas rising, with volcanic matter every now and then bursting forth through the vents or fissures with which they are traversed. We see other wide spaces slowly sinking without any volcanic outbursts, and we may feel sure that this sinking must have been immense in amount as well as in area, thus to have buried over the broad face of the ocean every one of those mountains, above which atolls now stand like monuments, marking the place of their former existence." [1] It was, however it amended Lyell, wholly Lyellesque, and indeed he thought it "perhaps the most interesting conclusion in this volume, that the movements must either have been uniform and exceedingly slow, or have been effected by small steps, separated from each other by long intervals of time, during which the reef-constructing polypifers were able to bring up their solid frameworks to the surface." [2] This, the deduction from the theory, seemed to mean almost more to him than the theory itself, having the more immediate relevance to problems of plant and animal distribution. The decision whether a specific area lay in a rising or subsiding area, roughly defined by the presence or absence of volcanic activity, "will directly bear upon that most mysterious question—whether the series of organized beings peculiar to some isolated points, are the last remnants of a former population, or the first creatures of a new race springing into existence." [3] Coral reefs were all very well, but *his* eyes were on larger game.

The *Zoology* was also now well on its way to completion. Waterhouse's *Recent Mammalia* had appeared first, in 1839,

[1] *Coral Reefs*, 148.   [2] *Ibid.*, 145.   [3] *Journal*, 569.

Owen's *Fossil Mammalia* in 1840, the *Birds* in 1841. This year Jenyns's *Fish* would be out, and next year the *Reptiles*. It had been a long and tedious task, and like all these early books held no prospect of monetary return, but it made this section of the results of the voyage at least relatively available, and was splendidly produced with many most pleasing coloured plates.

The Coral Reef proofs had exhausted him, but he was stronger now and recovered rapidly, and on following Emma to Maer in mid-May he turned with fresh vigour to his species notes and at long last set to work to sketch an outline of the theory which presented itself more and more insistently and definitely to him. He wrote as he rested, with a soft-leaded pencil, on the first paper coming to hand, still setting down notes rather than anything intended to be a finished essay, dropping words from his sentences, and on the backs of the filled sheets roughing out the course of the matter still to come.

It ran only to some thirty-odd pages, but was for all that extraordinarily ordered and complete, a recognizable first outline of *The Origin of Species* itself. Its first point was variation of species under domestication, the way in which mankind utilized inherited variation and selection to create new and presently stable varieties. From that he jumped to what had been at the back of his mind in dealing with his coral-reef theory: geological change of conditions causing variation in wild species. Malthus was also invoked to show a double struggle for existence bringing about a "natural selection" more vigorous than man's. He glanced briefly at sexual selection as a factor, then turned to obvious difficulties—sterility of crossed species, development of instinct, formation of eye or ear. . He admitted a completely satisfying answer as impossible, but urged that such problems diminished on examination, and were not of themselves sufficient to destroy his theory. Admit variation in nature. Admit inheritance of variation. Admit selection under subsistence pressure and change of environment, and—could we set a limit to variation? Yet, even so, have we *evidence* of such production of species? Much, he concluded, of a negative kind. Accept the genealogical system of classification for species, and the facts of the geological record and of geographical distribution begin to fall marvellously into place. "The affinity of different groups, the unity of types of structure, the representative forms through which the foetus passes, the metamorphosis of organs, the abortion of others"[1]—all these matters, "inexplicable to Creationists," now become intelligible facts. Earlier in the outline he had been cautious: "Be it remembered I have nothing to say about life and mind and all forms descending from one common type. I speak of the variation of the existing great divisions of the organized kingdom, how far I would go, hereafter to be seen."[2] But now exultation carried him away. Not only did he stress, as in 1837, the intensification of interest brought by his theory to scientific inquiry, but insisted also upon its truly religious quality. No longer was "the Creator" to be held directly responsible for the cruelties of parasites and the bloodier of the beasts, for false instincts, for Nature's incalculable waste: "From death, famine, rapine, and the concealed war of nature we can see that the highest good, which we can conceive, the creation of the highest animals has directly come."[3] He ended on a thought first set down in his notebook of 1837 and never thereafter absent from his reflections on the subject: "There is a simple grandeur in the view of life with its powers of growth, assimilation and reproduction, being originally breathed into matter under one or a few forms, and that whilst this our planet has gone circling on according to the fixed laws, and land and water, in a cycle of change, have gone on replacing each other, that from so simple an origin, through the process of gradual selection of infinitesimal changes, endless forms most beautiful and most wonderful have been evolved."[4]

So he first stated the great theory, scribbling hurriedly and only half legibly, finishing the sketch during three days spent at Shrewsbury, then thrusting it away where none would come on it. He told no one of it. Yet there it was, practically complete, the only essential omission a failure to grasp the means of constant divergence of species from the common stock, though even that was half-stated, had he but eyes to see it, in his comment on the manner in which a victorious species must "beat out other forms" and so people the varied (and so ever-further modifying) regions of the earth.

On June 18th he set out alone on a brief geological trip through North Wales, his first visit since that with Sedgwick eleven years

---

[1] *Foundations*, 50.   [2] *Ibid.*, 6–7.
[3] *Ibid.*, 52.   [4] *Ibid.*

earlier. He covered some of the same ground, and marvelled how blind he and Sedgwick had been in those days, partly from ignorance, partly because they were looking for other things. At Capel Curig a spell of bad weather was enlivened by a chance encounter with Charles Bunbury, botanist and Mrs. Lyell's brother-in-law, whom he had met before. All one wet day they talked agreeably together at the inn. He thoroughly enjoyed the excursion—fortunately, for it was the last of the kind he had strength to undertake.

III

Earlier in the year Emma had said they would not move that summer, but on July 22nd she and Charles went to view Down House, near the village of Downe, beyond Bromley in Kent. They spent three days considering the property on the spot. Neither absolutely cared for it. Emma liked the house and garden, but thought the surrounding countryside desolate and depressing. Charles found the country charming but the house ugly. It was open to the passing lane, and the garden and fields—there were eighteen acres in all—had not been looked after. Still, it had the merit of quietness and isolation. They doubted whether they could find anywhere else so admirable in this respect and yet so near to London—just sixteen miles from St. Paul's. Two hours to London Bridge, yet "absolutely at the extreme verge of the world." Walking up together under a dull sky and against a chill wind, from the small village inn where they were staying, they discussed prospects and alternatives. The lane could be lowered, a wall built, protective shrubberies planted. The grounds could be improved, even the house added to. It was cheaper than any comparable place they had seen—only £2200. And they must either find a place quickly or stay on in London some while, for Emma had been pregnant from the beginning of the year.

They must think it over and come and see it again, they decided, and meanwhile Charles would get his father's opinion. But he had even then little doubt of his own view, and the place was in fact very shortly settled on, and purchased for him by Robert. They moved in on September 14th. The expected child, Mary Eleanor, was born on the 23rd, but lived only three weeks.

It was a sad beginning, shadowed too by the onset of Uncle Jos's fatal illness. But here they were, and here they were to stay for forty years, the rest of Charles's life.

He had been, only a couple of years earlier, a tall, rather gaunt, pleasant-featured *young* man. His illness had made him seem suddenly older. His face had grown heavier, his hair was receding further from the always high prominent forehead, which now, with the heavy "beetle" brows, appeared to overhang the deep-set inquiring eyes. He bore in repose something of a grim, anyway a haggard, look. He was thirty-three-and-a-half years old.

BOOK THREE

THE MAN MAKING

CHAPTER XI

EARLY DAYS AT DOWNE

I

DOWNE village has always lain remote in a hollow of the low chalk hills of Kent, tranquilly distant from the main London roads to east and west, reached only by winding hedged lanes. It was, in 1842, a small, mainly self-contained community of about forty houses, largely cottages, grouped about a little flint church. The villagers were land labourers, dressing habitually in wide, coloured smocks, and in the summer evenings working their blossoming gardens or sitting contemplatively in the cottage doorways, listening to the cooing of the doves in the clustering trees and touching their foreheads respectfully to the passing gentry. The village innkeeper was grocer and carpenter too, and once a week the carrier went his horse's jog-trot way to London and back, "calling anywhere for anything and taking anything anywhere."

Down House stood above the village to the south, a quarter of a mile away. Bare and bleak it seemed that first autumn amid its untended fruit and fir, chestnut and mulberry trees, whose shed leaves the chill wind whirled about its shabby whitewashed walls. Comfort, though, soon prevailed inside. The downstairs rooms, if a little low, were large—study, drawing-room, dining-room, all won admiration. Upstairs was abundance of bedrooms, more than enough for all their visitors. Little could be done that winter, but all the following spring and summer workmen were busy on improvements both inside and out. At the south-west corner, facing the wide garden, a large bay was added on all three floors, the walls were freshly stuccoed, the roof examined for loose and broken slates, the lane lowered and a boundary wall built as further protection, banks made and planted for shrubberies, the garden set in order, and gravel put down upon the paths.

With that spring, as the trees everywhere broke into leaf and blossom, the charm of the house's situation, and the surrounding countryside, became manifest even to Emma. Grand walking country, Charles thought it, with field-paths giving access everywhere, following the smooth hill-crests between the streamless but fertile valleys. The abundance of singing birds, the soothing hum of roving hive-bees, the variety of twining plants and shy wild flowers along the hedge-banks, the scattered shady copses, perpetually delighted him. He was to become, just so long as health allowed, a constant wanderer by all these near-by ways, and, on pony-back, farther afield even to wooded Keston Common. As The Mount had been for thirty years, as No. 12 Upper Gower Street never was, so Down House was thenceforward to be his home, a living centre of gathering, growing, ever-deepening associations, a happy refuge which would never fail him though the wide world broke about his ears.

II

The first year, though, was unpropitious, all too full of distraction. There was Mary Eleanor's birth and death. Hensleigh Wedgwood for a while was seriously ill, and his three children visitors. Work began on the house and garden in March, continued into September. Uncle Josiah's illness dragged on, until in this July of 1843 he died. Erasmus Darwin went shopping with the Wedgwoods to buy the deepest mourning. Charles and Emma journeyed to Maer for the funeral, then returned with Robert for a few days at Shrewsbury. The loss was a sad one, and not least for Charles, to whose respect for a man of integrity was added not only affection but a deep sense of obligation to his "first Lord of the Admiralty"; only Aunt Elizabeth, her mind wandering in the far past or sunk in blank stupor, was beyond grieving. Emma was again pregnant, and on September 25th Henrietta was born. The Scotch nurse Brodie now entered the household, to become only less of a family figure than the butler Parslow, now in the third of his many years of service. In October, Charles was at Shrewsbury again, entertaining his father with tales of the fantastic claims of Homoeopathy; Robert was charitable—the world was full of fools! To Charles's own complaints he was less sympathetic. A numbness in the finger-tips? Neuralgia, he declared, as though the definition dismissed the disease. Spending too much on Down House? "Stuff and nonsense! Stuff and nonsense!" The family fortunes were sound enough; away with this worry about a few pounds more or less!

All this while, from October 1842, Charles had been working as steadily as possible on the second part of the *Geology of the Voyage of the " Beagle "*—the *Volcanic Islands*. No specific illness interrupted him, but he remained so nearly an invalid that he could work no more than two or three hours a day even when relatively well. He laboured, too, with a desperate sense of writing into a void, of compiling books to be issued at his own or the Government's expense and which no one—not even geologists—would ever read. Geologists *never* read each other's publications: "The only object in writing a book is a proof of earnestness." [1] This was, more than any other, *the* book he had so joyfully conceived, overwhelmed at the very thought of writing it, when basking in the sunshine nearly a dozen years ago upon the shore at St. Jago!

He persisted, however, and slowly the work took shape. He thought of it as a pamphlet, but it was as long as the main text of the *Coral Reefs*. It consisted of plainly geological description of the various islands he had visited; its interest was in its new observations, its new readings of known facts—and its stress upon the significance of small causes, for he remained steadfastly anti-catastrophic in treating even of the most catastrophic of earthly agents. His principal large conclusion referred to the distribution of volcanic islands and volcanic phenomena. He pointed out that volcanoes were generally found either under or near the sea, as though eruptions reached the surface "more readily through fissures, formed during the first stages of the conversion of the bed of the ocean into a tract of land." [2] He associated elevation and volcanic activity even more directly, conjecturing that "the forces which eject matter from volcanic orifices and raise continents in mass are identical." [3] They were views reinforcing rather than extending those of the *Coral Reefs*. A last chapter gave briefly his geological observations in New Zealand, Australia, Tasmania and the Cape, irrelevantly but in

[1] *LL.*, i. 334.   [2] *Volcanic Islands*, 126.   [3] *Ibid.*, 95.

order to leave him free to devote the projected third volume entirely to South America.

The writing was finished in mid-February, 1844, and the manuscript, with appendices by other hands on Fossil Shells and Corals, sent at once to the publisher, Smith Elder. Proofs came promptly, and Charles was able to correct them and to give time to his garden—some of the evergreens were not doing well and must be shifted—before spending a week at Maer and Shrewsbury in April. In late spring the book was out, price half a guinea. Charles had no hopes for its success.

In any case he was not waiting upon the event. Other subjects were in hand, and one laid down meant only another taken up. His species notes had never ceased to accumulate, and more than once in the past year he had turned aside from volcanic islands to ruminate upon his outlined theory. It needed, it must have, he felt, a fuller formulation. Its importance was obvious. Supposing he died, and the whole thing with him? Through May and June he was at work upon it, expanding the sketch of 1842 to 189 sheets. The earlier pages were before him as he wrote, and as each was done with he would decisively cancel it with a line from top to bottom. He wrote mainly from memory, with a minimum of reference to his books or the ever-accumulating folders of notes. The stuff was in his head, he had thought it over so often; now it flowed clear through his mind, freshly, abundantly.

He followed the plan of the earlier essay with exactness, adding only expansion. Variation of species under domestication seemed to him the absolutely essential starting-point, failure to appreciate which had made "veritable rubbish" of such other theorists as Lamarck and St. Hilaire. From that he turned to variation of organisms in a wild state under the natural selective agencies of the general and sexual struggle for existence. He did not pretend to explain variation under the influence of environment or the inheritance of such acquired characters; he took them for granted as evident facts. Other writers assumed a limit to variation; he could find no proof of it. He elaborated his comments on variation of instinct, always in his mind a point of special difficulty: he sought to show how by modification and inheritance the most complicated instincts might be slowly acquired, much as "the most complex and fine shades of expression" developed from infantile cries and grimaces! Perhaps the steps were not easily imagined—any more than those in the formation of that intricate organ the eye—but he insisted that there was no justification in that for rejection of his theory.

Next he proceeded, as before, to the evidence, most of it necessarily suggestive rather than concrete, negative rather than positive, but showing, he thought, that the production of exquisitely adapted species by modification was at least possible —and for the moment the admission of possibility was more important than that of probability. Admittedly there *were* difficulties, perhaps chief among them that of the imperfection of the geological record, the lack of "transitional stages" among known fossil remains. The point bothered him more then than later, for he had already glimpsed and was soon to grasp the key to that problem. For the time being he could only sum up the evidence in its despite. The common view premised the creation of organisms by distinct and separate acts. Very well! "It is impossible to reason concerning the will of the Creator, and therefore, according to this view, we can see no cause why or why not the individual organism should have been created on any fixed scheme. That all the organisms of this world have been produced on a scheme is certain from their general affinities; and if this scheme can be shown to be the same with that which would result from allied organic beings descending from common stocks, it becomes highly improbable that they have been separately created by individual acts of the will of a Creator."[1] The scheme *could* be shown to be such. He had shown it so in 1842; he demonstrated it again in greater detail. Study geographical distribution, the extreme dissimilarity of species in widely distant areas, lessening as the distance shrank. Observe the special case of slight dissimilarity between inhabitants of different islands in the same archipelago, combined with *general* similarity to those of the nearest continent: an absolutely crucial instance. Note the obvious relation between living and more recently extinct creatures in the various major continental divisions. On the creationist views all these facts were inexplicable—but they followed instantly and absolutely as a result of "specific forms being mutable and of their being adapted by

[1] *Foundations*, 133.

natural selection to diverse ends, conjoined with their powers of dispersal, and geologico-geographical changes now in slow progress and which undoubtedly have taken place. This large class of facts being thus explained," he continued, "far more than counterbalances many separate difficulties and apparent objections in convincing my mind of the truth of this theory of common descent." [1]

He set out as baffling on any other theory the difficulties and contradictions of classification, the plain fact of unity of type within the great classes, identity of embryonic structure, and the mystery of rudimentary organs. The embryo he suggested as "shadowing forth" the mature structure of earlier forms, thus indicating in its successive changes the order of its ancestry, but declared the evidence, though contributory, too incomplete to claim as positive proof.

He asked, finally, why the human mind was so reluctant to accept the idea of common descent, and suggested in answer that "we are always slow in admitting any great change of which we do not see the intermediate steps. The mind cannot grasp the full meaning of the term of a million or hundred million years, and cannot consequently add up and perceive the full effects of small successive variations accumulated during almost infinitely many generations." [2] For his own part he accepted it absolutely; it seemed to him as plain as gravitation, and as universal. All known organisms, he asserted, were descended from a few, probably "less than ten," [3] parent-forms into which life had been "originally breathed."

A "few" parent-forms, he wrote then, and a little later he was to hazard in a pencil note to the fair copy, "perhaps one only" [4] —and from that the entire branching tree of all the forms, dead and living, of past and present! The longer he considered it the more he was convinced, scarcely even realizing how the essential clue, the mechanism of divergence of species, still eluded him.

---

[1] *Foundations*, 193–4.   [2] *Ibid.*, 249.
[3] *Ibid.*, 252. He had written a year earlier to G. R. Waterhouse : " According to my opinion . . . classification consists in grouping beings according to their . . . descent from common stocks " ; and again : " I believe . . . that if every organism which ever had lived or does live were collected together (which is impossible, as only a few can have been preserved in a fossil state), a perfect series would be presented, linking all, say the Mammals, into one great, quite indivisible group."—*Memorials of C. Darwin*, 8.
[4] *Ibid.*, 254 n.

One incidental point had little immediate relevance, but some future interest: "According to our theory, there is obviously no power tending constantly to exalt species, except the mutual struggle between the different individuals and classes; but from the strong and general hereditary tendency we might expect to find some tendency to progressive complication in the successive production of new organic forms." [1]

The essay was finished early in July, probably on the 5th, for on that day he wrote a letter to Emma giving instructions for its publication "in case of my sudden death, as my most solemn and last request, which I am sure you will consider the same as if legally entered in my will." [2] A competent editor, who must be geologist as well as naturalist, was to be enlisted; Lyell would be best of all, then Forbes, or Henslow, or Strickland. Hooker or Owen would be "very good" if either would accept the task, which was the amplification of the essay by reference to Charles's marked books and portfolios of notes. Four or five hundred pounds, plus any profits from sales, was suggested as fair payment. Failing some such editor, "then let my sketch be published as it is, stating that it was done from memory without consulting any works, and with no intention of publication in its present form." [3]

At once he was off to Shrewsbury again, returning to settle down, on July 27th, to the third and final volume of the *Geological Observations*: on South America. He began it dubiously. Was it really worth all this time, labour, and expense of publication—the Government grant was all but exhausted—for a result rather like tossing stones into bottomless wells? Why exhaust oneself writing books no one wanted to buy or read? But his friends encouraged him and he went on—as he would have done in any case, for he never turned easily aside from a subject once he had set his hand to it.

III

He and Emma had by now quite settled to their country life. "We live like clockwork," he wrote to Herbert in 1844, and to FitzRoy in 1846: "I am fixed on the spot where I shall end my life." [4] A local directory figured him as "Charles Darwin,

[1] *Foundations*, 227.　　[2] *LL.*, ii. 16.　　[3] *LL.*, ii. 18.　　[4] *LL.*, i. 318.

farmer," and at least he did his best to supervize the working of his garden and care of his trees, and to help with hay-making in the big field. At the far south end of the property, overlooking a pleasant valley, he had planted a narrow belt of trees and shrubs, three hundred yards long, less than twenty wide, with a sand or gravel walk encircling it, and round and round this he would take his daily exercise, with dog for company, when he did not wish, or did not feel well enough, to go farther afield.

Nevertheless he did not want, he said, "to turn into a complete Kentish hog," and for several years he paid regular visits to London once or twice a month as health permitted (the ever-necessary proviso), to meet scientific friends and attend gatherings of the Royal, Geological, or Zoological Societies. It was, though, a tedious journey to and from the railway station in the horse-carriage behind the old gardener-coachman who cared as little for time and tide as they for him, and so sometimes he would stay several days, often at Erasmus's house, setting out each morning to breakfast with such friends as might chance also to be in town just then. But more and more he grew, unless absolutely well, disinclined for such trips; the thought of them could disturb him even more than the actual travelling.

It was better in some ways to see people at home where Emma, best of hostesses, made them all so comfortable. Wedgwood visitors were frequent, and Catherine and Susan would come from Shrewsbury when they could leave Robert, who, though he had visited his sons in London, seems never to have seen Charles in his new house at all, perhaps because he was in these later years too old, too ill, and above all too huge for easy travel.

Erasmus was always welcome, when he would come, for though he never tired of playing with the children and would even busy himself about garden matters with more vigour than Charles could muster, he loathed country life, and would return thankfully from what he called Down-in-the-Mouth to his overshadowed but comfortable "baby-house" in Park Street, casting up eyes and hands in mock horror as he murmured "Two such months, good heavens!" to Mrs. Carlyle's sympathetic ear.

An odd figure, Erasmus, of whom too little has been recorded; a quiet, passive and yet distinctive personality, loved by those who knew him best and discerned in him—despite the sedateness of his bearing, high-domed Darwin forehead, sparse hair, and trimmed side-whiskers—a Lambish charm, tenderness, and wit. Carlyle and his wife were both deeply attached to him, and welcomed him as one of their most regular visitors; they regarded him as "the type of *English* gentleman" and entrusted their family silver to him when they went out of town. Thomas was drawn not only by his thoughtfulness and sincerity but by something "sarcastically ingenious" in him; he actually "rather preferred him for intellect" to Charles, as Sismondi, similarly, to Carlyle. Jane was untiringly entertained by his epigrammatic acidity, sharp with satire though never sour with malice. No stronger than Charles constitutionally, he conserved his energies by what Carlyle called a "patient idleness," never attempting to practise as a doctor, but living alone and quietly, reading widely in literature and very little in science, secure in the assurance of a handsome income. For a long while he was very friendly with Harriet Martineau, family rumour hinting at a possible marriage, but the intention, if it amounted to that, came, like everything else in his life, to nothing.

Charles expressed constantly his deep affection for "poor old Ras" and "my dear good old brother," yet neither entered deeply into the other's life. They had been together as children and schoolboys at Shrewsbury, as students at Edinburgh, and briefly in Great Marlborough Street towards the end of 1836; after that more occasionally in London and at Downe. But they had so little in common that Charles found here an indicative proof that nurture was secondary, nature primary, in creating character and achievement. He said one day that he supposed Erasmus would read his *South America*. "Upon my life," cried Erasmus, "I would sooner even buy it!"[1]—and on another occasion declared that he did not give a damn for the whole kingdom of nature. Yet the mutual feeling between them, as between all these Darwins and Wedgwoods, was something very real, and Erasmus was, after all, the nicest person to have about one, often stimulating yet never disturbing.

Other more scientific visitors were not all so easy, largely because Charles took to heart his father's view that a host must never cease entertaining his guests, and always went about it so strenuously as very quickly to exhaust himself. He could scarcely concentrate upon writing even a letter while such visitors were

[1] *LL.*, i. 334 n.

in the house, and much as he delighted in scientific discussion it so excited and wearied him that half an hour of it, except perhaps first thing in the morning, when he was at his best, left him incapable of work. It was really best when such friends came to stay not for a week-end but for a longer period, bringing their work—and their wives—with them. The pressure was less then: they would meet at meals, talk in the study for a brief while immediately after breakfast, take exercise together on the sand walk. The Lyells, Henslows, and Horners came many times thus.

From 1844 one of the most welcome guests, and the closest of Charles's new friends, was Joseph Hooker, eight years his junior and the son of Sir William Hooker, then first Director of the Royal Botanic Gardens at Kew. Young Hooker had grown up to botany, but he and Charles had more in common than that. When they met passingly in Trafalgar Square in 1837, his highest aim was to emulate Charles. Captain Ross wanted to add to his Antarctic Expedition a naturalist not only "perfectly well acquainted" with his subject but also "well known in the world beforehand"—such a person, he insisted rather surprisingly, for even the *Journal* was still unpublished, as Mr. Darwin. Hooker, applying for the post, retorted justly enough that it was the *Beagle* voyage which had made Charles known, and that he hoped the voyage of H.M.S. *Erebus* would do as much for himself. He was appointed as assistant surgeon, and a copy of the *Journal* was his constant companion through a four years' voyage which carried him to many of the points described in it—Ascension, St. Helena, the Cape of Good Hope, New Zealand, Tierra del Fuego, the Falklands. Meanwhile, Charles heard much of Hooker from the Lyells, and wrote to him in December 1843, soon after his return, congratulating him on his work and sending some South American plants, suggesting a comparative study of these with European species. Other letters were exchanged, and they met early in the New Year at Erasmus's breakfast table. The first visit to Down House followed, and thereafter he was always welcome, Charles's affection increasing with the years for this tall, lanky, active, genial fellow who shared so many of his own interests, who had seen the sights of the world as he had seen them, and who, though so devoted to natural science that he named it his recreation as well as occupation, was unfailing in courtesy to Emma, in inventing games for the children, and in amiability towards all.

In the beginning visits were not frequent, for he was a good deal away, in Edinburgh in 1845, and from 1847 to 1851 in India. But letters linked them, and the correspondence was one of increasing importance, circling more and more closely about the basic problem of mutability of species. Up to 1844, though making no secret of his deep interest in the subject, Charles had kept his essential views very much to himself. Now he became franker, making them known to Lyell, Bunbury, Jenyns, Owen, and some others, but above all to Hooker, to whom he confided them even before their first meeting in his letter of January 11th, 1844, and who was permitted to read the sketch of that year within a few months of its writing. The first reference was diffident. He spoke of his "blind" collection of all kinds of relevant facts ever since his return from the *Beagle*. Now light seemed dawning, "and I am almost convinced (quite contrary to the opinion I started with) that species are not (it is like confessing a murder) immutable." He went on, revealingly: " Heaven forfend me from Lamarck nonsense of a 'tendency to progression,' 'adaptations from the slow willing of animals,' etc.! But the conclusions I am led to are not widely different from his; though the means of change are wholly so. I think I have found out (here's presumption!) the simple way by which species become exquisitely adapted to various ends. You will groan, and think to yourself, 'on what a man have I been wasting my time and writing to.'" He added, rather oddly, that "five years ago" so he too would have thought.[1]

Hooker's reply was encouraging, and further letters followed on geographical distribution in relation to the appearance of new species. But Charles remained diffident. Even in July, when the extended outline was completed, he could not allow himself the claim that "on my views of descent, really Natural History becomes a sublimely grand result-giving subject," without the instant addition: "now you may quiz me for so foolish an escape of mouth," [2] and in the following November he was writing that "in my most sanguine moments, all I expect, is that I shall be able to show even to sound Naturalists, that there are two sides to the question of the immutability of species." [3]

[1] *LL.*, ii. 23-4. [2] *LL.*, ii. 30. [3] *LL.*, ii. 29.

The subject was never dropped, however often set in the background by more pressing matters. When the anonymous *Vestiges of Creation* (actually by Robert Chambers, the Edinburgh publisher and bookseller) appeared towards the end of 1844, he read it with the closest attention. He thought it bad in its geology, worse in its zoology, and generally "unphilosophical," but well written and of more substantial value than most of his scoffing friends allowed. Of course it went too far for his cautious mind, especially in its attempt to set forth a family tree of organic life. Privately he asserted some such genealogy, but he thought "much too little known at present" to attempt to specify it. He received with mixed feelings, being both "much flattered and unflattered," the news that the work was being widely attributed to him.

It is very striking to note the range and variety at this period of his interests in geology, botany, zoology—the origin of coal, the vitality of seeds, cross-fertilization of flowers, glacial and other ice-action, cleavage and foliation, earth-movements, dust falling on mid-ocean vessels, and always, tirelessly, every aspect of that "noble subject of which we as yet but dimly see the full bearing," that "almost keystone of the laws of creation"— geographical distribution.

Once he asked anxiously if Hooker really meant some geographically indigestible statement (Hooker hadn't; Charles had misunderstood him), adding: "If so, it is a sickening fact." One wonders whether, as he wrote the words, his mind recalled the incident of the tropical shell in the Shrewsbury gravel-pit, and his youthful amazement that Sedgwick should regard as a geological misfortune what seemed to him a geological marvel. He had travelled far in fifteen years.

IV

The writing of the *South America* began well, and all through the fine August of 1844 he was hard at work on it. But writing —even the plainest statement of fact or idea—never came easily to him, and he groaned perpetually over his scored and blotted manuscript. He would painfully think his sentences over, set them down, reconsider, cross out, try again. What would he not give, he exclaimed, "to write simple English, without having to rewrite and rewrite every sentence." Practice, alas, made the process no easier.

He got two chapters done before his stomach became worse than usual. Late in September the fair copy of the species essay came to him, and he spent a week reading it carefully over, but making little emendation. He meant then to have a rest at home, going nowhere and doing nothing, but in October went alone to see his father at Shrewsbury. Robert was relatively well, but an old man now, approaching eighty, and burdened by his corpulence, so excessive now as to make it almost impossible for him to turn over in bed unaided. Dressed in his old-fashioned knee-breeches and heavy-lapelled coats, he spent much time in his wheel-chair in house or garden or greenhouse, living a good deal in his recollections of the past, though always eager to divert himself with Charles's anecdotes of scientific society. All his memories were not happy ones, for when Charles urged him, since he could not walk, to drive out as he had been used, for fresh air and change from the house, he refused, as has been told, because every road out of Shrewsbury was associated in his mind with some painful event. But on this visit at least he was cheerful, and Charles enjoyed the stay.

He returned to a rather tedious winter of slow if steady work on the *South America*, only setting it aside in April to prepare a new edition of the *Journal* for publication in John Murray's recently launched Home and Colonial Library. This he turned to with a will, for not only was he pleased to see his favourite work reappearing in a "popular" edition, but Murray was offering him £150 for it, his very first literary earnings from any source. He spent four hard months on the revision, tiring himself to exhaustion by the end of August, but rapidly recovering.

The changes were considerable, apart from the inclusion of new illustrations which Murray wanted and Charles didn't, and the exclusion of maps which Charles thought essential and Murray wouldn't have. The edition was actually shorter by more than ten thousand words, but still more had been omitted and much new writing substituted.

The first eight chapters revealed only occasional excisions, interjections, amendments, the one notable addition being a brief passage on a special breed of oxen found in the Rio Plata country, whose difficulty in getting food during droughts, owing to the awkward formation of their lips, seemed now to Charles a good illustration of how little we are able to judge from the ordinary habits of life, on what circumstances, occurring only at long intervals, the rarity or extinction of a species may be determined."[1]

There was more revision and some rearrangement from Chapter 8 onward, 9 becoming part of it, 10 and 12 uniting to make a new Chapter 9; 11 becoming 10, and 13-23 becoming 11-21. In the latter part of the extended Chapter 8 there was a rewriting of passages on the geology of Patagonia, the relation between living and extinct species, and the problem of extinction. On the second point he wrote with new certainty: "This wonderful relationship in the same continent between the dead and the living will, I do not doubt, hereafter throw more light on the appearance of organic beings on our earth, and their disappearance from it, than any other class of facts."[2] The matter of extinction was also given a new stress: "We do not steadily bear in mind, how profoundly ignorant we are of the conditions of existence of every animal; nor do we always remember, that some check is constantly preventing the too rapid increase of every organized being left in a state of nature." What that check was had been made plain to him by Malthus: "The supply of food, on an average, remains constant; yet the tendency in every animal to increase by propagation is geometrical." Nevertheless, it was not *just* as simple as that: "How rarely, if ever, we can point out the precise cause and manner of the action of the check! We are, therefore, driven to the conclusion that causes generally quite inappreciable by us determine whether a given species shall be abundant or scanty in numbers."[3] In the last resort we were forced to *assume* more or less favourable conditions.

Charles thought this brief discussion of real importance; he regretted, instantly, that he had not made it longer and "shewn by facts, as I easily could, how steadily every species must be checked in its numbers."[4]

Some of the minor omissions were interesting: a reference to certain plants as "created"; two sentences mentioning man as a recently introduced "adaptation to the existing condition of the world"; the already quoted passage on "two distinct

---

[1] *Journal* (1845), 147.
[2] *Ibid.* (1845), 173.
[3] *Ibid.* (1845), 175.
[4] *LL.*, i. 338.

Creators." Larger but more incidental omissions were the accounts of the visits to the Cape and the Azores, advice to collectors undertaking a similar expedition, and such of the Addenda as were not incorporated in the text. The coral-reef chapter was also considerably amended in the light of his further work on the subject, and in the pages on the final return to Brazil there was a sudden outcry against slavery, thanking God he would never again visit a slave country, inspired by indignation at Lyell's easier view of "this odious deadly subject" in his *Travels in North America*, which Charles was reading this summer.

The most radically revised chapter was that on the Galapagos Archipelago. The reasons are obvious. Not only had the specimens collected there now been thoroughly examined, classified, and described by experts, but Charles himself had grasped, as he had not in 1837, the cardinal importance of the collection. After the briefest descriptive opening, condensed from the *Journal*, he went straight to the point. In this geologically only recently upraised archipelago, stocked for the most part with "aboriginal creations, found nowhere else," differing between island and island, and yet with a marked relationship to American forms, he felt himself, "both in space and time, . . . brought somewhat near to that great fact—that mystery of mysteries—the first appearance of new beings on this earth."[1] He described the finches somewhat more particularly than before, adding that: "Seeing this gradation and diversity of structure in one small, intimately related group of birds, one might really fancy that from an original paucity of birds in this archipelago, one species had been taken and modified for different ends."[2] The prevailing dullness of colouring of most of the birds, insects, and plants, so unusual in tropic latitudes, led to speculation on the connexion between coloration and conditions of existence. Later he returned to the vital point of the distinctive inhabitants of the different islands: "The distribution of the tenants of this archipelago would not be nearly so wonderful, if, for instance, one island had a mocking-thrush, and a second island some other quite distinct genus. . . . But it is the circumstance, that several of the islands possess their own species of the tortoise, mocking-thrush, finches, and numerous plants, these species having the same general habits, occupying analogous situations,

[1] *Journal* (1845), 378. [2] *Ibid.* (1845), 380.

and obviously filling the same place in the natural economy of this archipelago, that strikes me with wonder." [1] So he led up to what was in effect his concluding declaration of astonishment at "the amount of creative force, if such an expression may be used, displayed on these small, barren, and rocky islands; and still more so at its diverse yet analogous action on points so near each other." [2]

He had said something, but how little, how cautious it was beside all that he might have written, or anyway hinted at, had he but had the full courage of his convictions. He hadn't. He could not, he dared not, declare himself. Hooker might share the secret, and Lyell, and even a few others, but no more. It was, considering everything, an extraordinary reticence, but also deeply characteristic and, more than that, at the very root of his ultimate achievement. Charles was no coward, or he would never have dared even to think so boldly and independently, but all his life long he was dogged by this diffidence, so extravagant as to be deemed by some plainly a neurosis born of infantile and youthful repression and directly responsible for his adult ill-health. He carried modesty to extremes, on one hand consistently depreciating his own worth and achievement, and on the other invariably praising the work of others in unmeasured superlatives. The effect on his own work was necessarily retarding, but made it, positively as well as negatively, what it was. Erasmus Darwin before him, and A. R. Wallace after, conceiving their evolutionary theories, were content to publish them with a minimum of supporting evidence. Charles was to go to the other extreme, seeking to test his view in the light of every sustaining or destructive fact, even, one suspects, would life have allowed, to infinity!

So, hiding its light under a very considerable bushel, the book went out, with its friendly dedication to Lyell, acknowledging Charles's great indebtedness to the *Principles*, and a new Preface dated June 1845 which hoped that the work had been improved for the general reader, and expressed Charles's intention of presently describing, in a series of papers, "some of the marine invertebrate animals collected during the voyage." It was, if a cheaper, a less attractive volume, bulkier and more poorly produced, printed, and illustrated. It sold well, however.

[1] *Journal* (1845), 397.   [2] *Ibid.* (1845), 398.

V

The state of his stomach compelling relaxation, he turned to gardening with Erasmus, making a new walk in the kitchen garden, and shifting evergreens, turf, and even banks of earth. Then, in mid-September, he went northward. Emma stayed at home. The birth of her fifth child, George, in July, had left her weak and in need of rest. He visited Lincolnshire, where he had—perhaps from some family association—lately purchased a small farm property. He also spent some time at York, and visited the famous eccentric and naturalist, Charles Waterton, at Walton Hall, to be entertained by a mixed company of Catholic priests and "mulatresses," and by Waterton's spirited account of how the very day before he, at sixty-three, had run down a leveret in a turnip-field. He went as well to Chatsworth, in Derbyshire, where he was entranced by Paxton's great conservatory, "a perfect fragment of a tropical forest" which renewed his happy memories of Brazilian ecstasies. He spent the last two or three weeks at Shrewsbury, and was back at Downe on October 24th.

Again he settled to "my wearyful *South America*," interrupting his labours only for an occasional day in London. Emma, in January, went alone to Maer to see her mother, whose strength was failing rapidly. In February Charles was at Shrewsbury once more; his father too seemed failing, but made an unexpected recovery. Mrs. Wedgwood did not. On March 31st her daughter Elizabeth heard her say: "Lord, now lettest Thou Thy servant depart in peace." It was not the first time, but it was the last; her prayer was granted. Even her daughters felt her death to be a happy release for everyone. The happy, the gracious, the spiritually and physically lovely Elizabeth Allen, who for nearly half a century had been the faithful wife, the watchful mother, "the gentle mistress" of Maer, had these ten years past lived more and more remote from all her world, lost in the distressing mists of a collapsing mind.

Now Maer itself was given up, sold to strangers. There was nothing else to do. Josiah, Harry, and Hensleigh were married and had their homes in or near London. Frank was in Staffordshire, but his home was at Barlaston. Elizabeth went to live near her sister, Charlotte Langton; old Sarah Wedgwood, last of her generation, came to Downe village. The trains ran faster than ever to Birmingham, and often enough Charles would have been "ready to steam down" upon them, but in vain. Maer, "dear Maer," belonged to the past, with all its enchanting associations and memories.

He went on working at Down House while Emma was off to Tenby with the children. He wasn't well—for three years, he told Hooker a little earlier, he hadn't had a single night without stomach trouble—but the *South America* was now going to the printers, and he could not keep pace with their demand for manuscript. "I am tired and overdone." Still, the business was nearly at an end. In July and August he was at Shrewsbury, his work with him. In September he and Emma had ten days at Southampton for the British Association meeting; the papers were dull, but many friends, old and new, were there, and it was all very pleasant. The final proofs were done with on October 1st.

It was a longer book than either the *Coral Reefs* or the *Volcanic Islands*, yet, like them, always succinct, never expansive. Its plan was simple. It dealt first with more recent elevations and surface consequences on the east and west coasts, then turned to tertiary deposits and older rocks. It endeavoured, and with marked success, to show every phenomenon as adequately accounted for only by acceptance of successive slow subsidences and elevations of the entire continent. It rejected catastrophism absolutely: to the gradual processes of nature through endless time all things were possible. The great Patagonian plains rising terrace-like one above the other, the huge mud-formed expanses of the pampas with their abundant fossils, the mighty Cordilleras themselves—these and a hundred lesser details fell into satisfying place in the scheme of his guarded yet decisive generalizations.

Only less impressive, indeed, than the actual conclusions is the spectacle of this human mind poring over these necessarily hasty records of observations, and with iron insistence wringing from their chaos a vision of orderly creation. There was in this man the power of consistent, dogged, penetrative thought. He would seize on an idea, brood over it till it formed itself to his purpose, till he could make plain out of the minute particulars new and unsuspected relationships, a pattern, a harmony. It was the most satisfying of mental tasks, but also the most exhausting. There need be little wonder that two hours' work upon it was all he could manage each day, for the close substance of the book itself shows the unremitting intensity of his concentration. A good deal of the volume, inevitably, was unexcited and unexciting drudgery, the setting forth of plain fact, nevertheless he was satisfied, knowing that there was much in its pages that was new and important, and as he wrote he seemed to *see* the activity of the geological processes by which the great continent had come to its present being, like the breathing and slow shifting of a giant in slumber.

One conclusion was subsidiary for the work itself, yet of great importance for his species theory. In writing his 1844 essay he had declared the absence of intermediate fossil forms the weightiest objection to his view, though he had never subscribed to the not uncommon notion that "the whole history of all departed time lies indelibly recorded with the amplest minuteness of detail in the successive sediments of the globe, . . . carefully preserving every created form and every trace of action."[1] On the contrary, he had held the geological record to be extremely imperfect, but only now, in a detailed examination of the matter as it applied to shells collected on the elevated South American coasts, did he realize how imperfect it was bound to be, in that preservation was the exception, not the rule, occurring in bulk only under conditions of slow, continuous subsidence. Without subsidence there was no reason why any "geological record" should be preserved at all. To demonstrate that was if not to abolish, then certainly to diminish one negative difficulty.

Decidedly it must have been with a deep sense of relief that Charles put the finishing touches to this concluding volume on the Geology of the Voyage of the *Beagle*. Four and a half years of "steady work," prolonged over almost twice as long by illness and other interruptions, had gone to the three works, and if geology had been his first scientific love he was ready by now to have done with it in the name of a better, one wider and more satisfying. On the title-page of the new edition of the *Journal* it had been natural history—not, as before, geology—which received precedence.

[1] H. D. Rogers, q. *South America*, 139 n.

The *South America* ended, it seemed, all his work upon the actual material fruits of the voyage, with one small exception. Fifteen years ago he had set forth on a two years' interlude in his career as country clergyman. The two had extended to five, and only now, after another ten, had he really (he thought) done with it. The zoological results had been made available in five large volumes (£9, 16s. the bound set, £8, 15s. in sewed parts), the geological in three, the more personal account was now arranged in a definitive edition and issued for popular reading, the collections themselves had all been given satisfactory homes. The plants all had been or shortly would be dealt with by Hooker, Henslow, and others; the insects by Walker, Waterhouse, and Newman. Bell was undertaking the crustacea, Sowerby the shells, White of the British Museum the Arachnidae. There remained for himself only the minor matter of some lesser creatures, a barnacle or two, mentioned in last year's preface to the *Journal*.

CHAPTER XII

BARNACLES AND MIDDLE AGE

I

It has been said that Charles never turned easily aside from a subject once his hand was set to it. That was true of his geology. It was truer of his work on cirripedes, or barnacles. Here was this specimen about the size of a pin's head, collected in the Chonos Archipelago late in 1834. He meant merely to describe it, but it had some odd features which necessitated comparison with many commoner forms, and so step by step he became interested in cirripedes as a class, until J. E. Gray of the British Museum, Mr. Stutchbury of Bristol and other collectors urged him to take up its systematic description, offering to lend their own collections, even for dissection. He was attracted, writing in 1844 to Hooker about his "beloved barnacles," and asking advice about microscopes, but in 1845 he intended no more than "to get out a little Zoology" (and then "hurrah for my species work"), and when he took up the matter in the autumn of 1846 he believed that "some months, perhaps a year" at most, would see it done with, when "I shall begin looking over my ten-year-long accumulation of notes on species and varieties, which, with writing, I dare say will take me five years." [1]

But one thing led to another. After the purely intellectual labour of the geological books, it was "delightful to use one's eyes and fingers again." Why not do as Gray and the others suggested—redeem the subject from its present careless chaos? Of course he wanted to get on with species, but . . . did he? He knew, or thought he knew, where that was likely to land him: "When published, I dare say I shall stand infinitely low in the opinion of all sound Naturalists." [2] Didn't he, he asked himself, *need* some such training as this monographing job would give him? Here he was setting out to treat of species from, if not an

[1] *LL.*, i. 351. [2] *Ibid.*

absolutely new, at least a revolutionary standpoint, and yet what did he *know* of anatomy, development, classification? He looked back upon his biological experience—almost he had had none. There had been enough dissecting and describing aboard the *Beagle*, of course, but how little he had known, how bad the instruments and conditions of labour, and how unsatisfactory the results. He could not afford to pass by this opportunity of systematic training. He consulted Hooker, who absolutely agreed and urged him on, suggesting that one had hardly the right to theorize on species until one had "minutely described many."

So he took up the task of a monograph on both fossil and living cirripedes.

Almost at once it began to extend itself. The dissection, though it gave him great pleasure, was terribly slow, and it was not long before he found himself bogged in difficulties about the names of species. The subject had hitherto been badly dealt with, made a hunting-ground of collectors more eager to enumerate varieties and establish priorities than to aid science. Indignantly he declared "a very wrong spirit" to "run through all Natural History, as if some merit were due to a man for merely naming and defining a species,"[1] and had health allowed he would have started "a crusade" against the habit of thrusting one's own name upon every apparently new variety one chanced to come across. As time passed, stretching into years, and years beyond years, and he still seemed but plunging deeper into the work, he felt compelled to defend himself from the actual or supposed reproaches of friends who seemed to wonder whether he would ever emerge from this new morass. *Et tu, Brute!* he may well have said to himself when Hooker in 1849 announced his preference for Charles on species to barnacles any day, and wished him back at work on the former forthwith. "This is too bad of you," Charles wrote, recalling Hooker's part in turning him to barnacles, and how he had already been able to show species and barnacles running in harness, his hypothesis that "a hermaphrodite species must pass into a bisexual species by insensibly small stages" causing him to spot precisely that process in cirripedes. Was that nothing? "Do not flatter yourself," he told his friend, "that I shall not yet live to finish the Barnacles, and then make a fool of myself on the subject of species."[2]

[1] *LL.*, i. 371.   [2] *LL.*, i. 376.

Henslow's protest against "unapplied" science struck even deeper home, for it attached no less to species work too. He really must demur, he retorted. "Would not your hearers infer from this that the practical use of each scientific discovery ought to be immediate and obvious to make it worthy of admiration? What a beautiful instance chloroform is of a discovery made from purely scientific researches, afterwards coming almost by chance into practical use! For myself I would, however, take higher ground, for I believe there exists, and I feel within me, an instinct for truth, or knowledge or discovery, of something of the same nature as the instinct of virtue, and that our having such an instinct is reason enough for scientific researches without any practical results ever ensuing from them." [1]

He had to confess, though, to frequent tedium, and even to doubt whether it was worth while "spending a week or fortnight in ascertaining that certain just perceptible differences blend together and constitute varieties and not species." [2] Yet it was, one feels, just that, the precise examination of "just perceptible differences" blending together to suggest, to demonstrate, the unreality of rigid species, which was the finally valuable fruit of these years of labour. Of his work on cirripedes he wrote to Hooker, still in 1849: "I have been struck . . . with the variability of every part in some slight degree of every species. When the same organ is *rigorously* compared in many individuals, I always find some slight variability, and consequently that the diagnosis of species from minute differences is always dangerous. I had thought the same parts of the same species more resemble (than they do anyhow in Cirripedia) objects cast in the same mould. Systematic work would be easy were it not for this confounded variation, which, however, is pleasant to me as a speculatist, though odious to me as a systematist." [3] Mere variation was not enough to denote species, for variation was universal. What he had perpetually to stop and ask himself was whether this or that variation was sufficient to make others deem this specimen a "God-created"—had he been a profane man he might have amended, a God-damned—barnacle, deserving its own specific name, or was it a variety only? And wasn't it the fact, as he had so long suspected, that there wasn't, save in degree, twopennyworth of difference between variety and species after all?

[1] *ML.*, i. 61.   [2] *LL.*, i. 379.   [3] *LL.*, ii. 37.

II

Inevitably the work was interrupted, as ever, but more than ever, by days, weeks, and longer periods of ill-health. He touched perhaps, in these years, his nadir of well-being. It was in 1845 that he said that for three years he had never been free from stomach-trouble for twenty-four hours at a time, and 1846 found him little better, and 1847 rather worse. Boils and other mysterious swellings bothered him, he was "almost continuously unwell," and towards autumn 1847 the least effort or excitement prostrated him on bed or sofa, incidentally preventing his farewell meeting with Hooker before the latter's departure to India in November.

He managed his usual visits to Shrewsbury in February and October of that year, but the latter found him very poorly, groaning and grumbling on the sofa in his bedroom, and trying, not too successfully, to divert himself with *The Last Days of Pompeii*. The most enjoyable vacation that year was to the British Association meeting at Oxford in June. Henslow, Hooker, Buckland, Owen and many other friends were there, and one "heavenly day," never to be forgotten, they all made a trip up the river to lovely Dropmore.

Through 1848 he became steadily worse. At Shrewsbury in May he was continuously weak, unable to forget his stomach for more than a few moments together, even though Robert was now seriously ill, looking forward to death and hoping only that it would come suddenly. In the intervals of his own pangs and sitting with his father, Charles wrote letters, tinkered with his notes on work in hand, and read an English translation of his old friend Mme. de Sévigné, with whom (he told Emma) he now fell in love. He also read Evelyn's *Life of Mrs. Godolphin*, and thought she needed all her beauty to compensate for her excessive virtue.

Apparently he was at Shrewsbury again in October, and saw his father for the last time, for the old man died quite suddenly, as he had wished, on November 13th, aged eighty-two. In the following March, Charles made the extraordinary statement to Hooker that "I was at the time so unwell, that I was unable to travel, which added to my misery,"[1] but "unable" must have

---

[1] *LL.*, i. 372-3.

been a slip for "unfit," for he undoubtedly went to Shrewsbury for eleven days. It was a great effort, however, and on the day of the funeral he could not accompany the procession on its slow journey to Montford, where on November 18th Robert was buried beside Susannah. Shrewsbury mourned the death of one of its principal citizens. The poorer cottages of Frankwell darkened their windows, and the children watched from the doorways, weeping. Charles even felt bound to refuse an executorship of his father's large property, but at the family gathering there was discussion of what Susan and Catherine were to do, and he approved their decision to stay at The Mount.

Robert's death was certainly a grief to him, for he had, like all the Darwins, great family affection, and of later years—one may say since the return from the *Beagle*, and perhaps even more since his marriage and independent family life—intimacy had grown as fear, or awe, diminished. Nevertheless the event had been expected; it did not come as a shock, nor could it, at Robert's age and in his state of increasing helplessness, be greatly regretted, least of all for his own sake. Yet it may, in some degree, have acted as a release for Charles. The memory and influence of parental dominance do not easily die while the parent still lives, and, more than that, the passing of the older generation has its own psychological effect. Now one inherits the earth. No longer that elder rank stands between oneself and death, to take, as it were, responsibility for life. Now oneself must be responsible, give of one's best, admitting no longer any other holding the right to say one nay. The lapping shadow has engulfed all who went before. One's own turn comes next, whether it be near or far. One must act, for better or worse, now or never. . . .

The strain of the whole occasion was great, leaving Charles practically an invalid. At the least undue exertion, or scientific argument, or even anticipation of the briefest journey his hands would tremble, his head ache and swim, his stomach retch and vomit; he could not sleep of nights for acute indigestion and the futile gadfly activity of a worried brain; depression and physical weakness would prostrate him on bed or sofa for hours and sometimes days at a time. All that winter he could work barely two days a week. He thought on his father and on Aunt Elizabeth, and believed himself to be dying.

Desperately, in March 1849, he transplanted the entire household to a rented house at Malvern, staying there till June to try Dr. Gully's water-cure. Some aspects of the Doctor's establishment disturbed him. It held all manner of cranks, and as for the Doctor himself, why, he believed in everything—clairvoyance, mesmerism, even homoeopathy! Yet the cure seemed to do him good, and he was while there able to prepare for the printer his first small *Monograph on the Fossil Lepadidae; or, Pedunculated Cirripedes of Great Britain*, which was to be issued by the Palaeontographical Society, though not until 1851. He continued the treatment—with lamp, shallow bath, douche, "dripping sheet" and exercise—after the return home, and seemed gaining strength, weight and the appearance of health (but he always *looked* well), all that summer and autumn. The giddiness, the trembling, the vomiting had gone, but he still tired very easily, and work went very slowly ahead; and at the Birmingham meeting of the British Association in September, which he attended with Emma and where he was a Vice-President, he was not well enough to go with the Lyells and Horners and others to Warwick. Very flat, he thought the whole occasion, compared to Oxford two years ago.

As Emma feared, the water-cure, though he recurred to it frequently, could benefit him only temporarily, and though throughout 1850 there was little sickness, and 1851 was an especially good year, the succeeding spring found him poorly again, upset by the least thing, the quietest visit to London—and when in October he declared himself "unusually well" it was now with the qualifications of the frank invalid who has to make his health his first consideration. For nearly forty years, it is said, "he never knew one day of the health of ordinary men." [1]

Yet still the cause of his trouble remained undiagnosed! Various explanations have been suggested, mostly since his death. There was the family constitutional weakness. There were the five years of intermittent but often severe seasickness on the *Beagle*. There was the unnamed illness at Valparaiso, and the marked ill-health at Sydney. There was the painful palpitation of the heart which so alarmed him at Devonport before the *Beagle* sailed, and again in London after his return. Robert thought the Valparaiso illness the main cause, and Charles inclined to agree with him. Francis Darwin disagreed, looking instead sometimes to Sydney for the beginning of the business,

[1] *LL.*, i. 160.

and at other times suggesting that "when we remember the ill-health of his brother Erasmus there is no need to seek for any cause beyond a hereditary taint."[1] Sulivan, of the *Beagle*, remained convinced that the "constant suffering" of his sea-sickness was primarily to blame, but Charles rejected the idea.

More ingenious solutions of the problem have been devised by American medical men. Dr. Kempf, of course, attributed all the trouble to his "anxiety neurosis," holding that the voyage itself left no permanent ill-effects, and regarding his repeated attacks as largely the workings of an unconscious delaying mechanism restraining him from completing his work and flinging it to the lions of public opinion before the last possible moment.

Dr. George M. Gould was sure it was all a matter of astigmatism. Even in youth, he noted, study depressed the otherwise cheerful Charles. He even questioned the nautical nature of the sickness on board the *Beagle*; eye-strain could cause nausea too! Charles wasn't seasick when he went to Scotland by steam-boat in 1838, for he wasn't then using his eyes excessively. Work and illness were in fact always associated; recovery came with respite, and relapse with return to his desk. Dr. Gould's diagnosis would have been considerably more convincing had he not, on the one hand, ascribed the varied sufferings of many of the great throughout the ages to the same single cause, and, on the other, minimized all the facts against his case, as when he asserted there to be no evidence of special fatigue during the *Beagle* voyage. There is indeed no evidence at all that Charles's sight was other than what Lyell declared it in 1838, excellent. In his sporting days he was a first-rate shot; which if not absolute proof against astigmatism makes its existence in severe form unlikely.

The most convincing of these commentators was in fact the earliest of all, Dr. W. W. Johnston, who in 1901 examined the evidence as given in the *Life and Letters*, and concluded the trouble to have been "chronic neurasthenia of a severe grade due first to the overstrain of the *Beagle* voyage and second to the life of hard intellectual work begun in 1837 and continued until 1882."[2] During the voyage Charles was, at a time of continuous physical exertion, subject also to incessant mental stimulation, and simply could not restrain himself. His work was always overwork.

---

[1] *FitzRoy and Darwin*, F. Darwin, *Nature*.
[2] *Ill-Health of C. Darwin*, Johnston, 158.

Aboard or on shore, amid whatever discomforts, he was ever at some labour. He landed in England with exhausted nervous system, only to plunge instantly into yet harder work, intensifying his exhaustion, and thereafter he continued to work busily, if interruptedly, all the rest of his life. There was definitely no organic disease, as the unchanging nature of his symptoms (eventually diminishing) made plain. Had he, on his return to England, only allowed himself a year or two of complete rest, his subsequent story might have been very different. As things were, "the continued overstrain of exhausted nerve cells" made his accomplishment of "the highest intellectual work," though never impossible, always "both painful and difficult." [1]

With the exception of Dr. Gould's astigmatism, there seems some substance in all these views. The family constitutional weakness was a plain fact, upon which the years of seasickness were bound to have their effect, and the illness at Valparaiso and low condition at Sydney only too probably were the first signs of a consequent breakdown in health, never overcome and soon intensified by the overwork following his return; while no great proof is needed to suggest that in so highly nervous and diffident an individual as Charles—desperately modest and desperately aware of the challenging nature of his most precious work, expectant that at best it would set him "infinitely low in the opinion of all sound Naturalists"—some form of anxiety neurosis, and conscious and unconscious desire to delay publication, may well have played effective part. The causes of his ill-health cannot perhaps be precisely defined, but there is no need to make a mystery of the matter.

III

It was with direct reference to his invalidism that the daily routine of his life was arranged. The exigencies of the water-cure determined some of his habits, continued long after the cure itself was discarded. The period of frequent visits to London, and attendance at scientific gatherings, was largely if not entirely over; the mere journey there seemed more and more of a burden.

He woke and rose early, taking a short walk before breakfast. In summer the grass would still be dewy, the air fresh, sweet

---

[1] *Ill-Health of C. Darwin*, Johnston, 158.

with the singing of many birds; in winter he would be out before sunrise, to see the foxes creep home in the grey brightening dawn. He breakfasted alone, then was at his desk from 8 to 9.30, often his most productive time. Next came an hour's rest, lying on the sofa, while he dealt with his post or lay idle listening to family letters or a novel read aloud to him, generally by Emma. Another hour or even two in his study would complete a good day's work in time for the invariable second outing with his dog, first about the garden to see how his seeds and experimental plants were progressing, then off to the sand-walk, stopping to speak to the children playing soldiers on the lawn, or along solitary country lanes and field-paths, ever attentive to the changing panorama of the seasons, missing no detail of new leaf or blossom, flighting insect or winging bird. Usually he walked, but sometimes, on medical advice or for a longer excursion, he would ride his placid cob Tommy.

After lunch he would retire to the sofa again, to read the newspaper in silence. Next, letters must be written, then he went upstairs to his bedroom couch for a placid smoke while a novel again, or a book of history or travel, was read to him. Often he would drop off to sleep, and "he used to regret losing parts of a novel, for Emma went steadily on lest the cessation of the sound might wake him."[1] Another walk, promptly at four o'clock, and then, were he unusually well, another hour's work, followed by more resting and being read to in his bedroom till dinner at half-past seven. That meal over, he would play backgammon with Emma, read some scientific work to himself, or listen to reading aloud, or to Emma playing the piano, the one they had tramped through the snow together to buy, when life in London was still a new adventure.

The children would be in bed by now, the house held in its country stillness. Lamplight fell softly over walls and furniture, rapt in a spell of music's loveliness. The candles in the piano sconces threw their kindly luminance over Emma's mature comeliness. No word would be spoken, only the notes sounding as her capable fingers moved deliberately upon the keys. He liked Handel, but he liked Beethoven too—the struggle of great forces, the bugle cry of the indomitable spirit of man defying the universe of pain and death and night. Out of the battle, out of

[1] *LL.*, i. 122.

the chaos, an order, a harmony. Deep was its appeal, for whether he was conscious of it or not, his own effort was a parallel one—towards the *beauty* of order not imposed but revealed in what had seemed but chaos. He called it truth, but it was the thing, not the name, that counted.

There was usually a final reading before he went to bed about half-past ten, but were there company and much talk, he would retire earlier, for the excitement of the most modest participation would keep him sleepless to the small hours, leave him listless all the following day. At best he slept ill enough, sitting up against his pillows to relieve his stomach's queasiness, turning over in his busy mind some trivial problem that swelled in the darkness to nightmare proportions. But always he would wake early and be up and out before breakfast. . . .

These habits changed but little in the last thirty years of his life. He found, in the true sense, rest in them. They conserved his energy, nervous even more than physical. They made a routine in which relative inessentials found a place, freeing his mind for attention to larger matters.

They ruled not only his but Emma's existence too, for she gave her life wholeheartedly first to him and then to the children. In her many of the most admirable aspects of nineteenth-century marriage were incarnated. Esmé Wingfield Stratford has remarked that "the married lives of the Victorians were, in an extraordinary number of instances, crowned with a happiness that no dreams of romance could have surpassed," instancing the Brownings, Gladstones, Peels and others.[1] The Darwins should have found a place in his list. Mutual love and respect were the basis of a common happiness that endured undisturbed from their marriage till Charles's death more than forty years later.

To that happiness Emma's contribution was not small. As she had been in youth, so she continued unperturbed into old age—sincere, frank, downright, yet generous, thoughtful, charming. In no way intellectually comparable to her husband, she had her own natural intelligence, and, added to that, unswerving integrity. She held her place at all times with the dignity and serenity of a true self-possession, pretending neither to herself nor to any about her, young or old. It was characteristic that when, despite her early resolves, she found it impossible to be interested in

[1] *Victorian Tragedy*, Stratford, 148.

Charles's studies, she said so frankly, yet with such transparent acceptance of her own limitation that he could never be hurt by it, and could laugh gleefully in relating to friends how one day, at a British Association lecture, he whispered to her that she must find it very wearisome, only to be told: "Not more than all the rest." In the period when admiration for Tennyson was universal, she said that she found his *Queen Mary* only less tiresome than Shakespeare.

Sentimentality she could not stand. She liked things simple, bare—direct as herself. Her piano-playing, it was said, showed all her inner nature: it had vigour, a crisp fine touch, intelligence . . . but never passion. She was too sensible for that as for other extravagance. She took things as they came, and things, and people, must take her as she came. If individuals bored her they bored her, and that was that. A good housekeeper, she would not be bothered by trifles. Charles had acquired a taste for tidiness on the *Beagle*: Emma taught him he must not be too particular. She simply couldn't, she said, keep up with the children's muddles, which had to accumulate till they could be borne no longer, then cleared away by nurse or maid!

In many ways she was the perfect foil for him—bright, happy-go-lucky, apt to brush aside useless self-commiseration, yet an admirable nurse, adequate to every occasion, and always, from first to last, deeply devoted to him. She did not care much for any poetry, and for Tennyson's less than most, but she copied into a little book kept for private reading three revealing verses of *In Memoriam*:

> I know that this was Life—the track
>   Whereon with equal feet we fared;
>   And then, as now, the day prepared
> The daily burden for the back.
>
> But this it was that made me move
>   As light as carrier birds in air;
>   I loved the weight I had to bear,
> Because it needed help of Love;
>
> Nor could I weary, heart or limb,
>   When mighty Love would cleave in twain
>   The lading of a single pain,
> And part it, giving half to him.

It made no difference that in this case *he* was rather giving half to her.

IV

Their lives were indeed in these years being re-oriented. The older generation was all but gone; Charles's own generation was scattering hither and thither as old homes broke up, new homes formed; the younger generation was growing fast. Of all the Wedgwood and Allen uncles and aunts, only Sarah Wedgwood, Jessie Sismondi, and Emma and Fanny Allen still survived, and of these the first two had little longer, Emma however living till 1866, and Fanny till 1875, when she died in her ninety-fourth year. Jessie died in 1853. Clear-headed, self-possessed, religiously sceptical to the last moment, she gave her final orders, said quietly, "I think that is all," paused, looked up as though at someone standing before her, cried suddenly, "Sismondi, I'm coming," and was dead.[1]

The younger Wedgwoods lived mostly in and about London, Josiah with Caroline in their handsome house at Leith Hill, Harry and Jessie at Woking, Hensleigh and Fanny actually in town, not far from Erasmus, with a country house later in Surrey. Elizabeth and Charlotte formed an outpost in Ashdown Forest near Tunbridge Wells, and Frank—alone of them all—in Staffordshire.

Willie, Charles's first-born, was fourteen in 1853, Henrietta ("Etty") ten, and George eight; Elizabeth ("Bessy") had been born in 1847, Francis in 1848, Leonard in 1850, and Horace in 1851. Annie had died, aged ten, three weeks before Horace's birth. She had not been well the previous autumn, and in March Charles took her to Malvern to try the water-cure. Henrietta, Brodie, and Miss Thorley the governess went with them. Hensleigh's wife joined them in April to look after Charles, while Fanny Allen was at Downe with Emma. Annie was worse. A vomiting attack turned to fever, there was a week of suspense, of recoveries and relapses, Charles beside her watching her "poor hard, sharp features" grow paler, harder, sharper. She died at mid-day on April 23rd, and was buried in the old Abbey graveyard in Malvern. Charles was stricken. She was, he said, his favourite child, and indeed wrote of her as though he had no other. "We have lost the joy of the household, and the solace

[1] *Emma Darwin*, ii. 152.

CHARLES DARWIN IN 1854

of our old age."[1] "A dear and good child," he set upon her tombstone, and a quarter of a century later wrote in his autobiography that the very thought of her could still bring tears to his eyes.

If death brought grief, survival brought problems! Charles was the fondest of fathers, thoughtful for his children in every way, and from infancy when he could study their dawning rudiments of expression to days when they could go hand in hand with him along the sand-walk, when he might play with them, read to them, tell them tales of his life on the *Beagle* or of his own Shrewsbury childhood, be interested in all their individual developments and hobbies. He refused to forbid them to enter his study even in his brief working-hours, and they so loved his company that one four-year-old offered him a bribe of sixpence to leave his desk and join them in a game. They loved the holidays, for then he was with them all day.

He never ceased to worry as to the possibility of their inheriting his burden of ill-health. A more immediate problem, about 1852, was that of education. He thought all the public schools gave far too much time to "the old stereotyped stupid classical education," which had profited him nothing, and which in due course he seemed to see shadowing William's mind, destroying the free use of reasoning and observation. He allowed, though, that a boy who could stick at Latin should be able to stick at anything! He also disliked sending his boys away from home so young, and though he thought schools less vicious (and more industrious) than in his day he wasn't easy about that aspect either. He applauded Fox for educating his innumerable children at home, but clearly couldn't follow that example, and so, lacking "courage to break through the trammels," sent Willie to Rugby ("not worse than any other school"), where the boy seemed happy enough.

He worried too about what the boys would do after school. The professions were (as ever) all in a bad way; one could only hope that emigration might radically improve the situation in the next few years.

He also feared the threatened French invasion. Again a Bonaparte, a Napoleon, was ruler across the Channel, a squabble over distant possessions had suddenly brought awareness to both

[1] *LL.*, i. 134.

nations of the superiority of the French Navy in both steamships and artillery, and it was said that in Paris there was open talk of landing an army in Kent to settle accounts with the old enemy. Charles visibly shuddered as he thought of "the French coming by the Westerham and Sevenoaks roads, and there enclosing Down." He wrote to Fox lamenting the old carefree times at Cambridge: "Ah, in those days there were no professions for sons, no ill-health to fear for them, no Californian gold, no French invasions." [1]

The "Californian gold" referred to his fears for the devaluation, due to gold-field discoveries, of his mortgage investments. In fact he had little need for worry on that score. He had been comfortably off, on an annual £1400 with expenditure of £1000, before Robert's death set his income nearer £5000 a year. But he remained uneasy, and would talk sometimes of emigration, generally to New Zealand.

But that was a dream, and meanwhile life went on, bringing its customary ripples of events—holidays, the visits of friends, new books, the flow of ideas. Swanage was visited one year, Ramsgate another. In July 1851 all the family had twelve days in London with Erasmus to visit the Marvel of the Century—the Great Exhibition in Hyde Park. Charles was enthralled, but the children found it rather boring after the first excitement, and preferred staying at home to scrub the back stairs; probably Emma shared their feeling, for she once said that she could see a cathedral in five minutes. In September 1852 he and Hooker attended the funeral of the Duke of Wellington, and the following August, after a jolly three weeks of sea-bathing at Eastbourne, went to Chobham to view, in Sulivan's care (Admiral Sulivan now), the unprecedented Army manœuvres of that year. After being thrilled by the jingle and thunder of galloping cavalry and almost caught between the two advancing armies, he possibly felt safer from the French.

He made few new friends, but again met Asa Gray, now an eminent botanist, at Hooker's house in 1851; they began a regular correspondence. The acquaintance with Huxley also began now, quickly becoming intimate on a basis of mutual interest and liking. One chilly winter day about the end of 1853 he was introduced to and had a few brief words with a

[1] *LL.*, i. 381, 382.

tall young man, a naturalist and great admirer of the *Journal*, and just home from four years on the Amazon. His name, Alfred Russel Wallace, meant nothing to Charles. There were occasional dinners, too, with Lord Mahon, whom he liked less at first than his charming wife and his entertaining father Lord Stanhope, who thought geology "all fiddle faddle," and would urge the visitor to give it up for some serious subject like occultism. Nearer home, at Downe itself, there were the Lubbocks, of High Elms, where the Darwins had been welcomed from the very first, Charles being something of a god-father in science to young John, now about twenty. His other village associations were social rather than scientific. He was the treasurer of local Coal and Friendly Clubs for the poor. For some years he was a county magistrate, and assisted in parish affairs. No longer a church-goer, he and the vicar, Brodie Innes, were close and amicable friends.

Now, in 1854, his health seemed better, he more active. Realizing how much out of touch with scientific circles he had become, he joined the Linnean Society and the Philosophical Club, a select private assembly of members of the Royal Society who met for dinner and friendly discussion. On June 10th he, with Emma and some of the children, attended the opening of the Crystal Palace, now removed from Hyde Park to Sydenham. It was an impressive sight. The crowd beneath the high arching glass was terrific. The Queen and the Prince Consort, the Emperor and Empress of the French, were there; the special galleries were bright with uniforms. Clara Novello sang *God Save the Queen*, and as the full sweet voice rang through the great building Emma was so moved that she broke into audible sobs, her cheeks wet with tears. It was the only occasion, even including Charles's death, that Henrietta ever saw Emma lose her self-control.

Charles, entering middle-age, was a notable figure, tall, broad, and despite his illnesses apparently well-built. His head was impressive, almost bald, the broad forehead sweeping dome-like up and back from the overhanging brows, the strong mouth grim under the rather prominent nose, the side-whiskers greying but the hair still thick above the ears and at the back. Always genial in talk, his features in repose seemed to betray a prevailingly sombre cast of thought.

V

Now, after eight full years, the cirripedes' work was drawing to a close. Illness and domestic affairs apart, there had been little interruption or diversion—two or three geological papers, an unsigned review of a book on natural history by Waterhouse, and—the only really ponderable item—an essay on Geology written early in 1848 for an Admiralty *Manual of Scientific Enquiry; Prepared for the Use of Her Majesty's Navy: and adapted for Travellers in General* (1849). Herschel edited the volume, and Sedgwick had been asked to write this particular essay, but, being unwell, had suggested Charles. It was purely formal, but was rightly praised for its breadth and freshness, derived perhaps from its recapitulation of personal experience.

Two cirripedes volumes had appeared in 1851, the one, already referred to, of 88 pages, from the Palaeontographical Society, the other, of nearly 400 pages, *A Monograph on the sub-class Cirripedia, with Figures of all the Species: The Lepadidae; or, Pedunculated Cirripedes*, issued by the Ray Society, which conveniently, for Charles said that he did not know how else the work could have been published, had been founded seven years earlier for the issue of valuable but specialized scientific works. Now in 1854 the study of the subject was completed by the addition of two similar volumes from the same sources, the larger first, of 673 pages, *A Monograph on the sub-class Cirripedia: The Balanidae (or Sessile Cirripedes); the Verrucidae, etc.*, and then the smaller, of only 44 pages, *A Monograph on the Fossil Balanidae and Verrucidae of Great Britain*. In his preface to the second large volume, dated July 1854, Charles mentioned his original intention to publish as well a small volume on his anatomical observations. But he thought the ground sufficiently covered, and he wished to give no more time to the subject.

Indeed, he had given enough. The period had not been wasted, but let it be no longer. By midsummer all the writing was done, and August and September were spent in tidying away his papers, returning the borrowed barnacles to their patient owners, and generally clearing his decks for a new attack upon the old yet ever-new subject of species.

Not once had his collection of information ceased, his attention to the subject slackened. The reading of "really numberless special treatises and *all* agricultural and horticultural journals" went on perpetually. He was in incessant correspondence with practical breeders, and Hooker in India was made to keep an open eye for Ornamental Poultry.

His busy mind, in fact, was never still. He found it at all times "delightful to have many points fermenting in one's brain," and if superficially he seemed to tire easily he had a basic persistence which permitted him to drop no topic till he had wrung something positive from it. Year after year casual letter-references or notes in books would show it still remembered, awaiting the moment when it might find some place in the larger scheme he was creating.

He speculated endlessly—it was like strong drink to him—while condemning all hypotheses not absolutely confirmed by every available fact. Between the dangers and fascinations of theorizing he swung from one extreme to another. Generalization was wholly evil, the speculator was a fool, the compiler alone "great." Not so! "Those who merely accumulate facts I cannot very much respect."[1] And again, more than once: "I am a firm believer that without speculation there is no good and original observation."[2] His abiding view, reconciling these extravagances, appears in two sentences: "I can have no doubt that speculative men, with a curb on, make far the best observers."[3] And again: "It is a fatal fault to reason whilst observing, though so necessary beforehand and so useful afterwards."[4] Speculation there must be, but bridled always by unprejudiced, unspectacled observation.

That was his constant aim, made effective in no small degree by his power of keeping many subjects actively in mind over long periods, and also by his insistence upon the facts which told against as well as for his views. Huxley, later, was to praise his intelligence, his memory, his imagination, and above all his "strict subordination" of all these to his love of truth. His son Francis thought one of his most striking characteristics "that supreme power of seeing and thinking what the rest of the world had overlooked."[5] These qualities he had, and also another, perhaps as important as any—his sense that, as he wrote to Lyell in 1849, "truly the schemes and wonders of Nature are illimit-

---

[1] *LL.*, ii. 225.   [2] *LL.*, ii. 108.   [3] *ML.*, ii. 133.
[4] *Nature*, January 3rd, 1931.   [5] *ML.*, i. 72.

able." [1] His experience on the *Beagle* had opened his eyes, and thereafter he could never close them wholly, to the infinite possibilities of the universe. At that most impressionable age he had perceived how inadequate our formulated knowledge was to hold reality in chains, and again and again he made implicit or explicit rejection of that knowledge as final. In the *Journal* he insisted on our ignorance of the conditions of life in relation to survival or extinction. In 1845 he wrote to Hooker that "we cannot pretend, with our present knowledge, to put any limit to the possible, and even probable, migration of plants. If you can show that many of the Fuegian plants, common to Europe, are found in intermediate points, it will be a grand argument in favour of the actuality of migration; but not finding them will not, in my eyes, much diminish the probability of their having thus migrated." [2] In 1858 he expressed both these points more broadly: "The more I think, the more evident it is to me how utterly ignorant we are of the thousand contingencies on which range, frequency, and extinction of each species depend." [3] Laws there were, but neither existing theory nor collected evidence necessarily compassed them.

Strictly speaking, the attitude was a negative one; in actual practice, it was truly positive—a rejection of limitation, an asseveration of mental freedom. It gave flexibility of thought. In its light, no domain of knowledge or thought was closed to his questioning.

And as beneath his superficially quick exhaustion there was phenomenal persistence, so underlying his surface timidity there was unique boldness. It has been said that it is almost as important to know the right questions as to know the right replies. Charles had that primary power. He felt now, in 1854, that he had been asking the right questions long enough. It was time to test the tentative—though ever more assured—answers in a full examination of all the relevant material. He had already —in or about 1852—joined the last evasive link in sudden comprehension of "the tendency in organic beings descended from the same stock to diverge in character as they become modified." [4] That was a step of cardinal importance, and he never forgot, to his life's end, "the very spot in the road, whilst in my carriage, when to my joy, the solution occurred to

[1] *LL.*, i. 378.   [2] *ML.*, i. 407-8.   [3] *ML.*, i. 117.   [4] *LL.*, i. 84.

me"[1]—the (once grasped) obvious enough fact that the conquering, increasing species must inhabit and accordingly become adapted to ever more and more diversified environments as it spreads from region to region.

Why should he longer delay? He could not suppose now, as does youth, that he would live for ever. Supposing, too, he were forestalled, his work wasted? If the thought of his father's displeasure had been a deterrent, it could be so no longer. The *Vestiges* had, however unsatisfactorily, made the subject one of public discussion. He must get on, at least put it to the test of his own examination. Sometimes he was despondent: "How awfully flat I shall feel, if, when I get my notes together on species, etc. etc., the whole thing explodes like an empty puff-ball."[2] More often he was jubilant, firm in his faith—"hurrah for my species work!"

[1] *LL.*, i. 84.   [2] *LL.*, ii. 44.

CHAPTER XIII

THE GREAT WORK

I

THE anchor was weighed, the last delaying ropes cast off, the destination clearly marked upon the map, yet still progress remained of the slowest. It may be unjust to say he pottered, but that is the impression. There were excuses. Anyone who has ever attempted the arrangement of large masses of scattered and diverse material for more than the most superficial and immediate ends will sympathize with his hesitant deliberation, for even in the most minor cases the impulse is always to probe a little further, to test a little more completely, that the conclusion, when at last it is reached, may be just as final and unassailable as possible. Charles had special reasons for obeying such an impulse. He expected contradiction, attack, abuse. His case, he determined, must have no Achilles' heel.

That mid-January, 1855, the whole family moved to a rented house in London—27 York Place, off Baker Street—seeking both change of air and diversion after a spell of children's illnesses. But suddenly the weather turned bitterly cold. Day after day unyielding frost made the pavements iron and the air ice. Even indoors one could not keep warm, and the streets, often, were thick with whirling snow. The children fell ill again, and Charles and Emma so coughed and sneezed and suffered from rheumatism, first one and then the other, that scarcely once were they able to go out together.

It was the dreadful Crimean winter, when the unfortunate British troops on active service died by the hundred day after day, for months on end, from hunger, disease, and cold. All through 1854 the war had gone on, and it was to continue into 1856, but not even the Charge of the Light Brigade roused Charles to any particular recorded interest in the matter. He lived to an amazing degree isolated from the wider world of public affairs.

The startling events of the great revolutionary year of 1848 found no reference in his printed letters. Public matters must be forced on his attention by direct personal relation—the actual visit to the Great Exhibition, the French threat *to Down House*—before he could care to take much note of them. The Crimea was, after all, like most British wars in the nineteenth century, a very long way off.

All the family were glad to return to Downe on February 15th, the road cutting through deep snowdrifts massed and moulded by the wind, the country-side stretching away white and dazzling on either side. Snow, feet deep, covered all the garden, frozen so hard that the children, to their delight, could walk upon it without breaking the smooth pure surface. Illness still dogged them; another epidemic appeared, and in March the house was noisy with whooping-cough.

It still seemed to Charles "far the greatest fact" that he had "at last quite done with the everlasting barnacles," but by now he was hard at work collecting his species notes and sending out personal appeals for further information, though with the actual writing of any book on the subject comfortably "two or three years" away. He could, at that distance, regard the project with equanimity; any suggestion of its nearer approach disturbed him instantly. He would begin to question whether he wasn't perhaps getting out of his depth, whether the subject was after all likely to do anyone good. All his boldness would vanish, and he would have to reassure himself that he was no advocate of any theory whatever, but just a neutral observer giving arguments for *and* against a just admissible hypothesis. Yes, that was it, he wrote to Fox, he was going to give all the facts he could master to see "how far they favour or are opposed to the notion that wild species are mutable or immutable."[1] In June and July he was still tentative, even to Hooker: "You ask how far I go in attributing organisms to a common descent: I answer I know not."[2] Only a year later did he return to his earlier assurance and confess to Asa Gray, by then a regular correspondent: "As an honest man, I must tell you that I have come to the heterodox conclusion, that there are no such things as independently created species—that species are only strongly defined varieties."[3]

[1] *LL.*, ii. 49.   [2] *LL.*, ii. 65.   [3] *LL.*, ii. 79.

Through all these months and for many more to follow he was constantly busy with multitudinous experiments directed to support or destroy that conclusion. Tests suggested in one field were painstakingly applied in others. "Horrid puzzles" were faced with patient persistence and that supreme honesty which must reject the most helpful evidence till positive proof speaks in its favour. Geographical distribution was at once his inspiration and his bugbear, in the large view both illuminating and illuminated by his theory, in detail presenting the most perplexing problems as to means of dispersal from common centres of development—the only alternative to multiple creation. There was much talk just then of vanished continents once joining lands now widely separated by deep oceans. It would have solved a great group of Charles's difficulties to have accepted the idea, and he "earnestly" wished he might do so, but could not, believing that the existing divisions of land and water were of the very greatest antiquity. Instead he gave himself up to the most exhaustive and exhausting experiments to discover the possibility of eggs or seeds being carried from land to land either floating in the sea, or upon or in the bodies of migrating birds. Specimens were kept in tanks of sea-water at varying temperatures for varying periods, then tested for germination. He supposed seeds swallowed by a fish, the fish by a heron, the heron flying far afield, the seeds being voided and falling upon fertile soil. Sometimes everything went well, and all his geese were swans; sometimes all went ill, and all his swans were geese—but well or ill he patiently went forward, testing every point, determined to accept nothing lightly. He was taking up botanical studies too, largely for amusement but with an eye on variation, and in 1855 he became, at Yarrell's urging, a pigeon fancier, the birds interesting him as a notable case of extraordinary variation from one known form. He wanted, particularly, to discover whether the young of the various breeds differed as much from one another as did their parents. He developed quite an affection for them, and they became, in the following years, as familiar a sight about Down House as half a century before at The Mount. He duly joined the Philoperistera and Columbarian Clubs of fellow fanciers, a strange set of odd beings with much of the single-minded intensity of racing touts, meeting in "gin palaces" to discuss with dark solemnity and "awful shakes of the head" the finer mysteries of successful breeding. They treated him with marked respect, addressing him formally (or perhaps informally) as "Squire."

The summer of 1855 he was, again, unwell, but attended the Glasgow meeting of the British Association. He thought the long journey more trouble than the event was worth, though he approved the gathering as a whole, and agreed that the Duke of Argyll spoke excellently. It was his last attendance at any of the Association's annual meetings. Returning, he and Emma had a pleasant day at Shrewsbury with Susan and Catherine.

II

With such slight interruptions the preliminary work still went on in 1856. It seemed indeed that it might well go on for ever without Charles ever coming to the actual point of printing CHAPTER ONE at the head of a sheet of paper and beginning to write. Sometimes he felt that time was slipping away and he growing old too fast, but always some point would arise which obviously—but *obviously*—must be attended to at once. It was Alfred Russel Wallace who, indirectly in 1856 as directly in 1858, was to spur him forward.

Wallace was fourteen years younger than Charles, a man of poor though literate parentage and indifferent schooling, thrust upon the world to make his own living (at land surveying in the company of an elder brother) at fourteen, and only by hard effort keeping alive his ever-growing interest in natural history. A chance meeting with H. W. Bates led to their joint expedition to the remote head-waters of the River Amazon, in prospect a most dubious venture but in achievement, despite set-backs, extremely successful. He spent four years in South America, from 1848 to 1852, then, following eighteen months in England arranging his collection and writing a book on his travels, he set out, early in 1854, upon an eight years' visit to the Malay Archipelago.

He had, he said, long been "bitten by the passion for species and their description," and as early as 1847, when he read Chambers's *Vestiges*, he inclined to the idea of a progressive development of animals and plants. Now, resting at Sarawak in February 1855, while Charles stood at the window of 27 York

Place, looking out at the driving snow, he wrote an essay *On the Law which has Regulated the Introduction of New Species*. It appeared in the *Annals and Magazine of Natural History* in the following September. Lyell read it there, early in 1856, and called Charles's attention to it, urging him to publish at least a sketch of his views, lest, delaying too long, he should be forestalled.

Lyell might well take that view, for every line of the essay revealed its author as running hard upon the track of Charles's ideas, a hare in pursuit of the tortoise. Here were the facts of geological change and geographical distribution, and the setting of groups—classes, orders, families, genera, species—in space and time, set forth to suggest an essentially genealogical classification ("the analogy of a branching tree"), and to declare the "law" that "every species has come into existence coincident both in space and time with a pre-existing closely allied species." It was a law which not only explained but necessitated the visible facts. Like Charles, Wallace stressed the imperfection of the geological evidence, and, like him again, he was cautious in the matter of alleged "progression" of species; he preferred to write only of change. There were minor differences and some omissions, but there could be no doubt that he was on Charles's trail.

It was, though the fact is often overlooked, Charles's trail by more than prior exploration. If Wallace presently caught up to Charles, he owed much to him in the first place. He had been studying the subject, he stated in the essay, for ten years past. The date itself is suggestive, for it was in the middle 'forties that Charles's *Journal* deeply interested him, presumably in its 1845 revised edition [1]; and in this essay of 1855 especial stress was laid upon evidence, as that afforded by the Galapagos Archipelago, clearly derived from that source.

Charles was perturbed. Lyell was pressing. He had begun to doubt whether Charles would ever write his book at all, and may even have wished to force his friend's hand, knowing that once committed he would have to go forward. He asked for something short—he did not mind how short—just to establish priority. Charles " hated the idea of writing for priority," though

---

[1] A letter written by Wallace in 1887 (*Life of C. Darwin*, 189) diminishes this interest and influence, but is contradicted by his other statements. (See *My Life*, Wallace, i. 256.)

he would certainly be "vexed if any one were to publish my doctrines before me." [1] He and Lyell had a long talk on the matter in London, early in May; Hooker was appealed to and supported Lyell, and at last Charles rather unwillingly agreed to give "a couple of months" to sketching out a very brief abstract of his views to be issued as "a *very thin* and little volume."

III

The writing began on May 14th, and for a while went steadily ahead. Bunbury saw him on June 20th, and there was talk of approaching publication. But he was as uneasy about it as a cat on hot bricks. Repeatedly he lamented—even twice in the same letter—how "dreadfully unphilosophical" it was to publish without full detail, protesting that *truly* he would never have dreamed of it but for Lyell's persistence. He so expected disapproval from his friends that he attributed it even to the faithful Hooker.

At last the old cautious impulse, the desire to delay before publishing, began again to have its way. By July the idea of the brief outline and the material so far written for it were abandoned, and he was deciding that any essay must be as complete as his existing material would allow. He would publish in, perhaps, a year, confining himself to his notes, rejecting new investigations. But even that he could not keep to, and soon he was back at his salt-water tanks and his skeletons of young pigeons and rabbits. He *must* verify his facts. If the task still got "bigger and bigger with each month's work" he couldn't help it. The job had to be done properly, for this was—he was quite firm now—*the* book that he was writing. He would call it *Natural Selection*.

Soon some of the material on Geographical Distribution was being approved by Hooker, but it was October before the chapter was even provisionally finished. The end of the year brought its private troubles. On November 6th, at her house Petleys in Downe village, Sarah Wedgwood died, aged eighty, last surviving child of the Josiah Wedgwood who had been Dr. Erasmus Darwin's friend. Charles, in black cloak and with crêpe tied about his hat, walked with the other Wedgwood nephews and

[1] *LL.*, ii. 68.

nieces to the grave, then back to Petleys, behind sad black standards, to hear the reading of her will. How things had changed since the old Tag Rag days at happy Maer; life, so gay, so easy, then, brought with the passing years its increasing burdens. Emma, now forty-eight, was pregnant once more, and exactly a month later, December 6th, her last child, Charles Waring, was born, mentally deficient, and never to learn to walk or talk.

Charles worked on all that winter at Variation. Then health became troublesome again. He would have gone to Malvern, but the memories of Annie would be overwhelming there, and he visited instead, in April, another water-cure establishment at Moor Park near Farnham in Surrey. The weather, as all that summer, was superb. He liked the place and people, he lazed, read idly, thought of everything but work. Nevertheless it was from here that, on May 1st, he wrote to Wallace for the first time, answering a letter and commenting on the paper in the *Annals* as showing their similarity of thought and views. He mentioned his own work now in progress, saying that he had written "many chapters" but did not expect to go to press for two years, and adding that while he could not explain his views on variation in a letter, still he held "a distinct and tangible idea" of how it came about. That was all; as Wallace later pointed out, he gave no clue even remotely indicating Natural Selection.

He was much more specific in another letter, written to Asa Gray four months later, and subsequently important as containing his only explicit statement to this date of the significance of that principle of divergence which had come to seem to him, next to the primary fact of Natural Selection, the keystone of his theory. It was a very typical letter. Gray, reading his views, would "utterly despise" him. Well, he realized the difficulties and even agreed with some of the objections. Lamarckian talk of habit producing delicate adaptation was quite futile. His own view was different and, he thought, more satisfactory. Human selection, he asserted, had been the main agent in forming our domestic species. Suppose, then, a *natural* selection working not on this or that one feature for a few brief years, but upon the entire organization, and unremittingly over millions of generations, only a minority in each generation surviving to propagate its kind, and constantly exposed to changing environment. "Considering the infinitely various ways beings have to obtain food by struggling with other beings, to escape danger at various times of life, to have their eggs or seeds disseminated, etc. etc., I cannot doubt that during millions of generations individuals of a species will be born with some slight variation profitable to some part of its economy; such will have a better chance of surviving, propagating this variation, which again will be slowly increased by the accumulative action of natural selection; and the variety thus formed will either coexist with, or more commonly will exterminate its parent form." [1] That was the main, basic point, but this matter of divergence was important too, for since "the same spot will support more life if occupied by very diverse forms," [2] divergence favours survival and is thus an intrinsic part of the life-process. Conversely, the slighter divergences conflict more than wider ones, so tending to mutual extinction and creating the branching tree of life now specifically declared by Charles as by Wallace. He couldn't, he admitted in conclusion, say much about the laws governing variation itself; these were unimportant save as providing the material that selection itself worked upon.

Variation had his attention all the summer. The Indian Mutiny, though its horrors filled the papers, was as remote as the Crimean War a year or two earlier. He was at Moor Park again in June. Henslow came visiting in August. Work went on. That autumn, probably through the winter too, he was considering Hybridism. In December he wrote to Wallace that the book was now half-finished, but publication still—*still!*—two years away. Wallace had asked: would the book discuss Man? Charles replied that he thought to "avoid the whole subject, as so surrounded with prejudices; though I fully admit that it is the highest and most interesting problem for the naturalist." And again he gave his characteristic wriggle away from responsibility: "My work . . . will not fix or settle anything; but I hope it will aid by giving a large collection of facts, with one definite end." [3]

On he plodded, two or three hours a day all he could manage. From Hybridism he turned to Instinct, and thence, apparently, back once more to Geographical Distribution. By April 1858 he

---

[1] *LL.*, ii. 124.   [2] *Ibid.*   [3] *LL.*, ii. 109.

had, he told Gray, completed "eleven long chapters," but had still to deal with palaeontology, classification, embryology, and some other difficult topics, so that he would hardly be ready for the printer before the spring of 1860 at earliest. That same month he was again at Moor Park, playing billiards between walks. He returned to Downe to send Hooker a manuscript on range and variation in large and small genera which had been bothering him for a year past. In early June boils were keeping him to his sofa. He had completed the work practically to the point corresponding to the end of Chapter viii. of the *Origin*.

IV

At some such pace, ever delaying, always with publication comfortably two years ahead, he might have gone on almost if not quite to the end of his life had not Wallace, quite unwittingly, intervened on June 18th, with his essay, written in the previous February, *On the Tendency of Varieties to Depart Indefinitely from the Original Type*. It came by the morning's post, an envelope containing no more than a few thin sheets closely penned over and a note expressing the writer's hope that "the idea would be as new" to Charles as to himself and that "it would supply the missing factor to explain the origin of species," and asking Charles, if he thought it worthy, to forward it to Lyell.

Charles read the essay instantly, and wrote at once to Lyell. His letter was odd, and revealing, not least in its tense restraint. He had received "the enclosed," he said, to-day. It seemed to him "well worth reading." (He must, as he penned the easy phrase, have felt its appalling irony: the essence of all his long labour set forth upon a feather-weight of foreign correspondence paper—"well worth reading"!) His feeling broke out irresistibly in the next sentence: "Your words have come true with a vengeance—that I should be forestalled. I never saw a more striking coincidence; if Wallace had my MS. sketch written out in 1842, he could not have made a better short abstract! Even his terms now stand as heads of my chapters." Well, the honourable thing must be done. Wallace didn't actually suggest publication, but Charles would "of course" at once offer to submit the essay to any journal. Again his strong emotion declared itself: "So all my originality, whatever it may amount to, will be smashed, though" (consoling afterthought) "my book, if it will ever have any value, will not be deteriorated; as all the labour consists in the application of the theory." Then, in the last sentence, control once more fully regained: "I hope you will approve of Wallace's sketch, that I may tell him what you say."[1]

It is a letter raising at once the question of Charles's real feeling about priority, about, to put it more widely, public recognition and applause. He commonly and sincerely protested his indifference to, even his distaste for, concern with such matters. All the "wretchedness" of scientific controversy sprang from desire for fame, he wrote in 1848, one eye on Owen: "the love of truth alone would never make one man attack another bitterly."[2] Yet he could not, for all that, deny altogether the taint in himself, specifically envying what he saw as Hooker's easier freedom. While on the *Beagle* the joy of breaking new trails, of being the very first to scale some hill or scan some view, was very strong in him. He rejoiced in the thought of priority in collecting in some remote area, was dashed when he heard that he had been preceded. In later stay-at-home, sedentary life this feeling quite naturally transferred itself to his mental exploration. It was, in fact, all quite natural. Men do inevitably desire priority. It is a fundamental human instinct. Life may be largely a matter of repetition, but it reaches out continually to newness; it exists for the sake of newness, to do the new thing, or, more deeply still, to *be* the new thing. Who has never created something unique, his alone, has never lived; who ceases to have that capacity for creation is already dead. To live so is, in its joy, its own reward. "Delight is to him—a far, far upward and inward delight—who against the proud gods and commodores of this earth, ever stands forth his own inexorable self."[3] The desire to walk ahead of other men is normal and good. It becomes evil, destructive, only when it is sought, not for its own sake, but that other men may acclaim and reward, and when, above all, the achievement of other men is denied in one's own self-interest. Wallace's essay put Charles to the acid test. He did care for "the bauble fame," he had confessed in 1857; he wished he could care less for it. Consequently some conflict in him was as inevitable as manifest. It expressed itself in his distracted waverings, his frantic appeals to Lyell and then to

[1] *LL.*, ii. 116–17. [2] *ML.*, i. 65. [3] *Moby Dick*, Melville.

Hooker, his conscience-stricken desire, expressed in a letter to Wallace half-written and then destroyed, to suppress his own work completely, leaving the entire field to the newcomer. The depth of the tumult going on in him is clear in a letter to Lyell, written a week after that already quoted yet still in obvious agitation. Desperately he wanted to accept Lyell's and Hooker's protests against his self-effacement, *if* he could feel he was doing so "honourably." (The word occurs three times in a few sentences.) He put the points for himself: that there was nothing in Wallace's essay not to be found, "much fuller," in his own sketch of 1844, that Asa Gray had his letter, "a year old," showing that he had anticipated *all* Wallace's essential points. He would "now" be "extremely glad" to publish a sketch of his theory, but since he had not done so before, since this eagerness sprang only from Wallace's intervention, weren't, he questioned, his hands tied? Back and fore he swayed between thoughts of such publication, and the conviction that even to dream of it was "base and paltry." He concluded in a final abasement by apologizing for troubling Lyell with such "a trumpery letter, influenced by trumpery feelings," upon so "trumpery" an affair. He was only sending it, he said, in order to be able to banish the whole subject from his mind for a time—yet next day was sending another note post-haste after it, declaring the same qualms in very much the same language. Desperately he was anxious to do the right thing; desperately he was anxious also that his years of work should find their recognition. The latter was an entirely natural desire, discreditable to him only by an impossibly high standard and inconsistent only with the extravagant laudations of over-zealous admirers. He wanted his due, no more. Basically his values were right. He spoke with self-knowledge when he said that without the incentive of fame he might work with less gusto but would work just as hard. The root of his effort was authentically "a sort of instinct to try to make out truth"; where the thought of fame was present, it was secondary.

V

One can scarcely doubt that in any case the only fair solution, some form of simultaneous publication, would have been arrived at. But at this point the whole matter was removed from Charles's hands. Scarlet fever had spread from the village to Down House, and within little more than a few hours the youngest child, his namesake, was dead. Henrietta was sickening, apparently for diphtheria, and very soon two of the attendant nurses were also ill. Charles was distracted, busying himself to get the other children safely out of the house, comforting his wife, reproaching himself more than ever for his "miserable" worry over priority.

Meanwhile Hooker was acting with decision. He had sent to Charles demanding Wallace's essay and the copy of the letter to Gray by instant return, and these were delivered to him by hand at Kew on the evening of Tuesday, June 29th, together with the sketch of 1844. Mrs. Hooker at once set to work to transcribe the relevant portions of both letter and sketch, and two days later, on the evening of July 1st, these, with the essay, were read to a meeting of the Linnean Society by the Secretary. Hooker and Lyell were present, and spoke shortly, but more to impress the audience with the importance of the matter than anything else.

The extracts from the 1844 sketch were read first, stating briefly but plainly Charles's essential conception of the working of natural and also sexual selection. Then came the passages from the Gray letter, adding the vital principle of divergence, necessary to be stated here since Wallace set it in the forefront of his essay, appearing even in the title, and dominating the argument. The opening point was the taking by some writers of evidence of rapid reversion to type in many domestic varieties as argument for permanent species. Wallace wanted to undercut that argument by suggesting a principle able to produce progressive divergence under natural conditions yet simultaneously explain reversion among domestic organisms. That principle was natural selection of favourable variation in the struggle for existence. In the natural state, his essay explained, only the favourable variation establishes itself, and reversion from that must mean extinction. Under domestication, on the other hand, variation has in the main no such vital value; it has no necessary correspondence to the natural conditions of survival. Wallace, like Charles, dismissed Lamarck, substituting another and more mechanical process: the feline species had not grown claws by inner volition, but by the survival of those having the greatest facilities for holding their prey. This, he thought, adequately accounted for all special attributes, for in fact "all the phenomena presented by organized beings, their extinction and succession in past ages, and all the extraordinary modifications of form, instinct, and habits which they exhibit." [1]

So, that evening of July 1st, 1858, the theory of Evolution by Natural Selection was publicly announced.

The audience, Hooker said, was intensely excited, but there was no discussion; at most murmurs and hesitant talk between individual members when all was over. They had come there that evening in anticipation of a paper by Bentham asserting the fixity of species, to hear, in effect, its total contradiction. They might have scoffed, but the plea of Hooker and still more of Lyell had made that difficult; they could only think their thoughts and go their ways, leaving the whole matter to secure burial in the pages of the *Journal of the Proceedings of the Linnean Society*. That in fact seemed likely to happen, for in the scientific world generally the event aroused scarcely more attention than Wallace's other paper in the *Annals* three years earlier. But one straw served to show that a wind *was* blowing—Bentham's immediate withdrawal of his essay, by no means in assent to these new views, but in instant recognition that the subject had been put upon an entirely new footing.

The meeting came out into the London streets under a night sky in which a comet, symbol of fate and change, burned whitely. Those who noted it did so purely astronomically. The days of destinies read in the stars were dead for ever now. Yet change, inevitable, irresistible, was at hand. They had just witnessed the release of a pebble which in the next few years would grow to an avalanche.

[1] *Linn. Soc. Journal*, 1859, Zool., III. 62.

CHAPTER XIV

CLIMACTIC

I

CHARLES was glad to hear the proceedings had all gone "prosperously" that evening. Doubts still worried him whether he had acted honourably, and wished that Hooker would "exonerate" him by writing direct to Wallace. But his mind was made up now on one matter. The thing was done. The Malayan feline had forced his cat out of its bag for ever. Delay belonged to the past. He itched to get on with a publishable abstract of that larger work which itself had begun as abstract also.

First, though, he and the family must get away from sickbeds and sickrooms, to recover health and vitality. They all went first to Elizabeth's at Hartfield, then to the Isle of Wight. It was there, at Sandown, on July 20th, two days after the death of his sister Marianne, that he received the Linnean Society *Journal* proof-sheets of his and Wallace's papers, the reading of which induced him to abandon his holiday intention. He started work on the abstract that very day.

Even then he was doubting whether it could be confined to the usual thirty-page maximum of *Journal* papers, and soon it seemed that the opening section alone would fill as much. He felt that either he would have to compensate the Society for extra printing expenses, or else offer the work as not one but a series of papers.

He was, for once, enjoying the writing, but it irked him to think of having to give his conclusions without the support of all available evidence, and he wondered whether such presentation might not destroy interest in the full work, untouched since the receipt of Wallace's essay but which he meant to take up again at the first opportunity.

He wrote daily, at Sandown and then at Shanklin, till August 12th, when the return to Downe interrupted him for more than a month. But with every day spent upon it the manuscript grew not only in fact but also in prospect. By early October the series of papers had become "a small volume" involving another four (in November five) months' steady labour. By the end of the year he had in view a volume of up to five hundred pages.

Once that autumn the drain on his energies sent him to Moor Park to recuperate. He amused himself in pleasantly renewing his memories of Cambridge, where William was now at Christ's—the coffee-drinking in Fox's rooms, the visits to the FitzWilliam, the walks along the college backs, the sweet-sad evening services in King's College Chapel, the glorious days of beetle-hunting. He got William to look up his old gyp, Impey, who asked blankly, of Charles: "Why, has he been long married?" He remained poorly all the winter, and was back again at Moor Park in February, but otherwise kept constantly at work from November till late March. By Christmas the first ten chapters, based almost entirely on material already put into shape for the Great Work, were finished, and he ready to embark in two chapters upon the wide topic of Geographical Distribution, writing almost entirely from memory. That was done by the end of February. Between March 2nd and 16th he wrote the thirteenth chapter, on Mutual Affinities of Organic Beings; Morphology; Embryology; Rudimentary Organs, and before the month's end the last of all, the "Recapitulation and Conclusion," was also completed, and he busily correcting the entire manuscript, to have it ready for the printer early in May.

He was tired, but content to have completed a full outline of the work which had engaged him so long, and which he had come (with Lyell) almost to doubt would ever reach an end. Now the whole field was charted, and with that as basis he surely could not fail to go ahead. The abstract and then the full exposition in print, he would be able to feel, he thought, his life's work done. He was happier about Wallace too, having had two cordial letters from him approving the Linnean Society proceedings and the idea of the abstract.

His hopes were rising. In December, Hooker had suggested a grant to assist publication, but Charles even then felt that it might "sell enough to pay expenses," and by April he was anticipating something like popularity and "a fairly remunerative sale," though he begged Hooker not to repeat his opinion lest the event falsify it and he be made a laughing-stock. John Murray next stepped in, on April 1st, to give colour to his expectation by unseen acceptance of the book on Lyell's recommendation and a list of chapter-titles. He even offered the terms—two-thirds of the net profits—he customarily gave to Lyell himself. Charles thought the decision precipitate, but by April 10th Murray had seen the first three chapters and given final confirmation. He questioned only the title, objecting to the word "abstract," which was accordingly deleted. He also queried the phrase Natural Selection, thinking it unfamiliar and likely to convey little save to the initiate, but Charles could not surrender that! The original title was *An Abstract of an Essay on the Origin of Species and Varieties through Natural Selection*. In the printed volume it became *On the Origin of Species by means of Natural Selection, or the Preservation of Favoured Races in the Struggle for Life*.

With publication settled Charles may have thought himself done with the book, but the book had not done with him. Although Miss G. Tollet, authoress-sister of Emma's closest friend, and "an excellent judge of style," found no fault with it, Hooker and his wife suddenly declared revision necessary to remove obscurities of statement. Charles was horrified, for on his life, he said, "no nigger with lash over him could have worked harder at clearness than I have done."[1] He was exclaiming how he longed to be rid of the whole thing, when, in mid-May, the punctilious Murray appeared with yet more perturbing proposals. He had been reading the complete manuscript, and his conservative Lyellism—he was a keen amateur geologist—found its essential theory "as absurd as though one should contemplate a fruitful union between a poker and a rabbit."[2] In doubt he had consulted the Rev. Whitwell Elwin, editor of the *Quarterly Review*, who shared Charles's qualms about presenting the theory without full evidence, and proposed—Murray passing on the suggestion—that the author should publish instead a full statement of his observations of pigeons accompanied by a quite brief account of his general views and the promise of a forthcoming larger work which would substantiate them for other creatures. Murray liked the idea, for "everybody is interested in pigeons." Naturally Charles would not consider it. He had done his best, and for

[1] *LL.*, ii. 157.   [2] *At John Murray's*, 170.

the present—till the big book was completed—was determined to stand or fall by the existing abstract.

Murray accepted his decision, but the very thought of reshaping his work, joined to the certain need for a running textual revision, came as a last straw, and on May 18th he collapsed and had to go away for water-treatment and complete rest. He read *Adam Bede* for relaxation, and was charmed by it. He returned early in June to find the proofs of the book awaiting him, and to plunge into a prolonged and wearisome task of correction which kept him hard at work (but for another brief hydropathic interlude in July) until September. He travailed, he groaned, he apologized. His emendations covered the galley-slips and ran over on to attached sheets. The style, he now agreed, was miserable—incredibly, inconceivably bad. He wrote to Lyell and Hooker what a splendid book "a better man" might have made of his material, how wretched his own feeble effort. Thanking Wallace for the opportunity of reading a new essay, he regretted that he could not profit by it; he must not add "one word" to his book—only try to improve the prose.

Doubtless it was not as bad as he felt it to be. He was in driven, tired, debilitated mood, when a man can look at his sentences and wonder whether they mean everything or nothing, fully reveal or totally conceal his thought. He became, towards September, progressively depressed, only impelled by "an insanely strong wish to finish my accursed book." His stomach bothered him, he slept badly, but day after day stuck at his desk till the morning session was done. Then he, too, was done till next day, much though it fretted him to spend his hours in idleness. On September 10th the last proof was corrected; he had only to run through the revises, and would then be free of "my abominable volume, which has cost me so much labour that I almost hate it." [1]

By September 30th he had finished. He had intended to leave the day before for a hydropathic establishment at Ilkley in Yorkshire, but was too ill—"in an awful state of stomach, strength, temper, and spirits." He started instead on October 2nd, and a fortnight later the family followed him, to stay until early December.

Most of the time at Ilkley his weakness kept him idle. He had

[1] *LL.*, ii. 168.

been paying the piper of over-exertion too long, and now must dance to its tune. He suffered from swellings in face and legs, a nasty rash, an unusually severe attack of boils; in addition, he sprained his ankle. These things, and the winter weather, kept him mostly indoors, doing nothing. But he enjoyed doing nothing, enjoyed the feeling of freedom from the task which had borne upon him so long. By God, it was good, to "hardly ever think about my confounded book." When he did think of it, he did so much more pleasurably than had been possible before. Though he could never wholly shed his diffident manner, he was convinced of the truth of his conception; it was not hypothesis but fact. Lyell, who had been seeing proofs, wrote in general praise but making reservations. Charles boldly held out, in reply, for the theory, the whole theory, and nothing but the theory. Lyell still felt the need to suppose a continued intervention of creative power, if only to account for Man. Charles would have none of it: "I would give absolutely nothing for the theory of Natural Selection, if it requires miraculous additions at any one stage of descent."[1] One had to assume, anyway in the present inadequate state of knowledge, some original creation, just as "philosophers assume the existence of a power of attraction without any explanation"; granted that, Natural Selection could account for every form of life from lowest to highest.

Soon he began catching up a little on his neglected correspondence. He wrote promising Huxley an early copy of the book, for "I shall be *intensely* curious to hear what effect the book produces on you. . . . I am very far from expecting to convert you to many of my heresies; but if, on the whole, you and two or three others think I am on the right road, I shall not care what the mob of naturalists think."[2] The "others" were, in fact, Lyell, whose verdict hung doubtful, and Hooker, to whom he wrote: "I remember thinking, above a year ago, that if ever I lived to see Lyell, yourself, and Huxley come round, partly by my book, and partly by their own reflections, I should feel that the subject is safe, and all the world might rail, but that ultimately the theory of Natural Selection (though, no doubt, imperfect in its present condition, and embracing many errors) would prevail."[3]

To yet others to whom advance copies were to go, he wrote more briefly and cautiously, even humbly. To Louis Agassiz:

[1] *LL.*, ii. 211   [2] *LL.*, ii. 172–3.   [3] *LL.*, ii. 175.

"I hope that you will at least give me credit, however erroneous you may think my conclusions, for having earnestly endeavoured to arrive at the truth." [1] To Henslow, as his "dear old master in Natural History": "I fear, however, that you will not approve of your pupil in this case." [2] To Lubbock: "No doubt I am in part in error, perhaps wholly so, though I cannot see the blindness of my ways." [3] To Jenyns: "I may, of course, be egregiously wrong; but I cannot persuade myself that a theory which explains (as I think it certainly does) several large classes of facts, can be wholly wrong." [4] To Fox: ". . . my weariful book on species." [5] To W. B. Carpenter: "When I think of the many cases of men who have studied one subject for years, and have persuaded themselves of the truth of the foolishest doctrines, I feel sometimes a little frightened, whether I may not be one of these monomaniacs." [6] To Hugh Falconer, who detested all treatment of the subject as mischievous, if no worse: "Lord, how savage you will be, if you read it, and how you will long to crucify me alive!" [7] Numerous advance copies went out in early November to all these, to Erasmus, to Sedgwick, Gray, Herschel, Kingsley, Owen, Quatrefages, Wallace ("I do not think your share in the theory will be overlooked by the real judges" [8]), and to many others in Britain and abroad. Charles meant that his theory should have at least the opportunity of consideration by those who, in his opinion, mattered most.

II

It has already been said that the *Origin*, with all its amplification, essentially followed the outline of the 1842 and 1844 essays. He made it an opening point that it was not enough merely to believe in and to assert the common origin of species unless the mode of modification could be shown. Having found his own first clues to that in his study of domesticated animals and plants, he retained this as his necessary starting-point.

Variation Under Domestication gave proof of the possible wide degree of hereditary modification by progressive accumulation of successive slight variations through selection, which,

---

[1] *LL.*, ii. 215.   [2] *LL.*, ii. 217–18.   [3] *LL.*, ii. 218.
[4] *LL.*, ii. 220.   [5] *LL.*, ii. 222.   [6] *LL.*, ii. 223–4.
[7] *LL.*, ii. 216.   [8] *LL.*, ii. 221.

though fully recognizing the complex nature of variability, and our deep ignorance of its laws, he believed to be the predominant factor. The pigeon, with its many varieties derived from one, was his prime example. Thence again the natural transition was to Variation Under Nature, asserting species as arbitrary, differing quantitatively not qualitatively, and denying the possibility of drawing any absolute line along the whole course from varieties through species, genera, orders and classes up to the greater kingdoms of organic being. Next the Malthusian Struggle for Existence. Accept inheritable Variation, posit the Struggle, and the notion, the necessary fact, of Natural Selection was irresistible. The fittest *must* survive. Here he introduced his newer realization of divergence, and the part of sexual selection was briefly dealt with too. Thus, he declared, grew "the great Tree of Life, which fills with its dead and broken branches the crust of the earth, and covers the surface with its ever branching and beautiful ramifications." [1]

So, in the first four chapters, the essence of the theory was stated; the rest, largely, was confirmatory evidence and argument. To begin with, the Laws of Variation—effect of external conditions not so much directly on the whole individual as (apparently) on the reproductive system; effect of use and disuse, of correlated growth or atrophy; the inexplicable innate tendency to variation, especially in highly developed parts. He could not explain the causes of variation—they were inexplicable—but as a fact he thought it indisputable, as also that "it is the steady accumulation, through natural selection, of such differences, when beneficial to the individual, that gives rise to all the more important modifications of structure, by which the innumerable beings on the face of this earth are enabled to struggle with each other, and the best adapted to survive." [2]

Again, of course, the Difficulties must be met, first some of the lesser, then the larger. Instinct had a chapter to itself, suggesting mental variation as no less subject than physical to natural selection, and so had the more negative problem of Hybridism, in which he sought to show the ticklish topic of sterility between species as much more fluid and open to exception than was generally supposed, and as an incidental rather than primary attribute.

[1] *Origin*, 130.   [2] *Origin*, 170.

The old poser of the Imperfection of the Geological Record had a chapter too, summed up in a passage interesting as presenting the customary packed intensity of his thought, and the number of aspects brought to bear on the settlement of a single, if vital question: "I have attempted to show that the geological record is extremely imperfect; that only a small portion of the globe has been geologically explored with care; that only certain classes of organic beings have been largely preserved in a fossil state; that the number both of specimens and of species, preserved in our museums, is absolutely as nothing compared with the incalculable number of generations which must have passed away even during a single formation; that, owing to subsidence being necessary for the accumulation of fossiliferous deposits thick enough to resist future degradation, enormous intervals of time have elapsed between the successive formations; that there has probably been more extinction during the periods of subsidence, and more variation during the periods of elevation, and during the latter the record will have been least perfectly kept; that each single formation has not been continuously deposited; that the duration of each formation is, perhaps, short compared with the average duration of specific forms; that migration has played an important part in the first appearance of new forms in any one area and formation; that widely ranging species are those which have varied most, and have oftenest given rise to new species; and that varieties have at first often been local. All these causes, taken conjointly, must have tended to make the geological record extremely imperfect, and will to a large extent explain why we do not find interminable varieties, connecting together all the extinct and existing forms of life by the finest graduated steps." [1] This was much more assured than anything in the 1844 essay, and in the next chapter, on the Geological Succession of Organic Beings, he turned from defence to attack, to show such data as did exist as far more consistent with his than any creationist view. The slow, successive appearance and modification of species, the extinction of old forms never to reappear, the diverging variation of dominant groups yet always with evidence of connection, even of unity, the general effect of progression in development—all these things the theory of descent with modification, and that only, explained satisfactorily.

[1] *Origin*, 341–2.

The great and "grand" subject of Geographical Distribution, so long and so carefully studied, had expanded to two chapters, embracing again the widest range of facts all directed to the same end—the plausibility of his view, the implausibility of creationism.

Finally he had brought together certain classes of facts, relating to internal structure of organisms, which he himself regarded as so decisive that he said he would accept his theory on their evidence alone, "even if it were unsupported by other facts and arguments." [1] Indicating mutual affinities in widely separated groups, he asked if community of descent was not *the* "hidden bond of connection which naturalists have sought under the term of the natural system." [2] Morphology, rudimentary organs, above all embryology—each bore its witness.

The closing chapter was Recapitulation, but also something more—"and Conclusion." The main objections to the theory were repeated, then the positive arguments, driving the case home with swift blow after blow till in its triumph it took wings, rising eagle-like above the reluctance of the older naturalists with their stuffed set minds to appeal to that younger generation in which his hope always thereafter lay. He could set no limit to the power of Natural Selection, and even dared to "infer from analogy that probably all the organic beings which have ever lived on this earth have descended from some one primordial form, into which life was first breathed." [3] Once again he foresaw the results of its acceptance—freedom from vain shackles, the opening of hitherto untrodden fields of inquiry, new classifications, new precisions in zoology and geology. Psychology would find a new basis, "that of the necessary acquirement of each mental power and capacity by gradation." [4] Last of all—a brief sentence, but one which had seemed to him to burn the pen in his hand as he wrote it—"Light will be thrown on the origin of man and his history." [5] His enthusiasm even led him to commit himself on one point he was as a rule by no means sure about—the relation of Natural Selection to Natural Improvement. Here he indicated the vast past of organic development and predicted a tremendous future: "And as natural selection works solely by and for the good of each being, all corporeal and mental endowments will tend to progress towards perfection." [6] So he swept on to his

---

[1] *Origin*, 458. [2] *Ibid.*, 449. [3] *Ibid.*, 484.
[4] *Ibid.*, 488. [5] *Ibid.* [6] *Ibid.*, 489.

last paragraph, declaring how all nature revealed these laws he had laid down, and how, "from the war of nature, from famine and death, the most exalted object which we are capable of conceiving, namely, the production of the higher animals, directly follows."[1] That sentence and the one which followed, the last of all, were taken, with only the slightest verbal change, from the earlier essays, from the earliest notebook of all, that of 1837.

III

Of course the book had its reticences, its omissions, its glosses, its limitations. It was written by man, not God. But it is by any standard a mighty product of the human mind. Question its validity, deny its truth, and still it stands a master-work, a synthesis of a whole section of knowledge such as only a handful of beings have achieved in the history of the world. Few books have so instantaneously and so impressively made their mark, and few have deserved to. The indicated fact (and fact it was despite Charles's assertion that up to 1859 he could never meet a single naturalist "who seemed to doubt about the permanence of species") that the idea of evolution—even, from Wells and Matthew to Wallace and Herbert Spencer, of Natural Selection—was "in the air," explained something of the impression it made, but by no means all. The most was in the book itself. "Abstract" perhaps, it was the fruit of more than twenty years of close observation and intensive thought, deriving from all the author's activities of that period with extraordinary comprehensiveness. Charles had that genius which is a capacity for taking pains; he had also that which consists in the power to draw, effectively, constructively, upon a vast range of thought and knowledge. All his work in South America, the Falklands, the Galapagos Archipelago, New Zealand, Australia; all his innumerable experiments from floating seeds in sea-water to comparing the skeletons of young animals and birds; all his wide reading from breeders' reports to accounts of old voyages; all his botanical, geological, zoological, palaeontological inquiries, all his speculations on rarity and extinction, areas of subsidence and elevation, processes of fertilization, action of environment, transmission of acquired characters, adaptation, correlation, the range and

[1] *Origin*, 490.

variety of species—these were but some, a few, of the preoccupations whose results were poured into these packed pages.

The book's power, its sheer compulsion, comes in no small degree from its character as a condensation, a selection of a far larger body of facts and views. The realization that behind each actual instance waits not one but a dozen, a score, perhaps a hundred, more, gives it a dynamic quality as striking as its pure logical exposition, which, for all that, is primary and very forceful, the more so for being logic applied with imagination. The reader is told: "He who will go thus far, if he find on finishing this treatise that large bodies of facts, otherwise inexplicable, can be explained by the theory of descent, ought not to hesitate to go further, and to admit that a structure even as perfect as the eye of an eagle might be formed by natural selection, although in this case he does not know any of the transitional grades. His reason ought to conquer his imagination."[1] But rather, in fact, is the converse true. The reader's imagination must conquer his reason, for—possibly a dangerous admission—the conception must be grasped in its entirety before it can be fully appreciated in its detail. Keen logical argument, massed proof, free imagination—these give the work its power. It can never be, in the common phrase, *dry*, for it is impregnated with a sense of the infinite variety of life, and yet, with that variety, of order and unity. Intrinsically, essentially, an assertion, a demonstration, of order amid apparent chaos, of harmony in heterogeneity, and as such, more than as anything else, irresistible in its apparent achievement, it yet never loses the sense of life's incessant, stimulating wonder. It has been charged, and not without justification, with being a confining work, enclosing all multivarious existence within the little prison of a wholly mechanomorphic process. One may even admit that, in the long view, so it does, and still claim on its behalf that it was, in itself and in its day, for Charles and for his fellows, a releasing book, inspired by the need and desire for liberation from an anthropomorphic imprisonment even more oppressive. Those who found it releasing were—the continual lot of men—perhaps but birds escaping from a little cage to an ampler one whose limitations, for the moment, were scarcely discerned, not seen as limitations at all.

Charles above all felt the book as liberating. Nowhere, save by

[1] *Origin*, 188.

a slip of the mind or pen, would he commit himself to anything like rigid assertion. Again and again he confessed his ignorance, refusing to fetter himself or any other by what he did not know. Facing the infinitely complex and sensitive balance of nature, he realized the necessity of preserving a basically open mind. It may be that he tended, almost too often, to use his ignorance as positive argument—to assert a thing as so simply because none could prove it *not* so—but the refusal to be bound wholly and solely by concrete proof was a valuable and necessary factor in establishing his theory at all. It admitted intuition. It admitted, to come back to the vital word, imagination. And where imagination is, there is life also.

He never pretended to have explained everything, or indeed more than a process: he specified that he was not treating of origins as such, and in that connection came quite quickly to regret his misleading title. He freely admitted that no change of species could be directly proved, that his case rested on what was ultimately circumstantial evidence. He only sought to claim that, if his theory raised difficulties, it solved still more; it afforded a more direct, economical and thereby harmonious explanation, better because simpler, than any Creationist view. It remains that to the present day; more, it remains, despite all, and some very serious, questioning of its adequacy, the most satisfying total account of its broad subject yet put forward. That is why it holds still, against all attacks, its essential place. Even the possibility that its limitations, enforced by vulgar misunderstanding, may make it a very dangerous book, cannot change that fact. We do perhaps most urgently need to-day another, a deeper-visioning Darwin: he has not yet arisen.

CHAPTER XV

RECEPTION OF A BOOK

I

THE small, green-covered volume appeared on November 24th, 1859, in an edition of 1250 copies, price 15s. each. Murray had originally proposed 500 copies only, but friendly encouragement increased the number. Charles, at Ilkley and on tenterhooks, thought 1250 a good deal too many. In fact the whole edition was disposed of at Murray's annual book sale a few days earlier, and on the morning of publication Charles was "confounded" —if also delighted—to receive Murray's demand for a revised second edition "instantly." He could do nothing at Ilkley, lacking his notes and distracted by the water-cure. The instant accordingly extended to a fortnight, but on December 7th he hastened home, abandoning an intended visit to his sisters at Shrewsbury, sleeping at Erasmus's house in London and breakfasting with Lyell the next morning, and reaching Downe that day. But even at his study table he revised very little, from lack of either time or energy (he was soon ill again), and most of his amendments, largely suggested by Lyell, were merely verbal, stressing, modifying, but not changing. He did, according to his promise, slip in another phrase linking Natural Selection to improvement. He made a more careful distinction between rudimentary and nascent organs. He quoted, briefly and anonymously, a friendly letter from Charles Kingsley ("a celebrated author and divine") in support of his new assertion that he could see no reason why his book should shock religious feelings. He omitted one or two, in Lyell's view, seeming-preposterous examples of transmutation of species. He also added, in more than one reference to the original breathing of life into some primordial form or forms, the conciliatory phrase: "by the Creator." By December 21st printing was in progress, and early in the New Year the second edition of 3000 copies was ready for the booksellers.

Charles reported the news to Gray, sending him sheets of the new edition for possible American publication. It appeared, however, that Harper and Appleton had both taken time by the forelock, and were printing unauthorized from the first edition. Harper withdrew on Gray's protest, and Appleton, publishing in January, promised "reasonable" recompense for the author in the event of "any considerable sale." Gray in acknowledgment offered the sheets of the second English edition for any further printing, and there was some suggestion of other special corrections; Charles, eager to get back now to the Great Work itself, demurred to that, but soon proposed sending a short "sketch of the progress of opinion on the change of species." This was completed and despatched, first outline of the Historical Sketch appearing in all subsequent British editions. To Charles's pleasure, a German translation was proposed and quickly issued in the form of three bulky pamphlets; it was not well done, though, and attracted little attention.

Meanwhile letters had been pouring in from friends and fellow-scientists, a few favourable, most at least preserving the decencies. Erasmus wrote fraternally: "the most interesting book I have ever read." Lyell, first unimportant questionings over, declared himself entirely for the book, save perhaps on the awful issue of Man. Charles wrote gratefully: "I rejoice profoundly.... I honour you most sincerely. To have maintained in the position of a master, one side of a question for thirty years, and then deliberately give it up, is a fact to which I much doubt whether the records of science offer a parallel."[1] Hooker of course was warmly with him, as the Introduction to his *Flora Tasmaniae* now publicly declared, and Huxley too, full of enthusiasm, warning Charles not to let himself "be in any way disgusted or annoyed by the considerable abuse and misrepresentation which, unless I greatly mistake, is in store for you,"[2] and promising to sharpen up his own "claws and beak" in readiness for the inevitable fray. Lubbock followed his "father-in-science" with certain loyalty. W. B. Carpenter the physiologist, Jukes, Ramsay and Geikie the geologists, announced their conversion. Wallace in far-off Malay could be taken for granted. Herbert Spencer, an evolutionist by pre-Darwinian conviction, doubted some aspects of the *Origin*, but found he

[1] *LL.*, ii. 229-30.   [2] *LL.*, ii. 232.

also had much to learn from it. Kingsley, though reserved, wanting time to consider so large and important a matter, was more than sympathetic. Gray, too, made theological reservations, but was enthusiastic.

A number of old friends and acquaintances—Henslow, Jenyns, Holland, Falconer, Bunbury—sat on the fence, waiting and watching in a kind of cautious conservatism. Like Henslow, they all went "a little way" with Charles, they were not "shocked" at him, but it did disturb them to see God their Loving Father removed from immediate intervention to almost infinite distance —for that was the effect, however Charles might protest his unintention of theological or metaphysical motive. Bentham was doubtful too, but soon came down on Charles's side, as, reluctantly, Jenyns and Holland too.

On the other side, though, against the *Origin*, were many weighty—perhaps the weightiest, because the eldest—names. Sir John Herschel was openly contemptuous. Gray, of the British Museum, damned it as Lamarckian "and nothing else." Whewell wouldn't have a copy in the library of Trinity College, Cambridge. Andrew Murray the entomologist and Harvey the botanist shook their heads politely but decisively. Philip Henry Gosse continued to contemplate his navel, or at any rate his *Omphalos*, in wintry disagreement. Louis Agassiz in America was as much annoyed as unimpressed, and Alexander Agassiz walked in father's footsteps. Carlyle snarled and sneered; it was the old *Zoonomia* over again, and not improved by time. "That the weak and incompetent pass away, while the strong and adequate prevail and continue, appears true enough in animal and human history; but there are mysteries in human life, and in the universe, not explained by that discovery." [1]

Sedgwick agreed: the poetry of the grandfather was superior to the prose of the grandson. He and Owen, strongest of the opponents, were also the bitterest. Sedgwick sought to subdue his feelings for old acquaintance sake, but thought the book no better than "a dish of rank materialism cleverly cooked and served up" *merely* "to make us independent of a Creator," and wrote to Charles in only too evident distress, declaring himself to have read it with far more pain than pleasure. Parts he could admire greatly, parts made him roar with ridiculing laughter;

[1] *Life of Carlyle*, Wilson, v. 517.

"other parts I read with absolute sorrow, because I think them utterly false and grievously mischievous"—because wild, based on assumptions incapable of proof or disproof, deserting true scientific induction, mistaking secondary consequence for primary cause. He leapt, thence, right to the heart of the matter in a prophetic passage whose insight should be the more apparent now, when the increasing brutalization and degradation of humanity are no more to be denied than detached from conceptions of Evolution and Natural Selection: "We all admit development as a fact of history: but how came it about? Here, in language, and still more in logic, we are point-blank at issue. There is a moral or metaphysical part of nature as well as a physical. A man who denies this is deep in the mire of folly. 'Tis the crown and glory of organic science that it *does* through *final cause*, link material and moral; and yet *does not* allow us to mingle them in our first conception of laws, and our classification of such laws, whether we consider one side of nature or the other. You have ignored this link; and, if I do not mistake your meaning, you have done your best in one or two pregnant cases to break it. Were it possible (which, thank God, it is not) to break it, humanity, in my mind, would suffer a damage that might brutalize it, and sink the human race into a lower grade of degradation than any into which it has fallen since its written records tell us of its history." His "moral taste," he said, had been greatly shocked, and he much disliked the last chapter for its confident tone and appeal to the future. He showed his realization of the implications of Charles's views by referring to himself as "a son of a monkey and an old friend of yours."[1] Charles thought the letter muddled and even childish, but replied placatingly: Sedgwick could not "possibly have paid me a more honourable compliment than in expressing freely your strong disapprobation." He was grieved to have shocked a man he had so long held in high respect, but he was not conscious that "bad motives" had influenced any of his conclusions, and "surely you will agree that truth can only be known by rising victorious from every attack."[2] Perhaps he had written too confidently—but then, others had complimented him on his restraint! The two did not quarrel, but soon, as Sedgwick attacked the *Origin* openly and Charles remained unrepentant, a friendship

[1] *LL.*, ii. 248–50.   [2] *LL. Sedgwick*, ii. 359.

already remote became remoter, though finally broken only by death.

Owen was a different matter. Charles expected his opposition, and, despite a first show of sympathy, got it. He was in fact one of the cleverest, oddest, and most repellent of eminent Victorian scientists, a man whose soul—such of it as existed—was perfectly revealed in what Emerson called his "surgical smile." Huxley at their first meeting was forcefully struck by his queerness, and found his bland and oily politeness almost frightening. Jane Carlyle said his sweetness reminded her of sugar of lead. No one liked him, for hatred and jealousy were too clearly his dominant emotions. Undeniably he was brilliant, but envy and bitterness corrupted his nature. He did not merely disagree with his rivals: he detested them. In earlier days he had been gracious enough to Charles, then but an obscure young naturalist, but the intervening decades had worked a change, and the publication of the *Origin* turned him from friend to enemy. Taking up pen and ink-pot—nineteenth-century substitutes for stiletto and poison-cup—he began to behave like an Italian Renaissance conspirator. Characteristically when he met Charles early in 1860 he cordially shook hands and discussed the book, no more than demurring that perhaps it tried to prove too much. Only later did it appear that at that very time he was busy upon an angry and anonymous attack—incidentally singing his own praises—for the *Edinburgh Review*. Charles never forgave him, and always thereafter delighted in any further proof of his "baseness."

II

Reviews also came pouring in, good and bad, for and against, learned and shoddy. The first was in the *Athenaeum*, five days before publication, a prominent and lengthy notice combining an appearance of fairness with an undercurrent of disagreement and derision, describing "the belief that man descends from the monkeys" as being "wrought into something like a creed by Mr. Darwin," noting his "evident self-satisfaction," and twitting him with "supremely loving his own offspring," and "selecting" it from among all others to be "the successful competitor in the struggle for existence."

The *Saturday Review* on Christmas Eve was antagonistic, and, Charles thought, unfair, though making some good points by the way. Two days later the *Daily News* attacked him for unacknowledged "stealing" from his "master," the author of the *Vestiges of Creation*, but the pain of that insult (worse than being called Lamarckian) was quickly lost in the pleasure of opening *The Times* to find the *Origin* dealt with there in a remarkable three-and-a-third-column article declaring in simple — even elementary—but forceful terms the basic Darwinian hypothesis. It was admirably and clearly planned, and still more admirably and clearly written, presenting in turn the known facts about species and the problems of species, the discredited Lamarckian solution, Charles's scientific qualifications, the working of domestic selection, and finally the idea of a natural selection. Judgment was suspended, beyond declaring the hypothesis superior to any other yet formulated: no more than an "active doubt" was recommended. "The combined investigations of another twenty years may, perhaps, enable naturalists to say whether the modifying causes and the selective power, which Mr. Darwin has satisfactorily shown to exist in Nature, are competent to produce all the effects he ascribes to them; or whether, on the other hand, he has been led to over-estimate the value of the principle of natural selection, as greatly as Lamarck over-estimated his *vera causa* of modification by exercise." Charles instantly suspected Huxley, believing, with some reason, "that there was only one man in England who could have written this essay, and that *you* were the man," his only doubt a wonder how Huxley could "influence Jupiter Olympius and make him give three and a half columns to pure science." [1] As presently became known, Huxley was the author, but by pure chance. *The Times* reviewer, Lucas, had lamented to a friend the labour of dealing adequately with the book, and had been recommended to ask the aid of Huxley, who leapt at the opportunity. He wrote the review at top speed ("faster, I think, than I ever wrote anything in my life"), and it was printed with the addition of no more than an introductory paragraph or two.

With this article Huxley took up his avowed labour as Charles's General Agent or, more metaphorically, Bulldog. He followed it up with essays in *Macmillan's Magazine* and the *Westminster*

[1] *LL.*, ii. 253.

Review, and between these lectured on the subject at the Royal Institution on February 10th: Charles was both present and pleased, thanked Huxley enthusiastically, then having gone away decided that as an exposition of Natural Selection it had been a complete failure! Meanwhile the *Gardeners' Chronicle* had graced the New Year with a "nice" notice which turned out to be by Hooker. Dr. Carpenter spoke favourably for the book in the *National Review* and the more severely professional pages of the *Medico-Chirurgical Review*. Even the *English Churchman* contrived to approve, if patronizingly.

Charles had said in the beginning that he cared nothing for reviews, but he was clearly anxious. (Later, as both approval and objection repeated themselves in seldom-changing terms, he grew more genuinely tired of them.) So far, so good, however. There was, these first few months of 1860, much cause for satisfaction. Other letters and reviews were less favourable, but there was considerable support and praise, far exceeding his first expectation, and the book still sold well. Bunbury noted in March: "Darwin's book has made a greater *sensation* than any strictly scientific book that I remember. It is wonderful how much it is talked about by unscientific people"[1]—both by those who had read it without understanding and those who had not read it at all! Owen also added his testimony, informing a Parliamentary Committee that "the whole intellectual world this year has been excited by a book on the origin of species," to such effect that visitors came to the British Museum demanding to be shown exhibits displaying "those phenomena that would aid one in getting at that mystery of mysteries, the origin of species." He added: "I must say that the number of intellectual individuals interested in the great question which is mooted in Mr. Darwin's book is far beyond the small class expressly concerned in scientific research."[2] John Stuart Mill in April expressed the view of many when he wrote to a friend that though Charles could not "be said to have proved the truth of his doctrine, he does seem to have proved that it *may* be true, which I take to be as great a triumph as knowledge and ingenuity could possibly achieve on such a question."[3] "Seriously," Charles wrote to Hooker in March, making out a list of complete or partial accepters of his

---

[1] *LL. Bunbury*, ii. 190.   [2] *Life of Owen*, ii. 39–40.
[3] *Letters of Mill*, i. 236.

theory, "I am astonished and rejoiced at the progress which the subject has made." [1]

But the battle was not yet won. Some of the enemy's heavier artillery was only just coming into action. Sedgwick, "rabid with indignation," contributed a "savage, unfair" and anonymous attack to the *Spectator* in March. The April *Edinburgh Review* had Owen's similarly unsigned 45 pages, malignant, clever and, Charles feared, likely to be damaging. Not till July appeared the most famous of them all, that in the *Quarterly Review*, signed by "Soapy Sam" Wilberforce, Bishop of Oxford, though Owen's part in it was alleged by more than a few. The essay did not doubt Charles's sound Christianity, but warned him against delusion and of the scientist's need to restrain his exuberant fancy. The author presented his own case against Natural Selection, without presenting Charles's case for it. Like most other attacks on the *Origin*, the review has to-day only historical interest. It, and they, looked backwards, not forwards, stood behind, not ahead of, the work commented on. In their own time they strengthened the reaction they supported, gave courage to obscurantism, and an appearance of reason to prejudice. The first dawn-glow of acceptance over, the barrage of that wider public opinion which is mainly prejudice closed down. Charles recorded in June that lately his lot had been more kicks than ha'pence, and in July that most of the reviews were now "bitterly opposed." The common spirit seemed that of the zoologist who said he would read the book but would never believe it. A casual survey of Charles's correspondence at this time gives clearly a quite wrong impression of the general feeling, since it deals so much more with the inner circle than the wider world. If the scientists declaring themselves for Natural Selection were a not inconsiderable proportion of his own friends, they were plainly a very slender minority among their fellows, and we have also, in looking back, to reckon with the fact that their very adherence to the winning side gives them in retrospect an added importance. When Huxley in the middle 'eighties recalled "a gale of popular prejudice," "outburst of antagonism," "attacks, instinct with malignity and spiced with shameless impertinences," and "the outpouring of angry nonsense," he may have been dramatizing a little but he was dramatizing on a basis of reality:

[1] *LL.*, ii. 291-2.

"On the whole, then, the supporters of Mr. Darwin's views in 1860 were numerically extremely insignificant. There is not the slightest doubt that, if a general council of the Church scientific had been held at that time, we should have been condemned by an overwhelming majority."[1] The *Origin* was attacked before the Royal Society of Edinburgh, and by Sedgwick at the Cambridge Philosophical Society, where Henslow, though still no convert, rose to defend it as presenting at least a fair case for inquiry. A long uphill fight it was going to be, Charles said in May 1860, but well worth while, and despite occasional pessimisms he stuck to his first faith that "if we can once make a compact set of believers we shall in time conquer." What sometimes did depress him was the number of people who couldn't even understand his argument. Systematic study blunted the perceptive faculties, he supposed; "the mob of naturalists" were soulless beings anyway, their minds closed to all but their own particular preoccupations. Well, he had already said that his hope was in the future, the young men. . . .

III

The one personally dramatic episode connected with the publication of the *Origin* was not least notable for the personal absence of both its real principals, Charles and Owen. It was the famous incident of the British Association meeting in Oxford at the end of June, 1860. Charles had meant to be there, but the usual trouble sent him instead scurrying off for a water-cure at Sudbury Park, Richmond. Owen was present for some of the proceedings, indeed sounded the first trumpet for battle, but thereafter vacated the lists in favour of his even more slippery colleague and agent.

The introductory gathering was placid enough. Lord Wrottesley's presidential address did not mention the *Origin*. Nevertheless the book was in the minds of many, and the next day, Thursday, June 28th, Oxford's amiable Professor of Botany, large-browed, squat-faced Dr. Charles Daubeny, delivered to the Zoological Section an address on "The Final Causes of the Sexuality of Plants, with Particular Reference to Mr. Darwin's Work on the Origin of Species," which he strongly supported.

[1] *LL.*, ii. 186.

Huxley was asked to speak by Henslow, in the Chair as President of the Section, but deprecated discussion before "a general audience, in which sentiment would unduly interfere with intellect"; but when Owen rose to attack Charles's theory on the ground of certain alleged differences between human and gorilla brains, Huxley stood up to deny the other's "facts" absolutely.

There this first incident closed, but keen antagonisms had been revealed, and Natural Selection became the topic of every informal gathering. Owen was reported to be plotting counterattack with the Bishop of Oxford, who was to speak on the Saturday, following the American Professor Draper's paper "On the Intellectual Development of Europe, Considered with Reference to the Views of Mr. Darwin and Others," and so large an audience assembled at the New University Museum that a move had to be made from the usual lecture-room to the Library. Even that was soon filled to standing, the Bishop's men a sombre phalanx massed in the middle of the room, while their ladies crowded the windows, white handkerchiefs ready to flutter at the slightest provocation. Owen was to have taken the Chair, but discreetly did not appear: Henslow held his place. The Bishop came late, pushing briskly through the crowd. Hooker and Huxley were both on the platform, though, unlike the principals, neither had meant to be present. The former, rather weary by now of Natural Selection, had accompanied a friend thither, meaning to turn away at the door, but had been lured inside. Huxley should have been on his way Londonwards, but meeting Robert Chambers in the street had been urged not to desert his comrades.

Draper did his best to soothe the excited gathering, even to send it to sleep. For an hour and more he droned on, strongly Yankee, stolidly professorial. The audience bore, if bored, with him, but impatiently shouted down three speakers who followed, the third irretrievably cooking his goose by chalking two crosses on a blackboard and beseeching: "Let this point A be the man, and let that point B be the mawnkey." One joyous whoop went up from the undergraduates present. "Mawnkey! Mawnkey!" they yelled, the high glass roof tingling to the sound.

The Bishop now was on his feet, to "spout for half an hour with inimitable spirit, ugliness and emptiness and unfairness," fluently, rhetorically entertaining, at first jovial, then more harshly scoffing, with twisted dagger-thrusts at Charles and Huxley too.

He ridiculed the former's "proofs," and denounced his conclusions as "an hypothesis, raised most unphilosophically to the dignity of a causal theory." It was a first-rate performance in everything but substance, and Hooker's and Huxley's spirits rose as they realized the weakness, and also the obvious source, of his speech. Only at the very end did self-satisfied valour get the better of his discretion in a frank appeal to prejudice, as he turned to Huxley with sarcastic smile to ask whether it was through his grandfather or grandmother that he claimed descent from a monkey—and thence slid instantly into his pulpit peroration declaring Charles's views contrary to Biblical revelation. Hands clapped, handkerchiefs waved, there was a sustained uproar of clerical approval.

Huxley, at the Bishop's taunt, had struck his hand upon his knee, whispering: "The Lord hath delivered him into mine hands." Now he waited, silent, the audience crying his name from all sides, till Henslow called on him to speak. He rose to his few friends' applause, his face pale under his thick actor's hair tossed blackly back, his wide, thin quivering lips pressed defiantly forward, his long creased frock coat open on his high waistcoat and stock. "I am here only in the interests of science," he asserted, "and I have not heard anything which can prejudice the case of my august client." Mr. Darwin's theory, he said, was much more than mere hypothesis; it was the best explanation of the development of species yet put forward. He stated a number of scientific points before indicating the Bishop's essential incompetence to treat of these matters and closing with his famous retort.

One strange fact about this occasion is the lack of any authoritative account of it. *The Times* gave a few inches to the opening and concluding ceremonies of the week's meeting, but its valued space was far too taken up by the loyal capers of the Rifle and Conservative Associations, and the tuneful performances of the French Orpheonists, to have space for the brayings of "the British Ass." The *Athenaeum* reports turned a deaf ear to personalities. So no contemporary record was made, and partisan recollection so tended to dramatize that not only the words spoken but their place in the speeches, and even their immediate effect, remain matters of contradictory account. According to some the Bishop's words were more, according to others less, personal. According to a few, Huxley jumped up instantly to fling the Bishop's insult back in his teeth. But however the reports of his words vary, their bitter essence is the same: that he would rather have an ape for an ancestor than an intellectual prostitute like Samuel Wilberforce [1]—and whatever individual listeners heard, the sense struck quickly home. There was instant commotion. The Bishop himself sat smiling fatly, but all about the audience men stood up, protesting and cheering. Lady Brewster achieved immortality by fainting and having to be carried out. The massed clergy raised their holy voices in unholy anger. They felt that the Bishop's cloth should protect him from the insults it was his privilege to hand to laymen. Others (laymen) thought Soapy Sam got only his deserts.

The occasion was dramatic in more ways than one, for when order was at last restored it was Admiral Robert FitzRoy who stood up to speak his last word on Charles. His life, since the New Zealand fiasco, had not been without honour. He had been superintendent of the Woolwich Naval Dockyard, commanded one of the new screw frigates in important trials, been promoted Admiral (retired), become a Fellow of the Royal Society and, in 1854, chief of the meteorological department of the Board of Trade, where he instituted the system of storm-warnings gradually developed into the present daily weather forecasts. He was a handsome man still, his hair greying over his half-Roman, half-Stuart features. He denied Huxley's claim for the *Origin* as a logical statement of fact. He regretted its publication; said how its reading had pained him; recalled (but perhaps only on the spur of the moment) how often he had protested with his once-shipmate for entertaining views contrary to the inspired word of God. Waving a Bible aloft, he asserted its unimpeachable authority. Cried down as irrelevant, he sat restively, nursing his thoughts. That he of all men should have hatched such a viper—a python, more like!—in his bosom. He was, as always, overworked, under severe nervous strain. It was just four years and ten months later, on April 30th, 1865, he took his own life, a bitter disappointed man in a world poisoned by false gospels.

Dr. Beale, following FitzRoy, was against the *Origin*. Lubbock spoke briefly for it. Last came Hooker, who, despite Lady

---

[1] " One who prostituted the gifts of culture and of eloquence to the service of prejudice and of falsehood." *Life of C. Darwin*, 239.

ROBERT FITZROY

By the courtesy of the Royal Naval College, Greenwich

Brewster, felt that Huxley hadn't made sufficient impression; *he* must save the day. So in his own estimation he did—though no other observer seems especially to have noticed it—"smashing that Amalekite Sam" amid, he told Charles, repeated applause. "I hit him in the wind at the first shot in ten words taken from his own ugly mouth; and then proceeded to demonstrate in as few more: (1) that he could never have read your book, and (2) that he was absolutely ignorant of the rudiments of Bot. Science. . . . Sam was shut up—had not one word to say in reply, and the meeting *was dissolved forthwith*, leaving you master of the field after four hours' battle."[1] Actually the *Athenaeum* report, longer than for either Huxley or the Bishop, suggested a soberer speech, asserting incidentally that science knew no creeds, only hypotheses, and that it was as the most fruitful hypothesis that he accepted Natural Selection.

That night the encounter was the talk of the Association, "almost sole topic" at Dr. Daubeny's reception.

Charles thanked heaven that the battle had been over his absent body. He would as soon have died, he said, as "tried to answer the Bishop in such an assembly." But he thought that the incident as a whole must do the cause great good, by "showing the world that a few first-rate men are not afraid of expressing their opinion"[2] even in the face of those chartered libertines of truth, the upper clergy.

IV

The subsequent progress of the theory of Natural Selection was steady if slow, not only in Britain but throughout the Western world. American interest was immediate, impelled by Gray's championship, until the Civil War came to offer the struggle for existence written in the book of life. Germany and France hesitated a while, partly, it seems, because of poor translations. For some years the French scarcely took it seriously; in Germany the fiery championship of Haeckel, the gentler advocacy of Fritz Müller, were needed to make it the predominating topic it eventually became. But by 1864 Charles could feel that progress was satisfactory at home and already "safe" abroad, and by 1872, when the sixth and final revision

[1] *LL. Hooker*, i. 526–7.   [2] *ML.*, i. 157.

appeared, there had been four French editions, five German, three American, three Russian, and at least one Dutch, Italian, and Swedish. Subsequent editions, and many in other languages, were of course to follow. Five new sentences in the sixth English edition serve to show Charles's view of the situation then: "As a record of a former state of things, I have retained in the foregoing paragraphs, and elsewhere, several sentences which imply that naturalists believe in the separate creation of each species; and I have been much censured for having thus expressed myself. But undoubtedly this was the general belief when the first edition of the present work appeared. I formerly spoke to very many naturalists on the subject of evolution, and never once met with any sympathetic agreement. It is probable that some did then believe in evolution, but they were either silent, or expressed themselves so ambiguously that it was not easy to understand their meaning. Now things are wholly changed, and almost every naturalist admits the great principle of evolution." [1] Increasingly throughout the 'seventies his theory seemed established. "Now," he wrote in 1878, "there is almost complete unanimity amongst Biologists about Evolution, though there is still considerable difference as to the means, such as how far natural selection has acted, and how far external conditions, or whether there exists some mysterious innate tendency to perfectibility." [2]

He himself, it might be said, was not by any means "unanimous" on many or most of these topics. Throughout the six editions of the *Origin* (1859, 1860, 1861, 1866, 1869, 1872), the main thesis of Inherited Variation, Struggle for Existence, Survival of the Fittest stood firm, but from the third to the sixth there was considerable change in detail as further consideration and new evidence suggested amendment, addition or deletion, and open-mindedness was continuously preserved. The special point of the causes of inherited variation particularly attracted him—he declared it the most perplexing problem of all—but his attitude was that of a true agnosticism. He was on the whole most dogmatic in earlier days, most strongly mechanist and not so much non- as anti-Lamarckian, belittling the direct effect of environment, the use and disuse of parts, and also the possibility of "sudden leaps" as a factor of any potency. *Natura non facit saltum*, he quoted. By degrees he broadened

[1] *Origin* (1872), 424. [2] *LL.*, iii. 236.

his view on all these matters. From "physical conditions seem to have produced but little direct effect," he modified to "physical conditions seem to have produced some direct and definite effect, but how much we cannot say." The power of use and disuse was also accepted, and the assertion that Nature "can never take a leap" softened to "can never take a great and sudden leap," though in this case beyond that qualification he would not move. The first two cases implied belief in inheritance of acquired characters, and to this he acceded even to the point of accepting the inheritance of mutilations as "conclusively" proved.

He was not absolutely satisfied even with the term Natural Selection, thinking it bad as seeming to imply something of conscious choice; he wished sometimes that he could set another, perhaps "natural preservation," in its place. Irked by charges of theological intention, he sought as early as the third edition to defend his personifying use of the term as purely metaphorical and also practical—"necessary for brevity"—and later, at Wallace's suggestion, coupled to it Spencer's impersonal Survival of the Fittest. He frequently sought to define it in non-personal terms, but never so neatly as Wallace's "the continuous adjustment of the organic to the inorganic world." In the first edition he claimed no more than that "Natural Selection has been the main but not exclusive means of modification," but even on that point later broadened his views, attributing more to sexual selection and to what he called, not quite seriously, "mere useless variability."

The one matter on which he wavered possibly least of all was his ascription of the main causes of variation to the organism itself as against the environment. "The organization or constitution of the being which is acted on, is a much more important element than the nature of the changed conditions, in determining the nature of the variation." [1] He was here in some degree setting up logical alternatives whose opposition a more organic view would have, and since has, largely dissolved.[2] He never took kindly to any idea of purposive or designed evolution or variation, though here again he could never arrive at final conclusion on a

---

[1] *Domestication*, ii. 291.
[2] " You can alter the character either by altering the gene in the constitution or by altering the conditions of the environment. . . . To believe that one alternative excludes the other . . . is to fall into an elementary logical and biological error." *We Europeans*, J. Huxley, etc., 85, 87.

subject which seemed to him to enter those realms of intellectual contradiction which left him always frankly confused. Chance, Design; Design, Chance—"Again, I say I am, and shall ever remain, in a hopeless muddle." [1] What is clear is that when he wrote of "chance" variation, he regarded the adjective as no more than concealing human ignorance.

Always, to the very end, he refused to commit himself on matters beyond his knowledge, his comprehension. Processes, not origins, workings, not causes, these were his whole quarry. He never absolutely closed his mind on any subject. Imaginatively limited he might become; imaginatively moribund, never.

[1] *LL.*, ii. 354.

CHAPTER XVI

THE GREAT GOD PAN

I

THE *Origin*, we recall, was only a sketch, in Charles's view a hasty and inadequate sketch, of the Great Work whose completion and publication he regarded as his life's aim. Now that the outline was in print, his whole inclination was to dismiss it in turning back to the fuller statement. That was not possible. Controversy, as has been seen, pursued him publicly and privately; and there were, too, the demands of new editions.

These came later, however; meantime, he would be about his business. As soon as January 1860 he was looking over his earlier manuscript, and resolving to concentrate his strength, "which is but little," upon preparing part of what had already been written for publication as a separate book, the first, he thought, of three volumes which would contain the whole. Naturally the plan would be the same as that of the *Origin*, opening with Variation under Domestication, then Variation under Nature, and so forth.

Finding, as he did, much of the material in "a pretty good state," it might be supposed that the task would have been accomplished with relative celerity. But to presume even that is to treat too lightly the physical and psychological obstacles he had still, had always, to contend with—the ill-health that plagued him, the inevitable elaboration of each subject that extended his chapters to books, his one volume to several, and delayed publication from year to year, almost from decade to decade.

It is not clear exactly what arrangement the division of the Work into three volumes implied, but the first was evidently meant to contain much more than the opening discussion of variation under domestication, for not till June 1860 did he find that material so spreading out as to occupy one volume "exclusively," at any rate with the projected chapters on Man and the expression of the emotions. Ultimately, of course, these subjects were to fill not one but five large volumes.

It has already been noted that most of his labours developed in this way, how the two years' voyage extended to five years, and how he spent the next ten in recording its zoological and geological harvest. A barnacle plucked from the *Beagle*'s keel occupied him for another eight years, and only then did he turn to develop theories conceived also in consequence of the voyage, imagining the presentation of his views in one work, large no doubt but such as a strong man might hold in one hand! Instead he was to write at least half-a-dozen volumes directly to his theme, and as many more relevant to it from one angle or another, but the Work itself stands still in the realm of the world's unwritten books. Charles in this matter was an idealist; he had had a vision of the Perfect Book, which would encompass its subject totally and leave nothing more to be said regarding it. He never wrote it because no such book can be written!

The very *Origin* itself began as an essay, became a pamphlet, then a small book, then a big book. He did indeed manage to finish that, complete from A to Z, under the incentive of challenge to priority, but that incentive no longer operated, and now he delayed what one can only call, even allowing for his continued illnesses, intolerably. Though it was doubtless only at his very worst times he entirely ceased the collection of relevant material, he seems in the five years 1861-65 to have given hardly more than a year and a half to consecutive work upon it, and even in 1866, when it was at last practically completed for the printer, his labours were interrupted and desultory.

In essence, all that he wrote subsequent to the *Origin* was—in the human and philosophical as opposed to the technically scientific sense—footnote to it: confirmatory, explanatory, divagatory, but all enclosed within its implied frontiers. This later work had its own originality, it made its own discoveries, but so far as Charles Darwin's fundamental significance is concerned, these are minor. Evolution, even Evolution by Natural Selection, stands independent of Pangenesis, of Sexual Selection, of Cross- or Self-Fertilization and the like. These may strengthen the case, they do not extend it—it is not dependent on any one of them.

II

The renewed work on Domestic Variation began on January 9th, 1860, but distraction was constant. Letters, reviews, visits and visitors. Henslow, friendly as ever, was at Down House in January, Lyell and his wife in early March, Hooker and Huxley a month later, they and Charles reading Owen's *Edinburgh* article and chuckling over its anonymous admiration of its author. But that night Charles lay long awake, turning over its deepest thrusts uncomfortably, wondering whether he should or should not reply to it.

A new source of controversy, foreseen by Erasmus, pooh-poohed by Charles, appeared in Patrick Matthew's letter to the *Gardeners' Chronicle*, indignantly claiming priority in exposition of the theory of Natural Selection. Hurriedly obtaining his work on *Naval Timber*, Charles combined apologies with the assertion that "neither I, nor apparently any other naturalist, had heard of Mr. Matthew's views."[1] Despite acknowledgments in later editions of the *Origin*, Matthew continued disgruntled, styling himself "Discoverer of the Principle of Natural Selection" on his title-pages and visiting cards, and denying any credit to Charles, who would not have been human had he not confessed sardonic pleasure on having his attention called to Dr. Wells's still earlier "White Female" essay. The fact was that the problem of priority still bothered him; unable to achieve complete indifference, he resolved never to defend his own claim. The Historical Sketch neither defended nor claimed, but it was brief, not to say brusque, in its treatment of both Lamarck and Grandfather Erasmus, the latter no more than creeping into a footnote as anticipating "the views and erroneous grounds of opinion" of the former, whose work, Charles had said elsewhere, "appeared to me extremely poor; I got not a fact or idea from it."[2]

Despite interruptions, however, the Domestic Variation work went fairly steadily on till May, when to his own ill-health was added the worry of Henrietta, who was seriously ill most of this year and the next, and for long after little more than an invalid spending much time in warmer climates abroad. In June he

---

[1] *LL.*, ii. 302 n. [2] *LL.*, ii. 215.

was himself so poorly that he went to Sudbrook Park instead of the famous British Association meeting at Oxford, and he returned home only to take Henrietta away to Elizabeth Wedgwood's house at Hartfield, where he amused himself in Ashdown Forest by examining the newly observed insect-catching Drosera.

Back at Downe he tinkered briefly with his main manuscript, then, "shamefully idle," turned back to Drosera again, finding, now as always, writing so much less enjoyable than watching and experimenting, to which he inclined more and more. Ever since *Beagle* days he had been an eager botanist, but his efforts in this field, mainly in 1838 and 1839, 1841 and 1844, and rather more seriously in 1856 and 1857, had been tentative and occasional, directed principally to problems of fertilization and structure in relation to visiting insects. He did not feel that he could give these matters sustained attention while the work on Natural Selection still hung on his hands and mind. Once that was out of the way, even if only in the unsatisfying form of the *Origin*, the case was altered, and though he often accused himself of idleness he did in his later years give increasing time to botanical subjects. Again largely for poor Henrietta's sake, the family that autumn spent two months at Eastbourne, and both there and back at Downe Charles "worked like a madman" to complete an essay on Drosera. Why, he said, he "cared more about Drosera than the origin of all the species in the world!"[1] Nevertheless, his results were so astonishing that he determined to postpone publication till further experiments could prove or disprove the plant "a disguised animal!" Actually the subject was set aside for nearly two years, taken up as a holiday amusement at Bournemouth in September 1862, and thereafter shelved, though sometimes thought upon, till 1872.

What intervened first was a new (third) edition of the *Origin*, for Murray this November had sold more copies than he could supply. Charles had been storing up cutting rejoinders to his critics, but suppressed them on Lyell's advice. December, January, into February, he was patiently adding, deleting, revising. The Historical Sketch was inserted, and a section to rectify what he felt to be his most serious previous omission, the failure to explain the persistence of simple unchanging forms.

The winter was a hard one. Henrietta, though better, could

[1] *LL.*, iii. 320.

get up only for an hour or two each day, and Charles was poorly too, and despondent. He preached idleness to Hooker, but nagged at his own work. He felt drearily old: "Never to look to the future or as little as possible is becoming our rule of life. What a different thing life was in youth with no dread in the future; all golden, if baseless, hopes." [1] He delayed reading Spencer, whom, like most people, he more admired than liked, in favour of Wilkie Collins's "wonderfully interesting" *The Woman in White.*

Variation under Domestication crawled slug-like on till June, as far as Chapter viii. In April he had paid a pleasant visit to London to attend a Linnean meeting and see his friends. Henslow had died, after some illness, on May 16th, still lamenting the *Origin*'s exile of the Almighty, and the question of a biography was canvassed. Charles shook his head, doubting whether any who had not known him personally could possibly be interested by his equable character and life. The more he thought of it the less he could imagine what might be written about him, nevertheless himself eventually contributed some pleasant pages of Cambridge reminiscence, of which two sentences revealed the debt he owed to his first effective tutor: "In intellect, as far as I could judge, accurate power of observation, sound sense, and cautious judgment seemed predominant. Nothing seemed to give him so much enjoyment, as drawing conclusions from minute observations." [2]

From early June, the Domestication work was scarcely touched for over eighteen months. Instead, holidaying at Torquay in July and August, enjoying health and good spirits and the company of Erasmus, he turned to a newer "hobby-horse," exercised occasionally at Downe during 1860, thereafter ridden almost continuously until the publication of *The Fertilization of Orchids* in the following May. He wrote this essay originally in brief form as a paper for the Linnean Society *Journal*, but in September 1861 sent Murray the manuscript for his opinion on making a book of it. Murray approved, agreeing perhaps with Charles that "the subject of propagation is interesting to most people, and is treated in my paper so that any woman could read it," [3] and the expansion was completed that winter. The author, as ever, doubted as publication drew near, but its reception

[1] *LL.*, ii. 360.   [2] *Memoir of Henslow*, 55.   [3] *L.L.*, iii. 266.

by botanists was highly satisfying, if the general public found it slightly esoteric in its technical detail. Lyell thought it, next to the *Origin*, his most valuable work.

He was also working on dimorphism in the primula, and a paper on this subject was read to the Linnean Society on November 21st.

The American Civil War had been in increasingly bitter progress most of this year, and Charles had been and was more deeply engaged by it than by any war his own country ever waged. The slavery issue set him whole-heartedly on the side of the North, and in June he could write to Gray in New England that he knew no Englishman who did not think with him. With passing months doubts of entire Northern disinterestedness began to grow, and though he saw only folly in the tension created between London and Washington by the Federal seizure of two Confederate envoys from a British steamship, his confidence waned, and his criticisms distressed the ardently patriotic Gray. The latter had agreed not to discuss the War with Hooker, but somehow he and Charles couldn't keep it out of their letters; they could only charitably "allow for longitude." But if, when they tried to write of Science, War kept breaking in, equally when they wrote of War, Science kept breaking in too, and so the correspondence, though sometimes strained, was never broken. After 1863 Charles's interest waned; the War was just something that went on and on, and he had little stomach for the "bloody details." Emma's keen Federalism annoyed Fanny Allen, who had given her heart to the romantic Southerners as completely as once to the Corsican.

All 1862 was plagued by illness. Emma's sister, Charlotte Langton (Charles's favourite once, in the old pre-*Beagle*, pre-Emma days), died in January. There was sickness at Down House. In mid-May, when the *Orchid* book appeared—"a flank movement" on the enemy!—Charles was taking a "much wanted" rest with Josiah and Caroline at Leith Hill Place. Back at Downe, he was wondering what would become of his Domestication work—but gave all his time to dimorphism. Throughout July two of the boys were ill with scarlet fever, Leonard critically so. He was better when Wallace, home from Malay, came to meet Charles for the first time since his return, an occasion of mutual interest and pleasure, but a relapse followed, with typhoid symptoms too. He was not expected to live, but survived the crisis. The other children had been hurried out of the house in Brodie's charge, and as soon as Leonard was well enough to travel all the family was on its way to Bournemouth, sleeping in William's house at Southampton *en route*. There Emma collapsed with scarlet fever. Charles stayed with her; the rest went on. It was the end of August before they were all together again beside the sea, Charles "squashier than ever" but hoping "two shower-baths a day" would prove a sure remedy. But he worked too hard at his "amusement," Drosera, and arrived home on September 30th "quite knocked up," having to abandon a proposed British Association visit. Bouts of vomiting and uncontrollable trembling afflicted him all the winter, wasting entire days.

It was practically Christmas when the *Domestication* manuscript was resumed, and though the appalling masses of material seemed driving him to the verge of collapse again, he settled by some effort of will to six months of steady productive labour, even taking his leisures gladly, enjoying the instalment and stocking of a hot-house. All the family, in February, had a week in town, he visiting Hooker at Kew while Emma and the children enjoyed the theatres; he read a paper on dimorphism to the Linnean Society. But the old ghost was not to be laid even by further visits to Hartfield and Leith Hill, and most of the spring he was poorly, turning for relief to the study of climbing plants, in which a paper by Gray had awakened his interest. The two chapters on Selection having been completed, *Domestication* was now set aside for, practically speaking, more than a year.

Incapacitating illness was settling upon him again. That August he revisited Malvern, despite its sad memories, desperately seeking recovery. It did him no good. He returned in deep depression, and through the winter there was no improvement. At the New Year he could neither read nor write, depending on others for these services, and about this time he began to grow, doubtless to rid himself of the intolerable burden of shaving, the beard he was to wear to his life's end. Getting worse, his stomach giving him little rest, he went to Malvern again in March, collapsed there—and began to recover. In April he actually had five days quite free from sickness, and occasionally he would trifle with the *Domestication*, though climbing plants, on which he was preparing a paper (completed and sent off to the Linnean Society *Journal*[1] in the following January), remained his main study through the summer. The *Domestication* was only seriously taken up in September, to be carried on, despite days of dizziness and nights of nausea, through the winter to April 1865; five more chapters were done. Doctors were repeatedly consulted, but remained as unhelpful as ever; they talked solemnly now of "suppressed gout," the fashionable ailment of the season.

III

For all the rest of that year and practically to the end of the next he was working, intermittently, upon a single chapter of the *Domestication*—a long and important one setting forth a hypothesis conceived as early as 1840 or 1841, long treasured and reflected upon in secret, and now regarded with an affection second only to that accorded to Natural Selection itself. Pangenesis, he called it—"the Great God Pan"—no less than an intended solution of the central mystery of organic reproduction. (He liked the very name of it, though Emma thought it sounded wicked, "like pantheism.") It was in May 1865 that he confided it to Huxley, who discouraged publication. Charles said he couldn't suppress Pangenesis, but would do his best to present it humbly. Whatever the objections, it was, he told Hooker in 1867, such "an infinite satisfaction to me somehow to connect the various large groups of facts, which I have long considered, by an intelligible thread,"[2] that he was sure he would never wholly abandon it till something more satisfying came to take its place.

The year 1865, the early months of 1866, went quietly by, health neither good nor bad, but never certain. Sometimes he worked on Pangenesis. Sometimes he painstakingly revised earlier chapters. From March to May 1866 he was preparing a fourth edition of the *Origin*, making numerous small changes and adding fresh matter on sterility, relative periods of modification and stability, and embryology. A new paragraph in the Historical Sketch remarked on Owen's apparent odd claim to have been a first prophet of Natural Selection. Later Charles expressed

---

[1] An off-print of the *Journal* publication was bound and sold separately.
[2] *LL.*, iii. 74.

his thankfulness that "others find Professor Owen's controversial writings as difficult to understand and to reconcile with each other, as I do," [1] and in fact what Owen's views on the subject really were no one, not even his own biographer, was ever clear.

While the new edition was in preparation, Charles and Emma had a week in London, partly to attend a soirée of the Royal Society, where he was presented to the Prince of Wales, who made a remark that Charles's reverent confusion prevented him from hearing, so that he could only bow as obsequiously as his elderly bones would permit and beat as hasty a backward retreat as etiquette allowed. He also met many old friends, who found his beard such a disguise that he had to introduce himself. Emma was there, but better enjoyed her visit to the theatre, even though the play was that tedious *Hamlet*.

Charles was in these years receiving official recognition of his work. The Royal Society of Edinburgh repented of its earlier intransigence and made him an Honorary Member, which pleased him for old time's sake. After rejection in the previous year, he was in 1864 awarded what he thought the highest honour of all, the Copley Medal of the Royal Society of London. A little longer, and the learned societies of a dozen countries would honour him in their various ways. Translations went far and wide, and foreign letters came in from the ends of the earth. He commenced collecting his correspondents' photographs—a notable gallery of wisdom and whiskers.

As 1846 had seen an end of Maer, so 1866 saw an end of Shrewsbury. Three years before, Catherine Darwin had married the widower Charles Langton. Relatives had shaken their heads dubiously, wondering as to the outcome, for both the couple were elderly—she was 53—and strong-willed, and neither in good health. Susan, the family beauty once, that tall, vivacious, happy girl, sixty now, an old maid not to be detached from her home since birth, alone remained at Shrewsbury and The Mount, bringing up there the children of dead Marianne. June, 1864, Emma Allen died, leaving of all her generation only the immortal Fanny. February, 1866, all Langton doubts were settled. Catherine died at The Mount, where she had gone to nurse the ailing Susan. Fanny Allen wrote of her sadly as one who had failed to use her capacities for either others' happiness or her own. "Sad, sad

[1] *Origin* (1866), xx.

Shrewsbury! which used to look so bright and sunny; though I did dread the Doctor a good deal, and yet I saw his kindness." [1] Eight months more, and Susan too was dead—"and now," Charles wrote, "after so many years there is no longer any connection between our family and Shrewsbury." [2]

The last act was the sale by auction of practically the entire interior furnishings of The Mount, a six-days' business, November 19th to 24th. Furniture, glass, old Wedgwood ware, silver plate, books, pictures, linen, and animals all came under the hammer. Among the portraits were an engraving of Grandfather Erasmus and coloured print of Robert (each in Gilt Frame); among the books copies of *The Botanic Garden*, Anna Seward's *Life of Dr. Darwin*, Robert's *Experiments on the Octra Spectra*, the *Journal of Researches* (twice), the *Zoology of the Voyage of the " Beagle*," the three volumes of the *Geology of the Voyage*, a copy of the *Origin*, and—relic perhaps of Charles's Cambridge days—the *Works* of William Paley in seven volumes.[3]

The house too was eventually disposed of, passing into the hands of complete strangers. To-day it is Government property —headquarters for North Wales of the Post Office Telephones. The bulk of the neglected garden has been purchased by a builder, whose "modern villas" are locally considered a great improvement. A near-by street bears Charles's surname, his statue stands outside the Public Library, once the Grammar School, and elderly inhabitants in the town have not yet entirely forgotten why these things should be, though apparently not sufficiently interested or informed to protest when a local historian unwittingly made all Robert's children bastards and he a widower before he became a husband.[4]

The children now were growing fast out of childhood, were almost if not entirely adult. William, in a banking business in Southampton, had long had his home there. He would soon be twenty-seven. Henrietta, twenty-three, much away on the sunny south coast of France, exercised her bright intelligence upon the work not only of her father but sometimes of Huxley too. George and Frank, twenty-one and eighteen, were both doing well at

---

[1] *Emma Darwin*, ii. 184. [2] Unpublished letter to Thos. P. Blunt.
[3] *The Mount, Shrewsbury : Catalogue of Sale*.
[4] "The house at the top of the Mount where Charles Darwin was born is modern, dating from about 1820 when the old Doctor married." (Statement in four revised editions, 1911-35.)

Cambridge. Leonard, sixteen, was at Woolwich, preparing for an army career. Horace, fifteen, was away at school. Yet none had married, and Down House remained home to all of them, the happy place where they had hung their childhood, and whither they returned to visit again the glimpses of that unblemishable moon of a serene and placid youth.

Charles never shattered their first affection for him as a good friend as well as father. He attempted no stern Victorianisms, did not presume upon his paternity, but behaved to them as a natural human being, winning in return their equally natural love and loyalty. His occasions of anger were so rare as never to be forgotten. There was in this same year 1866 a great to-do over the recall from Jamaica of the Governor, Edward John Eyre, following his suppression of a negro rising there with promptitude and slaughter. English residents and those with financial interests there were as grateful to him as those in India to General Dyer after Amritsar, but a commission of inquiry reported less favourably. His arrival in England divided the country between angry attack and bitter defence. Committees were formed to prosecute and exonerate. Mill was chairman and Huxley and Spencer supporters of a fund, to which Charles sent £10, raised to arraign Eyre for murder. "Nigger philanthropists!" spat Carlyle, subscribing with Ruskin, Tennyson and Kingsley for the defence. When the controversy was at its height, Charles and William chanced to dine with Erasmus in London, and William ventured some jest about the prosecuting committee spending funds on a banquet. Charles blazed into anger, telling him he had better get back to pro-Eyre Southampton, and the sooner the better. There was no reconciliation that night, but at seven next morning Charles appeared at his son's bedside, to say he had not slept at all, and to ask forgiveness.

IV

By December 1866 the exposition of Pangenesis and the revision of all the earlier chapters were done, the manuscript was in Murray's hands, and Charles was free to finish the work off with a relatively brief section of "Concluding Remarks," completed in January. Both he and Murray were taken aback by realization of the book's extreme length—more than twice that of the *Origin* despite the omission of the intended chapter on Man. A quite absurd size, he agreed, but he could not shorten it. Murray did not protest; it went to the printer at once.

With great relief he saw it go, yet almost at once found himself restless, unable to take any real recreation. Emma wished to heaven he would smoke a pipe or "ruminate like a cow." He couldn't; work, now, was all his life. Soon, however, he had proofs to keep him busy, damnable things (how much sweeter collection than correction) that took his time and energy for all that year, with occasional days in London and snatched off-hours in the garden or on the sandwalk. The more he pored upon the proofs the duller the book became, till he felt it, despite Lyell's encouragement and his own faith in Pangenesis, mostly unreadable and certain to be quite unread.

It should have been out in November, and at Murray's November sale nearly the whole first edition of 1500 copies was sold to the bookshops, but actual publication was on January 30th, 1868, a second printing of 1250 copies following in February.

In the main *The Variation of Animals and Plants under Domestication* was just what it had set out to be—an exhaustive statement of one division of the facts on which the theories of the *Origin* were based. Dogs, cats, horses, pigs, cattle, sheep, goats, rabbits, pigeons, fowls, ducks, peacocks, turkeys, geese, goldfish, moths, trees, plants were all in turn called in evidence. There was a further chapter on bud-variation, three on inheritance, five on crossing and the allied problems of sterility and hybridism, two on domestic selection as such, and another five on the causes and laws of variation. Here, in much more detail but with no more essential change than has been indicated in the references to post-*Origin* modifications of Charles's ideas, he hammered home his point of a universal and practically limitless organic plasticity able to be influenced not only by human agency but also by "incessant geological, geographical, and climatal changes" to almost any degree. More definite, as well as more factual, he might be, but he was treading no new ground until he reached the account of the Provisional Hypothesis of Pangenesis.

Perhaps *because* this was an unfortunate venture, never very enthusiastically received and all too soon to be thrust out of court by later and more satisfying views, one may see here even more clearly than in the case of Evolution by Natural Selection the naked need of Charles's mind for ideological order. No man can reign over mental any more than social chaos, and since reign he must or abdicate his humanity, and since abdication was not in Charles's stars, chaos was dreadful to him. In the introduction to these volumes he had asserted for Natural Selection that "in scientific investigations it is permitted to invent any hypothesis, and if it explains various large and independent classes of facts it rises to the rank of a well-grounded theory." [1] That in effect was what he claimed for Pangenesis—not that it was more than provisional, not that it was more than speculative, but that it did "to a certain extent" connect ("by a tangible method") "large classes of facts" bearing on inheritance and variation. For his own part, he said, he had been "led, *or rather forced*," to such a view; from others he asked only that it be permitted tentative currency till something better appeared. He was only too highly conscious of its "gratuitously assuming" nature—its presumption that each part of the body in each of its progressive stages of development casts off "free gemmules" or minute particles capable of self-multiplication, that these gemmules, in the first place "thoroughly dispersed" throughout the body, are, by a "mutual affinity," "collected from all parts of the system to constitute the sexual elements," a convenient "elective affinity" thereupon arranging them in their due order, and their subsequent development depending upon "union with other partially developed or nascent cells which precede them in the regular course of growth." [2] Incredible complexities—not to add incredible numbers of gemmules—were admittedly involved, and yet . . . "number and size were only relative difficulties." Look upon that picture, if you must, but also upon *this*! Observe how much may be explained that hitherto was inexplicable—reproduction generally, parthogenesis even, hybridism, regrowth of damaged parts, prepotency, reversion, transposition or duplication of organs, the laws of inheritance, variability itself. This was the irresistible appeal. "On any ordinary view it is unintelligible how changed conditions, whether action on the embryo, the young or the adult animal, can cause inherited modifications. . . . On our view we have only to suppose that certain cells become at last structurally modified; and that these throw off similarly modified gemmules." [3] The ultimate *how* of the process might

[1] *Domestication*, i. 8.   [2] *Ibid.*, ii. 374.   [3] *Ibid.*, ii. 395.

not be much clearer (for what caused the structural modification itself?), yet quantitatively at least the light of knowledge could claim an advance. Incredible it might be: "We cannot fathom the marvellous complexity of an organic being; but on the hypothesis here advanced this complexity is much increased. Each living creature must be looked on as a microcosm—a little universe, formed of a host of self-propagating organisms, inconceivably minute and numerous as the stars in heaven."[1] *Credo quia impossibile.* At all costs explanation was necessary to Charles.

Pangenesis was for him the book's gay blossom—the rest but the dull if necessary soil from which it, with the *Origin*, flowered.

The book made no reference to Man as such, but did assert, much more clearly than in the *Origin*, a view of the common descent of all mammals, birds, reptiles, and fishes. It also declared Charles's dissent from those who, like Asa Gray, would substitute for Natural Selection a kind of Supernatural Selection, variation guided along beneficial lines. He could not credit it. On the one hand "an omnipresent and omniscient Creator ordains everything and foresees everything." On the other, "if we assume that each particular variation was from the beginning of all time preordained, then that plasticity of organization, which leads to many injurious deviations of structure, as well as the redundant power of reproduction which inevitably leads to a struggle for existence, and, as a consequence, to the natural selection or survival of the fittest, must appear to us superfluous laws of nature." "Thus we are brought face to face with a difficulty as insoluble as is that of free will and predestination."[2] It was the blank wall which faces the logician, the ideologue, at every metaphysical turning! Facing the blank wall, Charles could only, then as always, shrug his shoulders and look away.

The reception of the book was, on the whole, quiet. It made, despite, or because of, Pangenesis, nothing like the stir created by the *Origin*. Most even of Charles's personal friends hesitated to speak out in its favour. Huxley remained dubious as ever, and Hooker was notably reserved. Owen's hand was suspected in a contemptuous *Athenaeum* review. Other notices were more

---

[1] *Domestication*, ii. 404.   [2] *Ibid.*, ii. 432.

favourable, however, and Gray thought Pangenesis as good a theory as could be postulated in the present state of knowledge. The two principal enthusiasts were Lyell and Wallace, who found it fascinating. It was cheering, anyway, that the first edition, for all the book's "horrid, disgusting bigness," sold so well.

CHAPTER XVII

THE QUESTION OF QUESTIONS

I

THE Great Work, clearly, was making very slow progress. Eight years gone, and no more than the first introductory topic accounted for. Nevertheless Charles remained apparently hopeful. In his Introduction to the *Domestication* he promised its continuation in a second work to treat comprehensively of Variation under Nature, the Struggle for Existence, and Natural Selection and its difficulties, and a third which would test his theory by showing its ability to explain varied classes of facts. But he had been so wearied by the drawn-out effort of his last volumes that he could not face the thought of immediate continuation. Besides, he probably justified himself, the subject he did want to treat was after all undeniably a siding—if no more—on the same system. Had there been space, it would have been done with in the *Domestication*. It ought to be treated and done with! It was—who could deny it?—"the question of questions." The proper study of mankind is Man. It was, moreover, a vital chapter, if not the next in logical order, in the projected proof of Evolution by uninterrupted natural process. Disrupt the idea of continuous process, introduce one supernatural act, one miraculous creation, and you might as well abandon it altogether. If one miracle, why not a million? Make an exception of Man, as some evolutionists, even among his intimate friends, were anxious to, and your case was yielded. The order of your *natural* universe was broken, gone; you were delivered over to something so far, so different, from the universal harmony which had enchanted Charles's mind, that it could only be regarded as chaos again.

Man *must* be dealt with.

He had glimpsed that necessity at the very beginning, back in 1837, but had kept it more secret than any. It was no more than implied in the essays of 1842 and 1844. In 1857 he had told

Wallace of his intention to "avoid the whole subject, as so surrounded with prejudices," and only his fear of the accusation of concealment had forced inclusion of the *Origin* one restrained sentence: "Light will be thrown on the origin of man and his history."

He had indeed been cautious, partly because that was his nature, partly perhaps because, though convinced that man and all the animals were alike flowers on the same family tree, he had up to the publication of the *Origin* thought "only vaguely" on the subject. But he knew that it still waited there for him, unless of course someone else could be persuaded to take it up. That seemed for a while his hope. From 1860 until *The Antiquity of Man* appeared in 1863, he looked to Lyell with quickening anticipation. At the beginning of the period he wrote to Hooker that "Lyell is going at man with an audacity that frightens me," and he soon came to expect "a bomb-shell," no less—smiling that the old cautioner should now himself need cautioning. The book itself, when at last an early copy came, was a dreadful disappointment. His tone, for a time, was almost acrimonious. Lyell, he protested, had been going about bearing himself with the bravery of an ancient martyr, yet what he wrote was as timid as an old woman! Why, he scarcely spoke out even on species, much less on man! Here he was, writing: "If it should ever be rendered highly probable that species change by variation and natural selection"[1]—*ever! highly probable!* He openly ascribed views to "Mr. Darwin" as though he himself had no share in them. Worst blow of all, he referred to Charles's ideas as a kind of modified Lamarckism! Charles wrote expressing pained disappointment; he had thought Lyell's statement would mark an epoch in the subject, but that was over now. In point of fact, Lyell in 1863 was not a very great deal bolder in the matter than Charles himself in 1859. He was able, however, a little to retrieve himself in his friend's eyes in a second edition and in subsequent public utterances, where he made his inclination towards transmutation, even in the case of man, markedly clearer. But he could not, ever, go all the way, for he had, as he said, "been forced to give up my old faith without thoroughly seeing my way to a new one."[2]

From Lyell, it seems, Charles transferred his hope to Wallace,

[1] *LL.*, iii. 8.   [2] *LL.*, iii. 29.

in 1864 offering him his accumulated notes on the subject, which he did not suppose himself ever likely to use. Wallace appeared rather embarrassed to be handed this human baby, a large and disturbing creature, inadequately nourished, he felt, by ascertained fact. He replied cautiously that he would gladly accept Charles's offer *if* he took up the subject. Eventually, even before the end of the 'sixties, his views on Man were proving less acceptable to Charles than even Lyell's reticence, for he came, and with a growing assurance, to support what irked Charles most of all, a setting of Man in a separate category from the beasts, and an assertion not only of the inadequacy of Natural Selection to account for him but the need to postulate guidance by a "superior intelligence."

It seems scarcely possible that Charles did not similarly approach Huxley, whose small but pregnant volume, *Man's Place in Nature*, appearing just before Lyell's unhappy work, had given him so much more satisfaction. But if the offer was made, Huxley had other fish to fry, and the notes remained in Charles's hands. He would have to write his book himself. The subject was less novel now; the worst of the prejudice had been dispersed; he thought it should be possible to discuss it in a calm, scientific spirit. Man, Expression, and, since it had a special human importance, Sexual Selection, should, he fancied, make up a pleasant, not too exhaustive or exhausting volume.

II

He had already, early in 1867, between sending off the last manuscript of the *Domestication* and receiving the first proofs, got his notes together, and even done a little writing on the matter. Now, a year later, on February 4th, 1868, he settled down to it in earnest, working busily on his notes and despatching inquiries to his friends on all three main heads of his subject. Practically the whole month of March was spent in London, working almost every morning—for the hour and a half which was all that he could stand—in the British Museum, and paying inquiring visits to the Zoological Gardens and Society. He also met many friends, and dined at the Hensleigh Wedgwoods' house in Cumberland Place, where he met the impetuous Frances Power Cobbe.

Through April and May he was making progress on Sexual Selection, but had to abandon all work for a month before the family left home on July 16th for Freshwater in the Isle of Wight. He grumbled there at this confounded interruption of his labours, and sought to relieve the tedium by copious correspondence with Wallace, by whose disagreement on "female choice" as a factor in Sexual Selection he was a good deal worried. But Erasmus was at Freshwater too, and their hostess, Mrs. Cameron, was exceptionally lively and intelligent. Tennyson, who knew the Camerons and had met Erasmus more than once with Carlyle, saw them several times, as guest and host. Charles and Emma thought him rather absurd, but better than his poetry. The travelling Longfellow also called, and there was talk of Agassiz, spiritualism and table-turning. Hooker came specially to read the presidential address—naturally a good deal about Evolution and Natural Selection—he was shortly to deliver to the British Association at Norwich; Charles was quite delighted by "such a eulogium." Most of the time, though, he was weak and shaky, idly jogging about neighbouring roads and fields on Tommy, brought for that purpose from Downe.

He was rather better that autumn, and Down House saw more visitors than usual—Wallace (and his wife), Edward Blyth the zoologist, and Jenner Weir one week-end, Mr. and Mrs. Hooker, Tyndall, and Asa Gray another, frequent relatives and Lyell a little later. Charles Eliot Norton of Massachusetts, living temporarily at Keston, often came calling, bringing fellow-Americans of varying degrees of intellectual eminence. It was all very delightful, but that, and the preparation of the fifth edition of the *Origin* (a matter mostly of minor corrections), and then botanizing, and visits to London, and a rather unnerving fall from Tommy on Keston Common, ate away the time practically through the winter and into the spring. He was disturbed too by the publication, in the *Quarterly* for April 1869, of an article by Wallace making their divergence on Man painfully clear. He had felt, these past years, so much at one with Wallace in most respects that it hurt him to find them so completely at odds. If Wallace could waver, who was to be depended upon?

Accordingly, though he worked spasmodically on the manuscript from February to June, it made but little progress. Then came a rather unfortunate holiday. The children were distinctly growing up now. First they had decided that they were too old to talk of Papa and Mama—Father and Mother it would be henceforth. "I would as soon be called Dog,"[1] Charles grumbled behind his beard. Then they declared that next year's holiday must be in North Wales. Again Charles grumbled, but to no better effect. Early in June they all set out for Barmouth, spending a night in Shrewsbury on the way. He wasn't well, and the journey had been tiring, but the familiar place stirred him, and he went with Henrietta to pay a visit—his last—to his boyhood home. The occupier of The Mount, honoured by such a caller, received them hospitably, and not only threw the house open to them, but went with them from room to room, disturbing Charles, who had wanted a few quiet moments with his memories. As he and his daughter went away down the steep slope towards Frankwell he told her, sadly: "If I could have been left alone in that green-house for five minutes, I know I should have been able to see my father in his wheel-chair as vividly as if he had been there before me."[2] That, it seems, was what he wanted most to see—not himself as child, not his mother, or Erasmus, or his sisters, nothing in fact of his youth, but the later days, when he was no longer afraid of his father, when they had been good friends together. . . .

Barmouth too had its memories. Caerdon, their holiday home, was quite delightful, its terraced garden beautifully situated on the slope between the green woods and the wide estuary, Cader Idris in full view beyond the opposite bank. Yet Charles's forebodings were justified. The younger people thoroughly enjoyed themselves, but Emma was not at all well, and Charles sadly depressed, feeling his physical weakness the more in recalling the Cambridge reading-parties, the tramps and talks with "Cherbury" Herbert, net in hand, eye watchful, thought a butterfly too. Forty years ago that had been, and with every one the burden of the flesh, of thought, of family responsibility, had grown heavier. To-day it utterly exhausted him to walk a bare half-mile, even slowly. And when he sometimes rode a little, it was fretfully, nervously, thinking all the while of the fall on Keston Common.

An unexpected diversion was the reappearance of Miss Cobbe, a summer visitor too, one who knew a great man when she saw

---

[1] *Emma Darwin*, ii. 192.   [2] *Ibid.*, ii. 195.

him and would make the most of her opportunity. They had many informal talks together, in house and garden and wood, he boring her with tales of Henrietta's dog Polly, she wearying him with bluestocking chatter on Mill and Kant. Had he read Kant on the Moral Sense? No, and didn't want to—though later he admitted himself interested by a Kantian translation she sent him, principally "to see how differently two men may look at the same points." (Hensleigh's daughter Julia also lent him Kant, but he quickly returned the *Critique of Pure Reason* because it "said nothing to him.") Mill he regarded much more warmly, though thinking that a course of physical science would have helped his case too. She had a gift for catching him on his rare high horse (he usually preferred his cob), for when they met later that year at Erasmus's house he talked of a German scientist who had been rude if scientific about the origins of the Catholic Mass, and preened himself "how much more *decency* there was in speaking on such subjects in England." [1]

The visit as a whole seemed to have effected only superficial improvement, and the rest of that year he achieved little more than some revision of the Sexual Selection chapters, a tedious labour which made him feel "as dull as a duck, both male and female." Murray in October actually announced forthcoming publication of the book with the title of *The Descent of Man*, but that was perhaps no more than *pour encourager l'auteur*. The New Year, however, brought good resolutions and better results, and through most of 1870 he worked with unusual steadiness, though delayed of course by the inevitable extension of his material, finally forcing him to hold Expression over for separate treatment. The manuscript was ready for Murray and the printer by the end of August. It was a big book again, only less huge than the *Domestication*, but Murray had now full faith in his author's capacity to sell, and demurred only on the point of one sentence which he thought disturbingly coarse, implying as it did feminine capacity to experience sexual desire, and the intention of animal courtship to provoke it. Charles promised to modify so alarming a suggestion. Later Murray sent proofs to Whitwell Elwin, who found them "little better than drivel," almost unreadable and certainly not worth refutation. Darwinism, he prophesied, was a bubble blown from the pride of mutual applause, and certain

[1] *Life of F. P. Cobbe*, 490.

to be pricked at the first appearance of "a really eminent naturalist."[1]

If Elwin found the proofs tedious, far more so did Charles, making his usual moan. He had never studied the craft of writing, he said. The most he could do was to strive for simplicity and brevity, but in these later years he had developed the bad—and expensive—habit of leaving his final stylistic revision to the proofs, which made a dreadful mess of them. Right through the autumn and on till the middle of January he stuck at them, until by the time he had passed the final revises he was so sick of the book that he had lost the least sense of its value.

He didn't, he may have felt, even want to think about it, and accordingly plunged direct into the writing of his next and continuing work, *The Expression of the Emotions in Man and Animals*. His material must have been in unusually good order, for despite the inevitable interruption of correspondence following the appearance in February of *The Descent of Man, and Selection in Relation to Sex*, the new work was written probably faster than any other, the rough copy being completed by April 27th, though attention to proofs was thereafter delayed till November by an intervening half-year's work on the sixth and final edition of the *Origin*. The latter appeared in January, the *Expression* in November, 1872.

III

The Descent of Man is clearly one of Charles's great books, but despite the intrinsic importance and interest of its subject it has not and cannot have the wide, and in a true sense the imaginative, appeal of the *Origin*. The later work presupposes the earlier, and the spell of that is a magic which will not work twice. The essential intention is no longer new, the horizons are narrower, and the space given to the contributory matter of Sexual Selection is strictly disproportionate, being more than two-thirds of the total work, and only an eighth or so of this having any direct reference to Man. It is, in fact, as Wallace said, two books mixed together. Primarily, Charles wanted to explain "how far the general conclusions arrived at in my former works were applicable to man."[2] His sole object, he said, was to consider "firstly,

---

[1] *At John Murray's*, Paston, 232.   [2] *Descent of Man*, i. 2.

whether man, like every other species, is descended from some pre-existing form; secondly, the manner of his development; and thirdly, the value of the differences between the so-called races of man." [1]

On the broadest issue all the evidence of both homological and analogical differences between man and the lower animals was massed from first to last to demonstrate the physical and mental differences between Man and the beasts as of degree only, not of kind. Even conscience, the moral instinct, was no more than the natural product of social living and habit. The very idea of God was evolved, not inborn or divinely revealed; there were races of men to whom the conception was unknown. There was *no* distinctive character in Man which was not both subject to, and the product of, Natural Selection.

The statement, it was true, needed a certain elaboration in relation to the second and third purposes of the book. To the broad workings of Natural Selection must be added the subtler effects of Sexual Selection, a matter not only of male victory but (this was where Charles and Wallace disagreed) of definite female choice in the struggle to mate and reproduce. But though of special importance in promoting racial and individual differences in Man, Charles believed it operated over the vast range of the animal kingdom wherever the male and female sexes were differentiated, and accordingly he had felt bound to indicate its effects in detail from the very lowest classes through insects, fishes, amphibians, reptiles, birds, and the lesser mammals up to Man.

There was throughout a growing tone of confidence. He now set forth for humanity a definite family tree, showing Man's close relationship to the Old World Monkeys. We should, he asserted, admit this community of descent; to do otherwise was to suppose that God had been playing games with us, making our structure, and that of the other animals, "a mere snare laid to entrap our judgment." "It is only our natural prejudice, and that arrogance which made our forefathers declare that they were descended from demi-gods, which leads us to demur to this conclusion." [2] Such objections, he repeated in his concluding paragraph, were beside the point: "We are not here concerned with hopes and fears, only with the truth as far as our reason allows us to discover it." [3] Still, if hopes and fears must be considered, was it not

[1] *Descent of Man*, i. 2–3.   [2] *Ibid.*, i. 32.   [3] *Ibid.*, ii. 405.

more hopeful to presume Man's rise from a lowly beginning, moving forward to ever greater material and spiritual destinies, than to take the contrary view of an aboriginal perfection turning to degradation, as among savage peoples? That rise was Charles's hope, his "infinite satisfaction." On his own view, the Ascent rather than Descent of Man would have been the fitter title. Yet in fact *The Descent of Man* expressed all too well the feelings of many at this removal of man from a place but a little lower than the angels to one but a little—and that not essentially—above the brute beasts. Even Emma regretted it as "again putting God further off."

The *Expression of the Emotions* was an altogether lesser, lighter work, subsidiary even in relation to the *Descent*. Following the latter, its purpose was to describe leading instances of emotional expression in man and animals, and to explain their purely natural development from instinctive physiological reactions. With the exception of Herbert Spencer's *Principles of Psychology* (1855), all previous works on the subject, as that of Sir Charles Bell, from an early dissatisfied reading of which Charles's interest in the subject had in part proceeded, had been essentially creationist, unable to progress much beyond the view that as things were, so they had been ordained. Charles had been making his own desultory observations for well over thirty years (not however, as report had it, by sticking pins into babies), and had also obtained, by personal questionings, letters, and printed questionnaires, much evidence from others in all walks of life and all parts of the world. His favourite subjects were infants, the insane, "savages," and animals—those in whom the expression of emotion was freest, least controlled—but he drew also upon art and photography. As a result he had conceived three main principles, which he thought explained so much that it would not be unreasonable to "hope hereafter to see all explained by these or by closely analogous principles." [1]

That was the main conclusion; it was only incidentally and at the end that he drew from his material certain wider deductions as to the source and stages of human development, all in accord with the views set forth in the *Descent* and adding no more than what he himself felt to be a "hardly needed" confirmation. He thought that the similarity of the main human expressions

[1] *Expression*, 350.

suggested a high degree of development before the period of divergence of existing races.

The sixth edition of the *Origin* was important principally as embodying Charles's final revision of the book. Some of its more vital modifications of original opinion have been referred to. One almost entirely new chapter, on Miscellaneous Objections to the Theory of Natural Selection, was inserted, much of it in direct retort to the criticisms of St. George Mivart, author of *The Genesis of Species* and one of the most persistent and—Charles thought—influential of Natural Selection's opponents. Even more than ordinary pains were taken with the revision, for Charles meant it to be the last; after this, he had decided, nothing would induce him to touch it again.

The sales of all three books were good, 7500 copies of the two-volume *Descent* being printed in 1871, and 9000 of the *Expression* practically upon publication (rather too many, for the edition took some years to sell out), and the *Origin* appearing in cheap, "popular" form.

Good too, on the whole, was their reception. *The Times* redeemed itself from the liberalism of its review of the *Origin* by six-column disapproval of the *Descent*, lamenting the publication of such morally—not to say intellectually—dubious speculations at the very moment of the Paris Commune; Mivart in the *Quarterly* attacked the work with learning and prejudice; in the provinces there was some angry dissent, but John Morley was gravely eulogistic in the *Pall Mall Gazette*, and on the whole there was far less abuse than Charles had anticipated. Haeckel was whole-hearted in his applause, and general German interest in it was immediate. The *Expression* too had its opponents, but found more of favour. The essential facts in regard to both books were, first, that Charles was now a fully established author, independent of particular reviews, a writer who must be read by all with any pretension to take his subjects seriously; and second, that neither one of them caused anything like the disturbance, either of enthusiasm or anger, aroused by the *Origin* a dozen years earlier.

IV

The time of sustained labour on these works, from 1870 till late in 1872, went relatively uneventful by. There were the usual visits here and there. Health stayed uncertain, but the water-cure was quite abandoned now. One pleasant trip, in May 1870, was to see George and Horace at Cambridge, walking again the "paradisaical" backs, fresh with spring leaf and blossom and the soft-flowing river; meeting Sedgwick once more on the old friendly terms, controversy forgotten for the sake of a yet remoter past; missing "dear old Henslow," without whom Cambridge seemed scarcely itself at all. Sedgwick was genuinely moved to see Charles's happiness with Emma and Henrietta and Elizabeth and the boys. He was eighty-six, but "half-killed" Charles with showing him the scientific sights!

That August they were at Southampton, reading daily of the Franco-Prussian War and hoping earnestly for German victory over "that vainglorious, war-liking nation" whose threat to invade Downe in the 'fifties had never been forgotten. Emma longed to see that hateful Louis Napoleon "kicked out" as he deserved.

The following summer—August 31st, 1871—Henrietta was married to Richard Litchfield, barrister, ecclesiastical commissioner, and one of the founders of the London Working Men's College. Charles attended the ceremony, but the excitement, and his weak stomach, made it a severe trial; he could scarcely sit—or stand—it out. He now "inherited" Henrietta's Polly, the terrier thereafter his close companion till his death.

He was, most of the next twelve months, working one step ahead of a breakdown. He grew sick of pen and ink and pencil and writing. He felt very old and weary. Whenever possible he turned eagerly away to the more enjoyable observation of Drosera and the cross-fertilization of flowers.

CHAPTER XVIII

BOTANY BAY

I

HE had, from 1872, ten more years of life. His health improved after 1870, and remained, at worst, relatively fair till within a few months of his death. He worked well and more or less continuously. Though the *Origin* was now quite done with, he prepared and published revised editions of the *Descent* (1874), *Domestication* (1875), *Climbing Plants* (1875), *Fertilization of Orchids* (1877), and—after more than thirty years—the *Coral Reefs* (1874). A new printing appeared of the *Volcanic Islands* and the *South America* in one volume (1876). The new works of the period were *Insectivorous Plants* (1875), *The Effects of Cross and Self Fertilization in the Vegetable Kingdom* (1876, second edition 1878), *The Different Forms of Flowers on Plants of the Same Species* (1877, second edition 1880), *The Power of Movement in Plants* (1880), *The Formation of Vegetable Mould through the Action of Worms* (1881), the short memoir of Grandfather Erasmus (1879), a number of miscellaneous scientific papers and communications, and the posthumously published autobiography (written 1876 and 1881).

Yet extensive and even important as much of this work may be, it has most of it a diminishing biographical importance beside his earlier writings. From 1872 he definitely set aside the intention of further fundamental theorizing. Much more than anything preceding it, the labour of these years was confirmatory, not originative. He had seen so many of his friends make fools of themselves by rash speculation in their later years, and feeling that "no man can tell when his intellectual powers begin to fail," he resolved to attempt no major speculation, especially on "so difficult a subject as Evolution," after the age of sixty or so. The Great Work itself was not avowedly abandoned till as late as 1877, but the last edition of the *Origin* excised many of the references to it, though others stand to this day. He realized that the *Origin* was to remain his central and master work, and he could account that now as securely established if not universally accepted. The storm of the high seas of controversy had been dared and weathered; he wished now, as well he might, to rest in smoother waters.

Work he must, of course, for long ago he had projected his whole being into it, had no existence apart from it, found it the only drug against physical discomfort. Man delighted him not, nor woman neither, save as the man was a fellow-labourer in his own field or the woman both pretty and able to listen intelligently to scientific small-talk. "When I am obliged to give up observation and experiment, I shall die." [1] His only relaxation now was a change of subject. Botany he had always enjoyed. He found plants and flowers restful, soothing, undemanding. He liked watching them, experimenting with them. Back in the old days of *Beagle* zoology and geology, when even species was dereliction of duty, he had turned relievedly to botanizing in his more exhausted vacations; and when, later, species sat like an ogre on his study desk, the new hot-house and the garden bloomed in ever-growing attractiveness. It was part of the pleasure that he never esteemed himself more than an amateur in this field, a gentleman-rider of hobby-horses which thrived on the mild wild oats he sowed amongst them. Gladly he wrote to Wallace: "I have taken up old botanical work and have given up all theories." [2]

II

The statement was true and untrue. He *had* done with new major theories and with main arguments for Evolution and Natural Selection, but he could no more totally detach himself from either than he could stop breathing—or working. In fact, the whole range of his botanical inquiry had at least a broad relevance to his larger work, if only as detailed verification of this aspect or that. It was, to begin with, all done in the light of evolutionary principle, serving, as he had written in 1861 of the *Fertilization of Orchids*, "to illustrate how Natural History may be worked under the belief of the modification of species," [3] and

---

[1] *Natural Selection*, Wallace, 472 n.
[2] *Life and Reminiscences*, Wallace, i. 278.   [3] *LL.*, iii. 254.

showing how that belief gave more, not less, of interest in making clear the significance, the use, of apparently useless details of structure. Three of the books, the *Fertilization of Orchids, Cross and Self Fertilization*, and *Forms of Flowers* (the last a collection of papers written in the 'sixties, and revised for book-publication), were concerned with establishing the universality of cross-fertilization, and its significance in relation to variation and, accordingly, selection. It was, he asserted, "hardly an exaggeration to say that Nature tells us, in the most emphatic manner, that she abhors perpetual self-fertilization."[1] (That being established, the advantage of division into male and female sexes needed no demonstration—one trod the border of the very deepest mysteries; but he was careful in these later days to keep well on the hither side of them.)

The *Fertilization of Orchids* had showed the adaptation of orchid structure to the end of cross-fertilization, and indicated a possible development from the simpler to the most complex forms: "It becomes far from incredible, if we had every orchid which has ever existed throughout the world, that every gap in the existing chain, and every gap in many lost chains, would be amply filled up by a series of easy transitions"[2]—transitions, that is to say, able to be accounted for by the normal workings of Natural Selection. *Cross and Self Fertilization* turned from the means to the beneficial effects of cross as compared with self fertilization, though its "most important conclusion," Charles thought, was "that the mere act of crossing by itself does no good. The good depends on the individuals which are crossed differing slightly in constitution, owing to their progenitors having been subjected during several generations to slightly different conditions, or to what we call in our ignorance spontaneous variation."[3] The *Forms of Flowers* dealt with some special and illuminating cases. No other discovery of his, he said in his autobiography, ever gave him the pleasure he got from "making out the meaning" of the heterostyled flowers.

Wallace was later to write that these three books revolutionized the science of botany by their clear demonstration of the significance of the many diverse forms of flowers and of the purpose inherent in every minutest variation. This latter was in no small degree too the achievement of the *Climbing Plants* and the *Power*

[1] *Fert. of Orchids*, 351.   [2] *Ibid.*, 323.   [3] *Cross and Self Fert.*, 27.

of Movement in Plants, complementary works, the first showing the advantages, means and development of climbing, while the second, written with Frank's collaboration, displayed movement as universal among plants, and described its various forms and the evolution of the more from the less intricate.

In subject the *Insectivorous Plants* stood apart—an account of the varieties of insect-devouring plants and of experiments made upon them—but it too was basically directed to making clear the evolutionary development even of these strange plants by increasing specialization of properties much more widely shared.

None of the books could be called large beside the *Domestication* or the *Descent*, but neither were they especially slender, most over 300, some over 400, closely-printed pages apiece, and each the fruit of prolonged and careful observation, admirable examples all of them of his capacities both for taking pains and for drawing large meanings from apparent trivialities. He enjoyed observation, he liked the verifiable fact, but his greatest pleasure was, as ever, in the generalization it evoked or supported.

Publicly, all were at least adequately successful, selling between 1500 and 3000 copies apiece within a year or so of publication, and meeting, even where they were not entirely approved, a milder reception than the biological works.

III

Through these years the personal life went on with, if possible, a smoother rhythm than before. Every now and then he would stick at some task just a little too long, and pay for his incaution with a period of idleness enforced if not by Emma then by his own tired brain, but even these crises tended to be less intense, as his efforts were directed to more modest achievement. (The revising first of the *Descent* and then of the *Domestication* possibly put him to severest strain; in each case the alterations were considerable and even important, though not to the purpose of these present pages.) He consented to take his holidays now with a better grace, and though one of his hobby-horses almost invariably travelled with him, he was less insistent than once upon galloping them daily.

There were now two houses in London where he was always welcome and quite at home — Erasmus's of course, and the

Litchfields' at 4 Bryanston Street, Portman Square. At either he could be completely himself, see people or not as he wished, invite his friends to meals just as though he were at Down House. He and Emma were always happy too with William at Basset, which they visited regularly summer after summer, making expeditions all about the district, once, in 1877, by train to Salisbury and thence by carriage-and-pair to Stonehenge, where worms were dug for, but found to have been sadly lazy. Leith Hill Place was unfailingly pleasant, and Charles in these years also made several enjoyable visits to Abinger Hall, the home of Sir Thomas, later Lord, Farrer, son-in-law from 1873 of Hensleigh Wedgwood; here he particularly liked to stroll about the neighbouring deserted common, a Wordsworthian wanderer with his broad-brimmed hat and long pointed staff.

Visitors to Down House inevitably increased with the growth of Charles's fame, and it was fortunate that he was now as able to cope with them as ever in his life. Some were complete strangers, wide-eyed students or eccentric wanderers who could not pass the house by without shaking the hand of the great man. Some—younger scientists—even genuinely sought his advice. More were distinguished foreigners, correspondents already or bearing notes of introduction. The Germans were the most tiring, bellowing their bad English across the dinner-table, but Haeckel at least had always the pleasure of going away completely convinced that Charles agreed with him in every particular. Alphonse de Candolle was met again after more than forty years. The Nortons brought in their travelling Americans, most of whom Charles found very much to his taste. Moncure Conway—whose strong selling-line just then was his own special brand of "post-Darwinite religion"—spent a night under the sacred roof-tree, full of Charles-worship and his own importance. From his bedroom window in the morning he ecstatically watched his hero walking in the cool of the garden, and thought that his beard was saying good-day to the bushes. Soon he was in the garden too, proclaiming a thrush's song a Vedic hymn to the rising sun. Charles said the bird was mating, but even this realism could not dash Conway's fervour, and he longed to clasp his master's feet and "try to tell him what he had been to me."[1]

Lubbock would also often bring friends across from High

[1] *LL. Galton*, iii. 471.

Elms, and one afternoon the skies fell when there appeared not only Lubbock, Huxley, Morley and Lord Playfair, but the unique, the incomparable, the garrulous Mr. Gladstone. Charles was struck dumb by such a visitation, but Gladstone was only too ready to seize the occasion to tell sad stories of the deaths of Bulgarians. His host, to whom a Liberal politician stood almost as high as a Prince, afterwards told Morley humbly: "What an honour that such a great man should come to visit me." [1] They subsequently corresponded, but Charles's hand, one feels, always trembled even thus to approach towards the altar of *real* greatness.

In these years many new acquaintances were being made in the wider scientific and social worlds, from Romanes and Burdon-Sanderson to Mrs. Ritchie (Thackeray's daughter), and from Anthony Rich—a wealthy admirer of Charles who wanted to leave him a valuable property—to Hans Richter.

But old friends were not forgotten or neglected. When in 1873 Huxley was overworked but financially unable to rest, Charles took active part in collecting £2100 and persuading him to use it for his health's recovery; and later, in 1880 and 1881, he was even more to the fore in pressing the bestowal upon Wallace of a Civil List pension. Hooker was more than ever his intimate companion, a frequent genial visitor, scattering bad puns and telling slightly blasphemous stories whose reception by Charles he waited with a slightly questioning eye. Gray had his last few days at Down House in 1881; their parting was elegiac, for they knew they could scarcely meet again in the flesh, nor did Charles at any time put much faith in the spirit. Carlyle, at Keston Lodge for the summer of 1875, came calling, and found him more likable than ever before; it was "a good while since I have seen any brother mortal that had more or true sociability and human attraction for me." [2] Francis Galton was sometimes met and often written to. Fox was still heard from, and so, at intervals, was Sulivan, now K.C.B., whose labour for missionary work among the Fuegians had compelled Charles to admit the error of his first feeling that savages so low in the human scale were beyond even heaven's help. His interest in the local natives remained kindlier and more immediate; and he always welcomed the walking-parties from the Working Men's College, sixty or

---

[1] *Life of Gladstone*, Morley, ii. 562.
[2] *New Letters of Carlyle*, ii. 314.

seventy men and women taking their tea on the lawn, strolling the gardens, and singing in the gathering shade of the lime trees.

Letters from strangers he treated in the same conscientious, amiable way. Many were from cranks or fools eager to waste his time by asking silly favours (one young man wanted an outline of Natural Selection, having to debate the subject but knowing nothing about it), but though he had a form printed to deal with these, he seldom brought himself to use it. When people sent him the most absurdly commonplace observations, he recalled how Henslow had once borne with him, and would say to Frank, now his secretary, "that they should all be pleasantly answered. It is something to have people observing the things in their gardens and barn-yards." [1]

IV

All who visited Down House spoke with pleasure, even affection, of its attractive situation and atmosphere. The former remained unspoiled, the latter never stiffened. No main roads approached to bring traffic and new building. Improvements to the house itself continued at intervals till quite late. A glass-roofed verandah, reached from the drawing-room by French windows and looking out over the large lawn across the low garden wall to the farther fields, was added about 1873, and another change was the transformation of the billiard room into a new and ampler study, where Charles might have two large tables, one for his writing and papers, the other for the plants and pots demanded by some current investigation.

It was part of your welcome there to have no standing on ceremony. Settled in the drawing-room, Emma would soon be back at her sewing-basket, Charles extended upon the sofa, both talking in friendly, easy, sensible fashion, seeking not to impress but only to set at ease, to discover your interest and to make you talk about it. Charles in particular liked to listen. He didn't as a rule talk either easily or quickly, though capable of both when excited; his passion for precision, for doing exact justice to each statement, then led him into long involved sentences, which only his animated manner, his readiness to laugh at himself, kept from ponderosity. For the visitor come to stay, books and

[1] *Autobiography*, Conway, ii. 325.

magazines lay about in profusion; there was the garden to sit or saunter in, always some member of the family to talk to, seclusion if one wished to work and communal gatherings after dinner in the drawing-room or, were the evenings warm, upon the verandah outside. There was music, more talk, sometimes reading, to make one forget the death of yet another day, one more, one less, in the lives of them all.

Charles was now, no more than in his sixties, definitely an old man, stooping markedly, his long beard and shaggy eyebrows quite white, his mild eyes sunk ever-deeper under dominating brow. The line of the lips was firm and straight, the cheeks deceptively ruddy, the smile animating his whole face, but the wrinkles grew deeper across his forehead and about his eyes, and the ugly nose coarsened. He liked to dress, as to live, informally in loose, easy-fitting clothes, a shawl over his shoulders in the house, a short black cloak (with soft black or straw hat) cast about them out-of-doors.

The routine of his days changed but little. There was the early rising and the walk before breakfast about the garden, noting the progress of the plants. There were the morning periods of writing and being read to, the so-many-times round and round the sand-walk with Polly or perhaps a visitor for companion, the easier afternoon, sometimes driving through the lanes in open carriage, then more resting and reading, an evening stroll with Emma, a short nap or rest after dinner to keep him lively till his final retirement at half-past ten or eleven. Nor were there any innovations in his reading—a certain, but in general a diminishing, amount of scientific matter, some reminiscences, some biography, especially of men he had known or whose work had interested him, and, for total relaxation, fiction. He delighted in Lewis Carroll, and the works of Mark Twain stood year after year on his bedside table to pass the wakeful midnight or early morning hours (he would press *The Jumping Frog* upon visiting Americans who had scarcely heard of and certainly did not esteem their humorous compatriot). But above all he enjoyed sentimental novels, compact of pretty heroines and of happy endings, than which nothing mattered more or could set a book higher. He would like, he said, to have a law passed against all other endings; and as for the other point, "A novel, according to my taste, does not come into the first class unless it contains some person whom one can thoroughly love, and if a pretty woman all the better." [1] *Her Dearest Foe, Fair Carew*, and *Denise* were all favourites, and so too was *Emma*, but Mrs. Gaskell's early problem novel, *Ruth*, was by no means approved of.

The whole family, without exception, retained all the appearance of singular, almost phenomenal, unity and happiness. Though he lamented that they should have to grow up at all, Charles's pride in his sons' successes was unending. He loved to write to his friends of how they were "all doing wonders," and the slightest accident would make him tremble for their safety and beg them to avoid all possibility of danger. When, not long before his death, they bought him a fur coat, he was overcome by emotion and told them that it would "never warm my body so much as your dear affection has warmed my heart." [2] He hated to see them going out into the world, away from home, and was glad to keep Frank, the only botanist among them, at hand as his assistant. (But they were all so willing, so helpful, in their various ways: no man, he was sure, ever had better children.) When Frank married in 1874 he went no further away than the village, and after his wife's sad death in childbirth two years later returned to Down House itself, bringing the infant Bernard to the delight of the grandparents. Bernard and Charles were frequent companions till the latter's death, and the child was never to forget his patient kindliness. William, in 1877, married a sister of the American Mrs. Norton; he was now a partner in his bank.

Fanny Allen, last of all her generation, died in 1875 at the age of ninety-four, asking that none should grieve for her or be summoned to her bedside. Lyell too died that same year, a few weeks earlier, an event Charles felt the less deeply since he had expected it and apparently feared for his friend the worse alternative of a weakening mind. Yet the death seemed the first stroke of a wider summons: "he is gone, and I feel as if we were all soon to go." [3] His thoughts went back to their memorable first meeting, and to all that Lyell's personality and writings had meant to him. How great was his debt: "almost everything which I have done in science I owe to the study of his great works." [4]

A happier return to the past, testimony to what, with Lyell's aid, he had accomplished, occurred two years later, in November

[1] *LL.*, i. 101.
[2] *Emma Darwin*, ii. 239.
[3] *LL.*, iii. 197.
[4] *LL.*, iii. 196.

1877, when he went to Cambridge to receive an honorary degree of LL.D. Most of the family were there: Emma, Bessy, Horace, Leonard, Frank, George. The Senate House was packed with undergraduates who cheered Charles, jeered an unpopular proctor, and dangled a monkey and a beribboned ring (the Missing Link) in mid-air. Emma thought the ceremony tedious, but was proud to walk with Charles in his enveloping scarlet gown beside the Master of Christ's. He stood the noise and excitement and the magnificent lunch in George's rooms unusually well, but returned that day to Downe, leaving Frank to attend and report upon the resplendent dinner at the Philosophical Club, where Huxley responded to the main toast first by accusing the University of waiting until its honour was quite "safe" and superfluous, and then by declaring his "deliberate opinion, that from Aristotle's great summary of the biological knowledge of his time down to the present day, there is nothing comparable to the *Origin of Species*, as a connected survey of the phenomena of life permeated and vivified by a central idea." [1]

V

Emma thought too much recalling of bygone days a morbid thing, but it was an inclination of age from which Charles at least was not immune. In this period as in no other the past seemed often in his mind. He had always liked talking about the *Beagle*, but he would go back beyond that now to his childhood, to his father, for whom he could never express a sufficient reverence; the wisest man, he would say, he had ever known. Even his work tended to look backward. He hunted up a diary he had kept in 1840 of young William's infant development, and made an article of it, published in *Mind* in July 1877, while the subject of his very last book went even further back to Uncle Jos's suggestion about earthworms and mould.

Before that, in May 1876, holidaying at Hensleigh Wedgwood's Surrey home, he had started writing his autobiography, continuing it a little each afternoon until he completed it in August—a pleasant if sometimes sad task intended to amuse himself and "possibly" to interest his family, for he thought sometimes how he would have enjoyed such an account written by Grandfather

[1] *LL. Huxley*, i. 481.

Erasmus. It was a leisure task, and he kept it cursory, attempting little re-creation of either characters or places. He had almost nothing to say of his mother, and little of his other relatives. Of his childhood he recalled principally some trivial misdeeds, of his schooling mainly its uselessness, of his youth the growth of undirected collecting habits. Edinburgh was represented by dull lectures, Dr. Grant, and the Plinian Society. Vacations, walking, shooting, burned brighter. Cambridge now appeared a place of friends: Henslow, Jenyns, Sedgwick. The voyage was treated very briefly: FitzRoy, activities, some striking memories. Personal life after that was quickly passed over; he recollected, rather, the famous figures he had known, the sequence of his books, detailing lesser points which had given him especial pleasure. He wrote of the *Origin* as his life's chief work, showing the Great Work abandoned for ever. Another book or two, he concluded, on work already done, and "my strength will then probably be exhausted, and I shall be ready to exclaim *Nunc dimittis*."[1] He viewed his life, it seemed, with a tolerable satisfaction; he knew that his children for whom he was writing would neither judge him harshly nor ask him to do so himself.

Most of his years his attitude to his grandfather had been one of disregard if not dislike. That, in regard to Erasmus's ideas, scarcely changed—he had no desire to make amends for what he never saw as a cavalier treatment; but he found himself more interested in his grandfather as a human being. So when he read an essay on Erasmus, contributed to the German periodical *Kosmos* by Dr. Ernst Krause, he not only obtained permission for its English translation and publication, but decided to preface it with a personal sketch, written from family knowledge, letters, notes and other documents—in order, he said, to refute Anna Seward's (long forgotten) "calumnies." He was, this spring of 1879, finding his work more tiring than ever before, and it should prove a pleasantly easy task before rejoining Frank to complete their *Power of Movements in Plants*.

Like every book ever written, it proved more trouble than had been expected, but was finished in time for an August holiday at Coniston, further afield than he had been for many years. Most of the children went too, and all delighted in the lovely Lakes scenery and the glorious weather. One very happy day

[1] *LL.*, i. 96.

was spent at Grasmere. Ruskin, recovering at Coniston from the first of his mental breakdowns, called friendly upon them, saluting his host as Sir Charles and warning the company of dreadful clouds that hovered threatening in the summer sky. On their return visit he showed some of his Turners to Charles, who thought them dreadful but did not dare to say so. An easier guest was Mark Twain, who not only deemed it the highest honour that his books should send Charles to sleep but was himself a humble admirer of the *Descent* and *Origin*. One regrets that no account of their meeting was set down.

The Lakes were to be revisited once again, in 1881, at Patterdale. Here, in this noble country, Charles rediscovered what he sometimes feared he had lost for ever, his delight in scenery. He and Emma would walk together along a cliff-edge looking down on the calm waters. Sometimes Bernard would go with them, his hand held fast against danger; sometimes he would stay behind with Henrietta or Frank or Bessy, and they would go alone, equally at one whether speaking or silent, an old man and an old woman strolling in the sunshine.

VI

The *Erasmus Darwin* was published in November. Charles had the usual misgivings, and the family generally was dubious of old Erasmus's interest anyway—doubts expressed by drastic cutting in the proofs and justified in the book's mediocre reception. Less than a thousand copies were sold.

Worse was to follow. Charles hated controversy. He once said that was because he conducted it so badly, and certainly it always demanded a quite disproportionate expense of time and nervous energy. The simplest contradictory letter to a periodical could keep him awake all night—the sort of thing Huxley would dash off between cups of tea and drink the better for. He had accordingly long since been glad to take Lyell's advice and to leave his assailants' bolts to melt in the thin air of unresponsive silence. A few trivial times he was forced to make some reply, but more often, as when he earned Miss Cobbe's dismissal and attack by his view of the Vivisection agitation of 1875 as hysterical exaggeration, he won peace if not victory by playing Brer Rabbit to the life.

The bother about the *Erasmus Darwin* tempted him only momentarily into the open, but led to the bitterest series of personal attacks ever made upon him by anyone. The incident really belongs not to Charles's but to Samuel Butler's biography, for though it gave the former "much pain" he never understood what caused the other's indignation, and after one letter to Butler himself, a couple drafted for but never sent to the *Athenaeum*, and a few to some friends, all written in January and February 1880, he gave the matter no more active attention, leaving Butler to spend year after year spitting like an angry cat or a woman scorned.

It is hard indeed not to see Butler as something of the disappointed suitor. There was a time when the rôle of beloved disciple to Charles would have delighted him, for on reading the *Origin* in New Zealand in 1860 or 1861 he had become the complete convert and even advocate in the local Press. Subsequent presents of Butler's own books—that of *Erewhon* accompanied by a disclaimer of intended disrespect to the *Origin*, "for which I can never be sufficiently grateful"—led to friendly correspondence and visits to Down House in 1872, and though he met Charles only once after that, at Erasmus's, he was for some years on informally friendly terms with Frank. In *Erewhon* he had written of machines as in effect additional human limbs; now he reversed the notion and began to look at limbs as machines, manufactured by ourselves at need by the action of habit, which was unconscious memory. The conception seemed to him, at least up to 1877, "an adjunct to Darwinism which no one would welcome more gladly than Mr. Darwin himself."[1] Only when he re-read the *Origin*, and with it Mivart's Darwinian criticisms, did he discover the incompatibility of his view, and the possibility of a faith in Evolution liberated from the "blind" working of Natural Selection. Charles denied the power of habit and memory; Butler stuck to it, and found his earlier loyalty still further sapped by a study of Lamarck. The latter chapters of *Life and Habit* (late 1877) openly advocated Lamarckism against Natural Selection, and were, in his own view, "downright disrespectful" to Charles himself.

But if Charles was bad at controversy, Butler was infinitely worse, being intrinsically egotistical and intensely suspicious.

[1] *Unconscious Memory*, Butler, 33.

His whole being was tainted, as he himself acknowledged, by a pervading "sense of wrong," and any act impinging inimically upon himself he could see only as deliberate. Following the appearance of *Life and Habit*, Frank Darwin wrote most amicably if a little surprisedly about it, but on a personal meeting Butler was sure that he was really very upset, and ever after he could never conceive any action on the part of any Darwin save as a move to conciliate or threaten him.

Meanwhile he was going on with his anti-Darwinian campaign —already seen as such—by reading all the evolutionary writers mentioned in the Historical Sketch, discovering in particular Buffon and Erasmus Darwin, and becoming more and more deeply suspicious. Why, he wanted to know, had Charles kept so remarkably quiet about his predecessors, scarcely as much as mentioning them in early editions, and dismissing them briefly and often scornfully in others? What could it be but a desire to steal their thunder, to take all their glory to himself? Clearly, too, this conspiracy of silence extended to his friends and supporters.

It was to dethrone the one god—Charles—and to elevate the three—Buffon, Erasmus Darwin, Lamarck—that *Evolution Old and New* was written. It was announced in February 1879, and appeared in May. Charles, working on the account of his grandfather, read the personal pages only and sent the volume on to Krause, then revising his magazine article for book publication, with the comment that it merited little or no attention. Krause shared or adopted Charles's valuation, and said in his manuscript that the recent attempt to revive Erasmus Darwin's ideas showed "a weakness of thought and a mental anachronism which no one can envy." [1]

Matters became intricate when *Erasmus Darwin* appeared with a preface by Charles seeming to guarantee Krause's essay as unchanged from its original printing, and noting the subsequent publication of Butler's volume. Butler spotted the reference, and incredulous to find it aimed at any but himself, hastily acquired a working knowledge of German while ordering the issue of *Kosmos* containing the original essay. Naturally interpolations and amendments were found, and instantly he saw the whole plot plain. The essay in *Kosmos* and the announcement of *Evolution Old and New* had appeared simultaneously. Charles, knowing

[1] *Erasmus Darwin*, 216.

that the book would be an attack, had decided to destroy it by falsely showing it as refuted before it was written. The volume, decked out though it might be with a grandson's seeming-innocent biographical study, had no other purpose! Charles's unscrupulous egotism had previously worked to bury the dead; now it was working to damn the living.

Instantly, on January 2nd, 1880, Butler wrote to Charles demanding an explanation. There was a very simple one. Charles in his original preface had stated plainly that "Dr. Krause has taken great pains, and has added largely to his essay as it appeared in *Kosmos*,"[1] and had that sentence only been retained the notion of conspiracy could never have arisen. But it was by pure chance struck out in the massacre of the proofs, and Charles in replying to Butler forgot that he had ever written it, only declaring the alteration of magazine articles for book publication to be so common that it never occurred to him to mention it. However, he added, *should* there be a reprint he would add an explanatory statement.

The letter to Butler seemed a cold dismissal, a continuance of the conspiracy, and late in January he stated his case in the *Athenaeum*, in a form Charles took as a charge of "lying, duplicity, and God knows what." He now recalled the wording of his first preface, looked up the proof, saw exactly what had happened, and drafted a letter of reply, making the facts clear though refusing further discussion. He would have sent it, and settled the matter then and there, but some of the family thought it beneath his dignity even to notice "so ungentlemanlike" a personage, and he was dissuaded. He drafted accordingly a second, shorter letter, but Litchfield disliked that too on the ground that "a clever and unscrupulous man" like Butler would twist anything to his own end, and Huxley too, when consulted, agreed decisively.

Silence pleased Charles, but Frank, who knew Butler best, regretted it, and undoubtedly it was an error, for it left Butler not merely unsatisfied but boiling angrily under the imputation of being "the only writer who has sought to gain notoriety for himself by offering personal insult to Mr. Darwin,"[2] and to vent a useless anger in new books and new editions of books, in periodicals and private letters and journals, beyond all proportion

[1] q. *Butler*, Jones, ii. 451.   [2] *Butler*, Jones, i. 412.

to the originating cause. He grew to detest Charles's very name as the incarnation of scientific arrogance, and his attacks upon this Pecksniff of Science, this very king of the Country of the People who are Above Suspicion, wrecked, he afterwards said, his literary prospects for years. His *Unconscious Memory* (1880) and *Luck or Cunning?* (1887) attacked Charles's views and his character; Butler even counted up the number of times the possessive words "my theory" occurred in the *Origin*. It was sadly stupid and wasteful, for Butler was assaulting an ogre of his own imagining, and meanwhile the documents which might have brought about reconciliation lay forgotten till after Butler's death. He had a case against Charles, in regard to both the latter's neglect of his predecessors and his actual theory, but bitterness exaggerated the one and obscured the other, so that the force of both was lost.[1]

[1] St. John Ervine, in his "The Centenary of Samuel Butler" (*Fortnightly Review*, December 1935), expressed the opinion that Darwin behaved badly in suggesting to Krause that *Evolution Old and New* was " really not worthy " of " much powder and shot." " A nod was as good as a wink to Dr. Krause.... A more scrupulous genius than Darwin would not have given Dr. Krause such a lead." But the whole idea of scrupulosity only arises in retrospect. One may assume that all Darwin meant was that while Krause should see this latest book on his subject, he need not give it much attention. Mr. Ervine's reading *presupposes* the fabulous conspiracy.

CHAPTER XIX

THE ULTIMATE HAVEN

I

IN March 1880 Cousin Josiah died, aged eighty-five, first of the four sons of Uncle Jos, and in November Elizabeth Wedgwood, aged eighty-seven, leaving only Emma of the daughters. Always small and bent from curvature of the spine, and growing blind with the passing years, Elizabeth had come to Downe in 1868 to be near her sister. She lived happily there, spending a good deal of time at Down House where she was always welcome, cheerful (despite her afflictions and the beggars who preyed upon her charity) until almost her last year, when she seemed suddenly to tire of life, existing more and more in memories of the distant past. Snow fell at her funeral, settling upon the black cape of Charles standing statue-like beside the open grave.

He was working now on the *Formation of Vegetable Mould*; it was written this winter and went to the printers in April—a relatively short account of the habits of earthworms and the part they had played in the world's history by hastening denudation, making new mould and burying ancient buildings. He expected little from or for the book, but people found it lightly and pleasantly written, and were attracted to have such importance and, even more, intelligence ascribed to this humble, familiar creature. "Go, proud reasoner, and call the worm thy sister," Grandfather Erasmus had declaimed in the *Zoonomia*, and the public delighted to see him obey. Besides, he was seventy, when all British writers are canonized. Six thousand copies were printed within five months, and correspondence poured in on the subject—to the surprise of Emma, who thought "training earthworms" a very poor sport.

It was his last sustained labour. He was subsequently to take up and "look into" some lesser matters, but he carried none to completion. He found it increasingly difficult to concentrate on new subjects at all; he was too old. A weary look had come into his eyes, and at times he felt his whole mind caught in the grip of a mental paralysis, and was glad to claim age's immunity "from being expected to know, remember or reason upon new facts and discoveries."

He had in 1877 and 1878 added to his autobiography fragmentary recollections of his father and brother, and now, in May 1881, he appended some final pages, describing the work he had done since 1876, and then, having no other personal life to tell, turning to speak of his manner of writing, and, at greater length, his mental qualities.

As a writer he admitted himself no stylist, by nature or nurture. As an appreciator he lamented his "curious" loss of "the higher aesthetic tastes." Once he had delighted in poetry: Shakespeare, Milton, Gray, Byron, Wordsworth, Coleridge, Shelley; once pictures had given him great pleasure (O Titian! O Venus!); once music had rapt his soul away. He was dead now to them all. He was worse even than Emma about Shakespeare, who bored him to nausea. Pictures—a few streaks of colour on canvas! Music?—aye, but he could no longer enjoy it, for it stimulated him too much, turning him back to his work just when he would forget it. He admitted his taste in novels to be deplorable. His mind, he regretted, seemed to have become nothing but "a kind of machine for grinding general laws out of large collections of facts." "If I had to live my life again, I would have made a rule to read some poetry and listen to some music at least once every week; for perhaps the parts of my brain now atrophied would thus have been kept active through use. The loss of these tastes is a loss of happiness, and may possibly be injurious to the intellect, and more probably to the moral character, by enfeebling the emotional part of our nature." [1]

He listed his more positive qualities: he was slow to apprehend, needing reflection; metaphysics and mathematics both baffled him; his memory was bad on detail though broadly retentive. Some invention he thought he had, and certainly (despite the adverse opinion of others) powers of reasoning, for the *Origin* was "one long argument from the beginning to the end, and it has convinced not a few able men." He thought himself "superior to the common run of men in noticing things which easily escape

[1] *LL.*, i. 101-2.

attention," [1] and to such matters he had applied himself with close and continuous industry, partly from desire for scientific glory, partly for the satisfaction of understanding, explaining, arranging. He thought that, considering his modest abilities, he had on the whole done well.

The need to understand, to establish order over the whole field of apprehension by whatever bold generalization might be demanded, the power of sticking at a subject until every aspect (or almost every aspect) had been explored, every exception or objection (or almost every objection) noted and dwelt upon and countered—these were cardinal. In 1871 he had written to Horace that so far as he could see the art of initiating discovery consisted in "habitually searching for the causes and meaning of everything which occurs. This implies sharp observation, and requires as much knowledge as possible of the subject investigated." [2] He had that habit, that faculty of observing, and the patience to acquire the rest.

The survey did not cover all his qualities—his almost strangely excessive modesty, his almost fearful caution contrasting oddly with his boldness, his lack, at least where his own work was concerned, of historical sense.

Two aspects he did not discuss at all, unless in those portions of the autobiography still unprinted: his religious and social views. In regard to either he probably had little enough to say. In the matter of religion, certainly, he never made up his mind in any final sense. It so baffled him that he could only set it in the category of the unanswerable ultimates (like the conflict between free will and determinism, only worse). "The more I think the more bewildered I become." [3] "I am, and shall ever remain, in a hopeless muddle." [4] Accordingly he was always reticent; if one could not speak with authority, and to console, silence were the better part. His general development was clearly one from simple faith to indifferent doubt, along, as he said, a path so gradual that he felt no acute distress. At Cambridge he had been definitely a believer, doubting only on minor details of doctrine, and Christian he sailed upon the *Beagle* and Christian came home. In the years following his return his disbelief grew until he saw that the theological life was not for him. He gave

[1] *LL.*, i. 103.   [2] *Emma Darwin*, ii. 207.
[3] *LL.*, ii. 312.   [4] *LL.*, ii. 354.

up Christianity, he said, when he was about forty. The *Origin*, he always insisted, was written without theological intention, and he disliked to admit its conflicting with Christianity; he was a theist when he wrote it, though later even that degree of assurance waned, until he had to admit himself, when questioned, an agnostic, one who did not assert the non-existence of a God, but simply did not *know*. He did not believe in religious revelation; he had no intuitive sense of a spiritual world or being; Natural Selection seemed to him to have destroyed Paley's argument from design; and as for a future life, that was a matter for each individual's personal judgment. He shrank from the term atheism mainly because atheists repelled him by their aggressive dogmatism—and their public preaching, for he questioned whether "the masses" were yet "ripe" for disbelief.

In the end he hung his doubt in the all-concealing cloak of a freely-admitted ignorance, refusing to commit himself in realms where faith was all and knowledge nothing, and wherein even his own feeling fluctuated. He frankly didn't know whether Design or Chance, God or No-God, ruled the universe. "My theology is a simple muddle; I cannot look at the universe as the result of blind chance, yet I can see no evidence of beneficent design, or indeed of design of any kind, in the details." [1] That was 1870; this 1881: "On the other hand, if we consider the whole universe, the mind refuses to look at it as the outcome of chance—that is, without design or purpose. The whole question seems to me insoluble, for I cannot put much or any faith in the so-called intuitions of the human mind, which have been developed, as I cannot doubt, from such a mind as animals possess, and what would their convictions or intuitions be worth?" [2] In the 'seventies he saw God as an evolved idea, and the animistic writings of E. B. Tylor helped to thrust him towards the completely naturalistic view.

Spiritualism he barely touched, and it had no influence on his views. In these later years Galton, Wallace, and presently Romanes sought to involve him in the subject, but Huxley's bitter scorn was sufficient antiseptic. He attended only one *séance*, when George hired a medium to perform at Erasmus's house. He, George, Erasmus, Emma, Henrietta, Litchfield, Galton, Hensleigh Wedgwood, G. H. Lewes and George Eliot made up the

[1] *ML.*, i. 321.   [2] *ML.*, i. 395.

party, but the manifestations unkindly commenced only after he had got tired of waiting and gone upstairs to lie down. He thought them all jugglery and nonsense, but George was more impressed, and it may have been to rescue him that Huxley came to a second demonstration. His analysis was devastating, and George accepted it. Charles and spiritualism after that were not on speaking terms.

His social outlook was scarcely less crude. Though interested in politics from the broadly liberal point of view, he was an onlooker only, taking his facts from the newspapers and attempting no individual or specific thought. His most consistent attitude was opposition to every Government for its indifference to science, and except for the mighty Gladstone he thought politicians as a whole "a poor truckling lot." When Wallace, in July 1881, urged him to read Henry George's *Progress and Poverty*, he said he would do so, but added that some years before a dipping into political economy had had no effect but to make him "utterly to distrust my own judgment on the subject and to doubt much everyone else's judgment!"[1] In the main he was a sturdy individualist of the *laisser-faire* school. In 1877, when asked to give evidence in a case brought against Bradlaugh and Annie Besant for circulating a book on birth-control, he wrote that "he disagreed with preventive checks to population on the ground that over-multiplication was useful, since it caused a struggle for existence in which only the strongest and the ablest survived."[2]

II

The final year went rather slowly by, with little work in hand. They were home from the Lakes in July, enjoying the warm summer days, and when visitors came calling he would sit with them in the garden in the cool shade of the limes, where the drowsy humming bees tumbled from the flowers drunk upon the smooth grass. There was another painter-fellow to sit to, a tribulation it seemed fame inevitably put upon one.

Fame!—one took the bad with the good. Earlier in the year there had been one thrillingly embarrassing moment when he had gone with Burdon-Sanderson to the Royal Institution to be

[1] *Life and Reminiscences*, Wallace, i. 318.
[2] *Autobiographical Sketches*, Besant, 136.

received by the entire gathering standing and applauding as he was escorted to his place upon the platform. Now, in August, he attended as a delegate the International Congress of Medicine held in London, actually dining with the President, none other than the Prince of Wales, who "talked familiarly"—and this time more audibly—to him.

Carlyle had died in February, and on August 26th Erasmus, after the briefest illness. It was a heavy loss. Poor dear Ras, so warm-hearted he had been, so affectionate. There had been an extraordinary clarity about his mind; everything he touched he illuminated with his charm and good humour. In youth he, not Charles, had been the family's bright boy, but he had done nothing, and of late years, though wealthy and with no apparent worries, had even lost the wish to live. Under his good cheer had developed—perhaps from ill-health, perhaps from idleness—fundamental melancholy. He was buried in Downe churchyard on September 1st, Charles's last funeral . . . until his own. Almost all Erasmus's money came to him; he tried to give some of it away, if only a little, to assist scientific projects. He said that he wanted to show his gratitude to science which had given him so much.

In October he paid his last visit to Cambridge, for a happy week with Horace and his wife, Lord Farrer's daughter, married in 1880; their first child was to be born that December, another Erasmus. He attended service once again at King's College Chapel, and the singing wakened as ever memories of those who had been all Cambridge to him, and now walked only ghosts along its streets.

That visit was in fact his last real holiday, for most of the winter his health was causing anxiety and he lived very quietly. He could scarcely work at all. Emma in December took him to London to see Dr. Clark about his heart, was relieved to be told there was no cause for anxiety, then perturbed when, calling one day on Romanes, he experienced on the doorstep a slight seizure, the first of that series of increasingly severe attacks which was to end only with his death four months later.

He seemed to recover, and except for a tiring cough was fairly well through January and most of February, though on the 13th, thanking a friend for birthday congratulations, he wrote that he felt a very old man: "my course is nearly run."[1] His pulse

---
[1] Unpublished letter to Lawson Tait.

troubled him again, and every afternoon there were distressing pains about his heart. He tired quickly, and could walk only very slowly about the garden and out to the sand-walk, leaning on his stick or the arm of Emma, his elder by nearly a year, yet looking a decade the younger. On March 7th he went to the sand-walk alone, had an attack, and reached the house with difficulty. He never went there again. The family was alarmed, and he distressed by a fatigue and faintness such as he had not felt for years. Doctors were summoned, and for days he lay in bed with scarcely strength or even inclination to look out of the window. Up again, he attempted a little sporadic work, but he was definitely an invalid now. He dared not go far from the house, or indeed anywhere without a companion. The weather was exceptionally fine, days of still bright sunshine laying a soft cloak of beauty over the Kentish countryside, and he and Emma instead of walking sat together under the bare boughs of the fruit trees in the orchard, watching with wistful eyes the fresh budding of flowers, the golden sunlight, listening to the clear singing of the birds. He must have known he had not long to live, and he tried to tell in adequate words all he felt of gratitude for so serene, so placid a long life. Never is human being so dumb as when he would declare his deepest feeling. The touch of hands, the meeting of eyes, means more. It was almost worth being ill, he would say, just to be nursed by her. She would answer gravely, tenderly, with only a hint of her brusque commonsense, that he spoke as if he had not done just the same for her.

On April 10th George was home from a West Indian holiday, and though Charles could talk little he enjoyed listening. He rejoiced to think how well his sons were doing. George, he was sure, would be "a great scientific swell" one of these days. William was as safe as a bank. Horace was happy in Cambridge. Leonard was making good progress. Frank might do much too, did he not waste too much time tending to his old father's affairs.

He was well for nearly a whole week, apart from a sudden attack of giddiness at dinner on the 15th, when he fainted in trying to reach the sofa. But he seemed to recover, and Emma noted April 17th as a "good day, a little work, out in orchard twice." April 18th he was up and about, writing a little, sending a subscription to Cambridge University.

But he was the grandson of that Erasmus who said that he

generally looked well before he became ill, and that evening, just before midnight, a severe attack came on. After taking medicine twice he woke Emma, saying that he could bear it better having her with him, then fainted and only with difficulty was restored to consciousness. Doctors and children were sent for. Through the night and into the morning he suffered repeated bouts of nausea and pain. He knew his state, and told Emma: "I'm not the least afraid of death." Later he said that his children must think only how good they had been to him, and after one very bad attack, to Emma again: "I was so sorry for you, but I could not help you." [1]

The morning became the afternoon, and those about him could do little but wait and watch while he passed from bout to bout, growing gradually paler and with ever-dwindling strength. Romanes afterwards said of him, rhetorically, that "he died as few men in the history of the world have died—knowing that he had finished a gigantic work, seeing how that work had transformed the thoughts of mankind, and foreseeing that his name must endure to the end of time among the very greatest of the human race." [2] It may be doubted. He had not notably lived in that conviction, and even if some of his modest protestation be discounted there seems no special reason why it should have welled up in him in that moment when of all moments a man must look his achievement in the eye and wonder whether it can be accounted more than winds whispering over desert sands. The truth was, of course, that he had never finished his work at all. It remained, beyond that one bare outline, a series of fragments. Certainly the broad theory seemed to have been established: *The Times* in a day or two would write of it as "one of the accepted generalizations of science. It is not too much to say that there is no man of real scientific eminence in Europe or America who does not hold to it in the main." [3] He had, it might be said, changed the outlook of educated opinion in a generation. That was much—or little. Everything—or nothing. All he could really say was that he had worked to his best ability. The rest lay in the hands of—God, Fate, the Eternal Indifference. He had laboured long, and now he was an old man, dying. Life, that had seemed endless, was over, no more than a brief flicker

---

[1] *Emma Darwin*, ii. 253.   [2] *LL. Romanes*, 131.
[3] April 21st, 1882.

of bright memories before the darkness. Wheels white foaming water. His sisters in their gay dresses laughing in the sunshine before the tide came in. Maer on summer evenings, windows softly aglow, Charlotte singing in the stillness, the moon rising yellow over lake and garden. Crisp early mornings on the moor. Fishing boats along the Forth. The quiet beauty of Cambridge in spring. The high cool peak of Teneriffe, seen with the eye of fancy, then in fact. Blue waters of the South Atlantic. The sting of the salt spray, the crying gulls. Vista upon vista—the bay at Rio, the dark Horn, the incredible Andes. The whole world a memory now, yet in its beauty, its power, its glory a sustaining magic. The brightness blurred, like the snowflakes beating soundless upon the windows of Macaw Cottage. A child crawled crowing across the carpet—William was it, Leonard, Horace ... Bernard? The Downe years had passed like a dream, one melting into another, measured by book or child. The boys growing out of the nursery, out of school, marrying, having children of their own. Some had died—Annie, poor Charles, little Mary Eleanor. Ras, Marianne, Susan, Catherine, Charlotte. He too, now, come to this favour. Only Emma, Emma, remained. . . .

He stirred, opened his eyes, murmured her name. Saw Etty, Bessy, Frank. They told him she was lying down, resting. He whispered that he was glad. They mustn't call her; he only wanted to know she was still there. They too, he said, were the best of dear nurses.

Later, she was with him again. They were all here, all who could get to Downe in time. He was quite conscious until nearly half-past three, when he grew suddenly worse, then sank peacefully into unconsciousness. Quarter of an hour later he was dead. It was almost four o'clock. Young Bernard, escaping from the strange, strained atmosphere indoors, was picking lords and ladies along the deserted sand-walk.

III

The world broke out in acclamation. Even *The Times*, surrendering past attacks upon him, had two days later four columns of dignified eulogy, setting him "beyond rivalry among the men of to-day, and side by side with two or three great discoverers of the past whose names are household words. *Punch* too saluted in judicious verse this "Recorder of the long descent of Man And a most living witness of his rise," and it quickly became clear that no less than Abbey burial could do him justice or his country honour. His achievement transcended any question of whether he was or not formally a Christian, though in fact on Sunday, April 23rd, two sermons had been preached in Westminster Abbey, and one in St. Paul's, making it quite clear that, despite earlier misunderstandings, God and the late "Professor" Darwin were at one on all essential points.

The funeral took place on the 26th, the coffin having lain the previous night in the small Chapel of St. Faith. Entrance was by black-edged invitation card through the Poets' Corner, and there was no admission except in mourning. It was a grand occasion, but not, Galton said, particularly touching. But then Abbey funerals never were, being too much like "the ceremonial of giving a University Degree"—which, of course, was just what it was. "The feeling promoted by the ceremony is *not* a solemn one but rather the sense of a national honour and glory." [1] The congregation, as *The Times* said, included "leaders of men and leaders of thought, political opponents, scientific co-workers, eminent discoverers, and practitioners of the arts." [2] France, Germany, Italy, Spain, Russia, the United States, and most of the major and many of the minor scientific societies of Great Britain, were officially represented. The pall-bearers were: Lubbock, Hooker, Huxley, Wallace, James Russell Lowell, Canon Farrar, an Earl, two Dukes, and the President of the Royal Society, thus merging, as Galton noted, all shades of opinion and rank. Herbert Spencer laid aside his sense of the absurdity of ecclesiastical ceremony to attend, and among other distinguished persons present were: Bates, Morley, Frederic Harrison, Max Müller, Carpenter, Ray Lankester, J. W. Judd, E. B. Tylor, John Murray, and—inevitably—Mr. Moncure Conway, lyrically envisioning "the throng of marble statues" as "a cloud of witnesses gathered to receive the newcomer in their Valhalla." [3] Thomas Hardy was also there, possibly the greatest figure among them and the most significant inheritor of Charles's legacy of thought, but naturally no one noticed a mere storyteller amid all these captains and kings of Truth.

---

[1] *LL. Galton*, ii. 198–9.   [2] April 27th, 1882.
[3] *Autobiography*, Conway, ii. 328.

There were so many Darwins, Wedgwoods, Allens, and other relatives present that they could not possibly all walk in the procession. The five sons and two daughters were there; Emma was not. The body was brought from the chapel by way of the cloisters, the choir preceding the wreath-hidden coffin. A sudden stir of anticipation, of sadness, swept over all as it appeared. Huxley, Hooker, and Lubbock were visibly moved. The words of the first anthem were: "Happy is the man that findeth wisdom, and the man that getteth understanding"; and beside the grave the choir sang: "His body is buried in peace, but his name liveth evermore." The noblest temple of the Church of England echoed the praises heavenward to assure the long-translated vicar of St. Chad's—where bells this evening rang a muffled peal—that his prayer of seventy-three years and seventy-four days earlier had not been entirely disregarded.

IV

William was chief mourner in the Abbey: Emma was chief mourner at home. For so long Charles had been the very centre of her life, more even than she of his. His work, his health, had been her first consideration, and of later years especially her days and her nights had been arranged to be with him whenever he wanted her. She always longed to travel, would have delighted in it, but denied her desire entirely for his sake. There had been and of course still were the children, but they were grown up now and living their own largely separate lives. Charles alone remained depending upon her as ever. He had been the beginning and he had been the end, and now the beginning and the end were gone. Vainly in the first hours after his death she tried to make little notes of the things she wanted always to keep fresh in her mind—his words of kindness, of thanks, their happy days together, their close solitary moments. She wished to forget the thought of his last few hours of suffering.

She stayed calm, keeping her grief for her loneliness, yet was grateful when Leonard's distress forced her to break down too and find relief in tears. It was better when immediate things were over, the funeral, the condolences of friends and, especially, neighbours, when she could retreat once more into the routine of regular daily living, fill her days with constant activity, reading, writing letters, playing patience or the piano, being with Bernard, perhaps her greatest consolation now. She presently became a little used to *not* having Charles at Down House, but it shocked her still to visit places she had last seen with him—she kept "making contrasts." She feared the loneliness of the winter too, and when the summer ended took a house in Cambridge to be near George and Horace, and where Frank too could carry on his botanical work. She could even enjoy the pure gossip of the dinner-parties, though something seemed missing, "a want of real interest."

That was characteristic. She went on living. She enjoyed life. Almost she seemed to take on a new lease of energy and interest. At Cambridge she entertained, made new friends, studied and discussed current affairs—with a novel Unionist inclination of her own—and back at Downe each summer got full pleasure from the sunshine, the garden, the flowers, looking after her poor women in the village, talking and writing with no diminution of her old liveliness. She read widely—a volume of fiction a day; lives of Maurice, Clough, Richard Jeffries, Severn, Dean Stanley; letters of Coleridge (a detestable being), Swift, De Quincey, Jane Carlyle (marvelling that her pen could be so "disloyal" to her husband); Rousseau's confessions, Greville's memoirs, Emerson's essays (but Transcendentalism was useless to Wedgwoods who had no intuitions!), Henry James, Stevenson, A. J. Balfour, Leslie Stephen, Morley, Milton, Voltaire.

Yet with all this activity something lacked. Life, she wrote within two months of Charles's death, was not flat to her, only "all at a lower pitch," and that in essence remained true to the end of her days. But it was an advantage, she added even more revealingly, no longer to grudge the passing years.

In 1885 she read Grant Allen's book on Charles, and thought it "prancing." Two years later appeared at last the *Life and Letters*, on which Frank had been busy ever since 1882. Emma was never happy about it, fearing the publicity, the revelation of intimacies, and asking for the excision of passages likely to give pain to more orthodox friends. But Frank's personal chapter on his father delighted her, and in the end she found too many pleasing "characteristic bits," even in the scientific letters, to wish the work undone.

Leonard had married in 1882, Frank again in 1883, George in

1884. In 1885 both Caroline and Harry Wedgwood died, in 1888 Frank Wedgwood, and in 1891 Hensleigh. Her friend from childhood, Ellen Tollet, died in 1890, and no one was left now to share her memories of Maer and Shrewsbury. The youngest of all Uncle Jos's children, she had outlived her generation.

Her own health, after 1889, was not good, and a growing deafness closed her the more within herself, but she remained mentally alert and sane, and even cheerful, to the end. That came unexpectedly, when she seemed recovering from an illness. On the evening of October 2nd, 1896, she wound up her watch as usual, sank back on her pillow, fell asleep, and never woke again. Her watch went on ticking long after her heart had ceased to beat.

She had been part of Charles's life as none of the children could ever be. With her death something was ended, something which had had its beginning close upon a hundred and forty years before, when an aching leg brought together in a little house at Burslem two young men who had no more notion than had "Doctor," placidly munching his oats outside, of all that was to result from their casual convergence.

BOOK FOUR

COMMENTARY

CHAPTER XX

THE FRAGMENTARY MAN

I

CHARLES DARWIN gave birth to Darwinism, a large and sometimes various conception, but implying primarily the evolution of living things, and secondarily the evolution of living things by more or less of natural selection. By bold hypothesis, by persistence and skill in the accumulating and ordering of more facts bearing on the subject than anyone had ever brought together before, by suggesting not only that it had taken place but also the means by which it might have taken place, and by the patience and honesty with which he sought to take every objection by the forelock, he established in less than a generation Evolution as accepted fact in the minds of men. Others had written of Evolution, and even of Natural Selection, before him; none had been able to achieve as he did, for he alone presented adequately what someone has called their biological credentials.

The effect was truly tremendous. Almost by the mere statement of a new principle of approach, dynamic not static, he revolutionized every department of study, from astronomy to history, from palaeontology to psychology, from embryology to religion. Dr. Charles Singer's remark that "the whole of modern biology has been called a commentary on *The Origin of Species*" has in essence a much wider application. The intellectual fermentation begun by Darwin's work has increased rather than diminished its activity with each subsequent decade. To-day it is almost impossible for us to return, even momentarily, to the pre-Darwinian atmosphere and attitude. The evolutionary conception is the oxygen in the air we breathe; it has woven itself into the very fibre of our thought; it shapes the world we see before our eyes, one in which permanent stability is unknown, in which life is change and change is life. Our interest is not in being but becoming, in process, and what is everywhere looked for is everywhere discovered.

There is good in this. To look no further, one great effect has been to charge the universe, but especially the world of organic life, with new meaning, displaying significances invisible before. In earlier days Darwin could consider an odd botanical specimen as "merely a case of unmeaning variability," and while it might be excessive to say that he eventually came to realize that no such thing as "unmeaning" variability existed, he did immensely reduce the area of the meaningless by his demonstrations that all kinds of characters, previously ascribable only to the idle whims of a Creator, were absolutely and essentially functional. In this as in other respects he offered to the world a large and profoundly satisfying extension of that law which redeems chaos to order and harmony.

II

As a scientific method adequately applied, Darwinism has worked well and fruitfully. As a social method, more loosely utilized, the results have been, perhaps, more dubious. Were not the consequences in some cases so potentially disastrous, it would be almost comic to see how Darwinism has been seized upon by all parties as a strong bulwark in defence of their contradictory preconceptions. On the one hand Nietzsche, on the other Marx, and between them most shades of Aristocracy, Democracy, Individualism, Socialism, Capitalism, Militarism, Materialism, and even Religion. (Marx is said to have thought so highly of the *Origin*, as giving him "a basis in natural science for the class struggle in history," that he wished to dedicate his *Capital* to Darwin—who for his part steadfastly denied connection with any form of Socialism.) Always you could find some aspect to serve your need, the Struggle for Existence were you a good Individualist, the moulding power of environment did you wish to display men victims of their circumstances. To-day Darwinism is part of the official creed of Communist Soviet Russia, while Sir Arthur Keith, equally on Darwinian grounds, lauds Capitalism and intra-specific competition as necessities of racial survival. You pay your money and, if you can find a pin of difference between them, you take your choice.

The disaster is that there is so little. Whatever aspect of Darwinism you choose to stress, Darwinism it broadly remains, with—in the popular view at least—Natural Selection, the Struggle for Existence, never very far from the foreground. Think Darwinianly, and you will act Darwinianly. What that means in human terms has been, in my opinion, most shrewdly analysed by Mr. Gerald Heard, passingly in several books but notably in *The Third Morality*.

Not idly, Mr. Heard would say, did Wallace term Darwin the Newton of Natural History. In the very deepest sense, Darwin was Newton's successor, as Freud his. And all three were the children of Galileo, who laid the foundation of modern science in divorcing primary from secondary characters and seeking to present reality in terms of mass and movement, of quantity isolated from quality. "Newton banished God from nature, Darwin banished him from life, Freud drove him from the last fastness, the soul. It was all latent in Newton, in Descartes, in Galileo: mechanism would conquer all, once it had conquered nature, for man's body was sprung from nature and his mind from his body." [1] Sedgwick in 1859 saw with strange insight the consequences of that process; what he failed to see was that he was already standing half-way down the slide, and that what was begun must perforce be ended. The link between material and moral, which he charged his old pupil with seeking to shatter, had been broken long before. His prophecy is fulfilled to-day with a strange and terrible fatality. "Humanity, in my mind, would suffer a damage that might brutalize it, and sink the human race into a lower grade of degradation than any into which it has fallen since its written records tell us of its history." Now, nearly eighty years later, every newspaper, any morning or evening, will cry out from its headlines what degree of brutality and degradation has come upon the world like a shadowing cloud. It is the type-activity of our civilization to prepare frantically in every continent and every country for the supreme brutality of modern war.

Some will question the connection between Darwinism and that. Mr. Heard makes it clear. "As we think, so in the end we must act." More than that, to act morally we must acknowledge a morality, and that is something with which the mechanist-

---

[1] *Third Morality*, Heard, 33.

materialism of Darwinism cannot adequately provide us. Morality depends on a more than personal purpose. Individual self-belief is not enough, nor is talk of Nation, Society, Mankind and the like, for what is any of these but one damned individual after another? Some faith there must be, however ill-defined, even indefinable, in a reality higher than material humanity. The anthropomorphic religions, all the way (and it is a long way) from Voodooism to Christianity at its best, cannot, whatever their power (that is, their beauty) as metaphor, serve us any longer as fact; we see them too plainly now in their static definition as but man casting his own huge shadow on a cloud. Yet while they lasted they at least could sustain an authentic morality. Mechanist-materialism could not, its only sanction, in denial of the supra-material, that of individual physical survival, the anarchy of the unabated struggle to sustain immediate existence. That is the condition of the world to-day—the struggle of individuals and of nations to survive at whatever price of brutality, since survival *is* the only ponderable value. It was one of Europe's unnoted historic moments when, at the conclusion of the Austro-Prussian War of 1866, a member of the Austrian Reichsrat opened an important speech: "The question we have first to consider is whether Charles Darwin is right or no."[1] Yet others had jumped swifter to the mark, for within six weeks of the issue of the *Origin* and the writing of Sedgwick's letter, Darwin told Lyell: "I have received, in a Manchester newspaper, rather a good squib, showing that I have proved 'might is right,' and therefore that Napoleon is right, and every cheating tradesman is also right."[2] It has been ever since the habit, if not of cheating tradesmen at least of our Napoleons, from Bernhardi to Mussolini, to justify themselves by Darwinian appeals and in Darwinian terms. The cook's intentions may have been of the highest, but the proof of the pudding, for those who must perforce sit at the supper-table, is in the eating.

III

Must we sit at it? Is Darwinism established, fixed, for ever? It may and must be questioned on more levels than one. There is the point of how rigidly or otherwise Darwinism itself is to be conceived. There is the problem of where Darwinism really

[1] *Long Life's Work*, Geikie, 129. [2] *LL.*, ii. 262.

stands to-day in the light of recent inquiry. There is the larger question of the adequacy of science to describe the universe objectively.

The essence of Darwinism, as popularly conceived, and acted upon, is the "gladiatorial" Struggle for Existence and, by Natural Selection, the Survival of the Fittest—of, that is, those who, by whatever means, do survive. Darwin himself was much more elastic in his view of these conceptions than most of his interpreters—one may even say than most of his interpretations. He said himself that he too often overlooked the extreme subtlety of the working of Natural Selection, and he certainly did that of his own definition of the Struggle for Existence, which he premised using "in a large and metaphorical sense, including dependence of one being on another, and including (which is more important) not only the life of the individual, but success in leaving progeny."[1] The effect of his treatment of the Struggle is infinitely more one of competition than co-operation, and his bald opposition to the alleviating consequences of birth control, even in human society ("over-multiplication was useful, since it caused a struggle for existence in which only the strongest and the ablest survived"), is extremely revealing of his instinctive attitude.[2] It might almost be said that if Neo-Darwinism is Darwinism without the nuances, so is Darwinism! Popular Darwinism may be a crude but it is scarcely an unfair or inaccurate presentation of the broad effect of Darwin's basic writings. The appeal to Darwin himself may open a window upon broader horizons, but scarcely one large enough for the prisoner to escape through.

The problem of the standing of Darwinism to-day is more complicated and more controversial. Dr. Singer sums up "the general position, after seventy years of research," thus: "Evolution is universally accepted as a general description of the history of organic forms. So far Darwin has completely conquered. When we come to the mechanism by which evolution acts, we have made little progress."[3] You may in fact, by accepting one learned "authority" against another, hold almost any belief you wish regarding the machinery of evolution. The one thing certain is that there is no thing certain. Every statement, every

[1] *Origin*, 62. [2] See also *Descent of Man*, i. 180; ii. 403-4.
[3] *Short History of Biology*, Singer, 323.

claim, is contradicted by others. Professor Berg, of Leningrad State University, lists ten fundamental Darwinian points, and asserts the opposite of every one of them. There is no aspect of the process of evolution which is not in question, whether Natural Selection, Sexual Selection, character of the Struggle for Existence, causes of Variation, inheritance of acquired characters, effects of use and disuse, influence of environment, even divergence and monogenesis. The pro-Lamarckists and the anti-Lamarckists, the Mechanists and the Vitalists, toss the balls of their controversies to and fro. A good time seems to be had by all, even though the spectator's inclination may be to come out by the same door wherein he went.

The main battle has raged round about Natural Selection, whose standard is on the whole discovered still to brave the breeze, though allowed less power (down to vanishing-point) by some than others. Sometimes it gains a little ground, sometimes it loses a little; at least no other hypothesis of comparable force has been set against it, though its proof, like its working, remains more negative than positive. A middle position may be stated in the words of J. B. S. Haldane: "It would seem that Natural Selection is the main cause of evolutionary change in species as a whole. But the actual steps by which individuals come to differ from their parents are due to causes other than selection, and in consequence evolution can only follow certain paths. These paths are determined by factors which we can only very dimly conjecture. Only a thorough-going study of variation will lighten our darkness." [1]

Sexual Selection also stands, a little blown upon, a little less important than Darwin seemed to think, but by no means without its place.

Variation was the great mystery in his day, and, as Haldane indicates, remains so still, despite the re-discovery of Mendel, and all the labours of Weismann, de Vries, Bateson, and many others. The gene and chromosome mechanisms (incidentally sweeping Pangenesis quite out of court) have almost as much complicated as illuminated the matter, and while we see further to-day, and while mutation has been established somewhat against Darwin's preference for "very short and slow steps," the causes of such sudden variations are still almost wholly

[1] *Causes of Evolution*, Haldane, 143.

mysterious apart from a few seemingly successful experiments with radium and X-rays and a hypothetical glance at cosmic rays. The effects of use and disuse and inheritance of other acquired characters, vital to Lamarckism and which Darwin took for granted, are asserted by some and regarded as dubiously by as many others. What seems largely agreed is that variation generally is much less miscellaneous, much more restricted, than Darwin believed, tending to occur not in all but only in certain directions, following definite though evasive laws, and the possibilities diminishing in mathematical retrogression as a species develops. The phenomenon of degeneration, "far commoner" than progress and less striking only because it tends to lead to the oblivion of extinction, is also much more widely appreciated than eighty or even fifty years ago.

It is in relation to the Struggle for Existence, particularly, that Darwin and those who came immediately after him are felt, despite his reservations, to have scanned their evidence too narrowly and jumped to conclusions too hastily. They overstressed the intensity of the struggle, fixing attention upon biological competition and disregarding biological co-operation. Seeing some organisms as predatory, they ascribed such motives to all. Noting the great tusks and talons given to species extinct and living, they supposed that it was by weight of these they held their places. They drew for humanity an ancestry in which the form of the ape was joined to the ferocity of the tiger. By tigerishness the ape had raised himself to man; by tigerishness would man raise himself to whatever higher form evolutionary progress held in store for him.

Thomson and Geddes, in their monumental work on *Life: Outlines of General Biology*, state the case against this view: "What has got into circulation is a caricature of Nature—an exaggeration of part of the truth. For while there is in wild Nature much stern sifting, great infantile and juvenile mortality, much redness of tooth and claw, and—even outside of parasitism—a general condemnation of the unlit lamp and the ungirt loin, there is much more. In face of limitations and difficulties, one organism intensifies competition, but another increases parental care; one sharpens its weapons, but another makes some experiment in mutual aid; one thickens its armour, but another triumphs by kin-sympathy. It is realized by few how much of the time and energy of living creatures is devoted to activities which are not to the advantage of the individual, but only to that of the race. Not that this is deliberate altruistic foresight, it is rather that in the course of Nature's tactics survival and success have rewarded not only the strong and self-assertive, but also—and yet more—the loving and self-forgetful. Especially among the finer types, part of the fitness of the survivors has been their capacity for self-sacrifice. . . . The fact is that the struggle for existence need not be competitive at all; it is illustrated not only by ruthless self-assertiveness, but also by all the endeavours of parents for offspring, of mate for mate, of kin for kin. The world is not only the abode of the strong, it is also the home of the loving." [1]

Kropotkin saw much of that fifty years ago, and since then realization has increased. It is no fundamental law of life that organisms should live at each other's expense, and competition between species is by no means on the same footing as that within species, for in the latter case, as has often been pointed out, the advantage of the individual may come to mean the disadvantage of the group. There is so much violence in the natural world only because there is so much failure. The angry ape and the contentious caveman no longer stand in the direct line of human ancestry; they in their violence (as perhaps we too in ours) are but the perishing offshoots of a better because more sensitive and adaptable ancestor. Adaptability, awareness, intelligence are the faculties by which the light small-boned but better-brained tree-shrew, tarsier, gibbonoid being and their descendants have survived while the great armour-plated beasts and the "gorilla-like" dawn-men have gone their way to extinction. The more highly specialized the creature the less readily he can make the necessary adaptation to changing circumstance. Resort to violence may be the final paroxysm of the dying species, and heavy armament has again and again shown itself the last defence of the beast already marked for extinction—a sweet thought for citizens of this modern world to wake to in the chillier small hours.

Natural selection still functions, but changing its ways with the ways of the creature selected. Intelligent birth-control is as much a factor for survival as co-operation. The brute looks

[1] *Op. cit.*, 1317.

backward and will perish with the brutes. Modern war, even Sir Arthur Keith might be expected to perceive, is less a pruning-hook than a universal and quite undiscriminating lethal chamber.

The further question of the relation of science to reality must be dealt with more briefly, indicated only. One answer has already been hinted above in the suggestion that science from Galileo and Newton forward has been based essentially upon abstraction. Spengler has commented: "How superficial, how glib, how mechanistic the conception with which the science of the Darwinian age worked. In the first place, this conception groups an aggregate of such grossly palpable characters as are observable in the anatomy of the discoveries—that is, characters that even a corpse displays. As to observing the body qua living thing, there is no question of it. Secondly, it investigates only those signs which very little perspicacity is needed to detect, and investigates them only in so far as they are measurable and countable. The microscope and not the pulse-sense determines." [1]

Even more widely it is incumbent upon us to realize the subjective, selective nature of perception and still more of all statement whatsoever, scientific perception and statement as much as any. Herman Melville somewhere remarks that all classifications, all systems of thought, are arbitrary, have a conventional basis; change the convention and you change your conclusions, which may yet have equal validity. Even the seemingly solid world we see about us, Gerald Heard would say, is in large degree not objective reality but subjective creation, shaped for us mainly by our dominant emotions (too often those of greed and fear) out of an infinitely wider range of ultimately incalculable being—"a construction made by the highly selective capacity of the partial objective mind." [2]

The subjective elements in the most careful scientific theory have accordingly to be looked for, and in fact were frankly recognized in Darwin's views nearly thirty years ago by even such relatively orthodox writers as J. A. Thomson and Patrick Geddes. In a semi-official Darwin centenary volume (1909) we find Thomson quoting Geddes as follows: "The substitution of Darwin for Paley as the chief interpreter of the order of nature is currently regarded as the displacement of an anthropomorphic

---

[1] *Decline of the West*, Spengler, ii. 124-5.
[2] *Source of Civilization*, Heard, 318.

view by a purely scientific one: a little reflection, however, will show that what has actually happened has been merely the replacement of the anthropomorphism of the eighteenth century by that of the nineteenth. For the place vacated by Paley's theological and metaphysical explanation has simply been occupied by that suggested to Darwin and Wallace by Malthus in terms of the prevalent severity of industrial competition, and those phenomena of the struggle for existence which the light of contemporary economic theory has enabled us to discern, have thus come to be temporarily exalted into a complete explanation of organic process."[1] A year or two later the same writers recurred to the theme in an elementary volume on Evolution.

IV

It has been pointed out by Thomson and Geddes that such derivation does not necessarily invalidate a theory, but it surely does suggest a dependence of the latter's validity upon the individual conceiving and authenticating it. In that belief this study of Charles Darwin was partly though by no means solely carried out and written. Disinterestedness is the condition of all truly creative achievement, in whatever field. There must be, to the utmost degree of one's capacity, total submission to the object: no bias, no self-assertion, or the consequence will be, as always in politics, as too often elsewhere, vision cancelled by blindness, positive balanced by negative, and the end equal to the beginning. But only the complete man, fully conscious, aware of himself on every plane, can achieve such disinterestedness. It is never achieved, for the complete man—unless in such figures as Jesus and the Buddha, whom the world calls divine—has never existed. Nevertheless only as man approaches to such completeness does he become capable of conceiving and establishing a faith, a philosophy, an outlook by which men may live in fullness of being.

Was Darwin disinterested? Unbiassed? Two outstanding facts make one question it. First, the driving force of all his imagination's waking life was his "intolerable desire not to be utterly baffled," to find for every phenomenon capturing his attention some satisfying explanation. He found facts "dull"

[1] q. *Darwin and Modern Science*, 15.

without a theory to connect them, he said, and the discovery of even a tentative explanation was "an immense relief." This need "led, or rather forced" him to adopt his quite fabulous Pangenesis theory, and to determine to hold to it till something better took its place. It distressed him acutely to turn his back even on factitious order and yield himself again to chaos. Second, it is very important to realize that with all the credit showered upon him for so patiently collecting facts over twenty years without definite speculation, without, as Galton said, "allowing his imagination to play on them," the truth was that over all that period, from 1838 at latest forward, he had his theory practically complete, full-formed in his mind save for the relatively minor matter of divergence. Even Sexual Selection and Pangenesis were very early conceived. In each case the theory preceded the *main* business of collecting facts which either confirmed it or else unconsciously became "difficulties" to be considered and countered. He deserves all credit for his patience and his honesty in that process, but it is absurd to suppose that he had, with his strong desire to establish order, no determining leanings to this side or that. Alexander Agassiz, on reading the *Life and Letters*, wrote to Murray: "I was somewhat surprised in Darwin's Life to see the element of wishing his cause to succeed as a cause brought out so prominently. The one thing always claimed by Darwin's friends had been his absolute impartiality to his own case. Certainly his correspondence with Hooker, Huxley, and Gray shows no such thing."[1] This applies especially to the period of the writing of the *Origin*, when, except in his moments of excessive modesty, his anxiety for his cause is unabashed, while the closing pages of the work itself have, as Houston Peterson says, "a veritable religious glow." Despite the common view, there are many elements in Darwin's life and work to which Kretschmer's account of "the paranoiac thinker" is by no means wholly inapplicable: "The paranoiac thinker is usually a man of tenacious and deep emotionality who, through some acute experience, is forced into a definite line of thought. He then pursues the line of thought relentlessly and with the greatest consistency, so that his spiritual life becomes more and more tyrannically and one-sidedly controlled by it. With the fixation of such an over-valued idea in the emotional life of the individual

[1] *Letters of A. Agassiz*, 228.

there generally enters a systematized delusion, *i.e.* an increased power of combining impressions, extending even to the smallest and most negligible matters of daily life, into a support for the original belief. Ultimately the controlling system of thought develops a whole array of tributary ideas, whilst everything that is no use to it is shut out of consciousness and overlooked with complete, passionate blindness. Such over-valued ideas . . . can bring those who entertain them into the deepest misfortune or the lunatic asylum, or drive them to the greatest deeds, according to the favourableness of circumstances and the intellectual power of the individual." [1]

There is no need to discuss and differentiate between emotional and imaginative; in this context the two are one, and Darwin's "acute experience" is clearly the voyage of the *Beagle*, the period in which he not only first imaginatively awakened but also discovered in the pursuit of natural history the possibility of reconciling his own interests with his father's hope and demand that he become in some kind a useful citizen. The "line of thought" followed fast, to be pursued both relentlessly and consistently till not only his spiritual but his physical life became completely and certainly one-sidedly controlled by it. The "increased power of combining impressions" and the development of many "tributary ideas" is very evident, as also the growing blindness to all irrelevant issues.

V

One paradox of Darwin's life is that while he lived so generally secluded a life, isolated always at least from the industrial struggle typical of the time, he yet stated a theory not to be detached from and singularly expressing the spirit of the century. In the machine age he established a mechanical conception of organic life. He paralleled the human struggle with a natural struggle. In an acquisitive hereditary society he stated acquisition and inheritance as the primary means of survival. He participated actively in none of these things.

The explanation is simple. Cut off though he might be from his times, he lived mentally entirely within them. His whole life in fact was really extraordinarily limited. He was, to extreme

[1] *Psychology of Men of Genius*, Kretschmer, 138.

degree, a family man, a Darwin and a Wedgwood. Both lines were profoundly middle-class, rising by prudence and steady worth from yeoman to gentlemanly rank, their typical members sturdy individualists, men of serious, even solemn, turn of mind, independent in opinion, not disregardful of culture and spiritual matters, but intent on making their ways in the world against whatever odds. They exhibited a plodding persistence, triumphing by patience and application over illness and difficulty. Even Erasmus, the most original of them all (and to whom Charles owed most intellectually if not physically), was very much a man of the new age then dawning. Darwin grew up essentially accepting. He had as boy and young man little initiative or ability to grasp principles, and this cannot be set down wholly to unfortunate teaching, for other men fared worse and did better. Simply, he had no intellectual avidity. He absorbed the values of his environment as he absorbed the basic ideas of the *Zoonomia*, unconsciously. His collecting, if persistent, was much a matter of fancy, of rivalry with others, and though his Plinian Society activities have been cited as exhibiting intellectual precocity, nothing more has been demonstrated than a very moderate natural capacity and—again—inclination to adapt himself to any tolerably congenial atmosphere. Through Edinburgh and Cambridge he continued imaginatively sleeping. Family interest in natural history, boyish collecting, the *Natural History of Selborne*, the *Wonders of the World*, the *Zoonomia*, sailing on the Firth of Forth, Grant, the Plinian Society, Fox, beetle-sticking, Henslow, Paley–Herschel–Humboldt, a little of Sedgwick—these represented the line of his development, but they were not yet added together, no balance struck.

Had he not sailed on the *Beagle*, he might have gone on sleeping —finished his theological studies and entered the Church. He would doubtless have continued his entomology, perhaps even been drawn into definitely scientific study. But never in his "quiet parsonage" could he have had thrust on his virgin mind as in South America those vast and varied masses of material whose impact *forced* him, as he said, into comparison and generalization and ever more venturesome hypothesis . . . along lines already absorbed and forgotten, dormant in his consciousness. Here he saw the range and wonder and beauty and savagery and intricacy and illimitable fertility of natural life, its variety and similarity, its adaptability and its persistence. Here he learned to think for himself, the pleasure, the satisfaction, of such thinking. He sailed, psychologically, a boy; he came home a man, holding the confidence of others, and with work done and work to do.

Yet, with all that change, little had been added save impulse and awakening. He brought back ideas, curiosities, living in his mind, but they had been there in essence before he went away. Evolution by transmutation of species, variation, the struggle for existence, sexual selection, survival of the fitter, the benefits of cross-fertilization, the oneness of organic life and man's relation to the monkey—Erasmus had declared them all. South America had illustrated them all, made them vivid, real, but Erasmus had not found them in South America, nor Malthus his views on population. And when Darwin himself came to test them nearer home he could not but perceive their application. Erasmus of Lichfield, Robert of Shrewsbury, Josiah of Etruria, all these had struggled for existence and, as the fittest, survived and begotten . . . himself. Even he was only a little removed, watching his investments with anxiety. The process was in his blood, and he looked out at it through a glass window.

He shared its crudest prejudices. "That such a great man should come to visit me." He believed as absolutely as any Kipling in the Anglo-Saxon's Burden, rejecting birth-control not only for its wider effect upon the struggle for existence but also because had it "been in action during the last two or three centuries, or even for a shorter time, in Britain, what a difference it would have made to the world, when we consider America, Australia, New Zealand, and South Africa! No words can exaggerate the importance, in my opinion, of our colonization for the future history of the world."[1] He also feared, in this connection, the "great danger of extreme profligacy amongst unmarried women"[2]—without a word, however, concerning male participation.

On all such issues, indeed on everything outside his own work, he spoke and thought with the perfect simplicity of a leading article in a respectable Victorian newspaper. He was a specialist getting on with his job, in a sociological, political, and religious vacuum. It meant great achievement on one side, and on the other an increasing desiccation, till he ceased to have any vital

[1] *Scientific Meliorism*, Clapperton, 340-1. [2] *Ibid.*, 341.

life outside his work at all, was just a nice, a very nice, old man.

Werner Sombart has written of "the fragmentary man" who is the type of the capitalist executive, seeing all in acquisitive terms, subordinating the whole to the part, making the quantitative aspect his total consideration till "all else within him dries up" and "everything about him becomes a wilderness, all life dies, all values disappear."[1] The resemblance of this fragmentary man to Darwin is evident. The business man lives for the acquisition of wealth; the Darwinian scientist for the acquisition of knowledge—each irrespective of human consequence. The scientist may be thought the better man. Nevertheless he in this case produced the perfect philosophy for the business man, and both are alike in recognizing, in the last resort, material survival as the only ultimately effective value.

The fragmentary man can only manifest a fragmentary truth. Darwin was incomplete, and Darwinism accordingly inadequate as a philosophy by which men may live. It was releasing, but only as Lincoln released the American negro slaves—from one prison into another, and in effect in each case the same, that of naked industrialism, an unrestrained materialism. That is the inevitable fate of every generalization. Its truth first releases; finally its falsity imprisons. Darwin delivered his generation from the Egypt of an oppressive anthropomorphic theology. Men did not understand in the moment of their liberation that a mechanomorphic materialism might prove even more confining. What they need to realize is the transitory nature of every vision of the universe: scientific, poetic, prophetic—all are symbolic, not real, and reality alone endures. It is ill to make, as Huxley said, the heresies of one generation the superstitions of another.

VI

With all his limitations, however, Charles Darwin was nevertheless a great man—perhaps because of his limitations, the representative figure as well as the greatest naturalist of the century his life so centrally spanned. None lesser could have changed the minds of men as he did. That, for good or ill, was an amaz-

[1] *Quintessence of Capitalism*, Sombart, 351.

ing achievement, beside which his more specifically botanical, geological and zoological labour, most of it still standing firm to-day, and nearly any item of it enough, it has been said, to make a lesser man's lasting reputation, looks almost trivial.

His supreme service to mankind may be a different matter, and many think it so. Henry Fairfield Osborn would state it "that he won for man absolute freedom in the study of the laws of nature." Henshaw Ward finds it in his teaching that "nature is utterly beyond our thinking" and that we are foolish to seek to enclose it within the narrow confines of our childish logic, a rather bold seizing upon Darwin's more private than public doubts "whether the convictions of man's mind, which has been developed from the mind of the lower animals, are of any value or at all trustworthy." [1]

A third alternative is his devotion to the fullest truth he could conceive. That, in effect his honesty, these pages have never meant to question. His bias was outside his consciousness, his flaw only that his consciousness was not broader. As an individual, his integrity was as complete as one is entitled to expect in any ordinary human being; he displayed it in that open-mindedness, his care, his patience, his attention to many sides of each problem, those reservations he was so careful to make even when their application so frequently evaded him. To-day, we need such integrity more than ever, both as becoming the more necessary as we recognize our only knowledge of the universe as coming ultimately from the subjective vision, built if not of our open interest and free curiosity than of our greeds and fears, and also because not for many years has truth, over the greater part of the world, been so stifled and disregarded and denied as to-day in the intensifying bitterness of the Darwinian struggle between individuals and nations.

I must confess to a very great liking for Charles Darwin the man, in himself, in his life, in his family circle, and even in his labours detached from their consequences; and finally it is to the man rather than to his scientific achievement as a thing in itself that I would look for his enduring value.

> If Luther's day expand to Darwin's year,
> Shall that exclude the hope—foreclose the fear?

---

[1] *LL.*, i. 316.

Not, one is forced to say, an atom.

> Then keep thy heart, though yet but ill resigned—
> Clarel, thy heart, the issues there but mind.[1]

The basic human problems are eternal, and the explications of science touch them but superficially and incompletely; the true solutions must be sought always in men's hearts, in an individual bravery, a personal courage, whether of the body, the mind, or the spirit.

[1] *Epilogue to "Clarel,"* Melville.

BIBLIOGRAPHY

OF WORKS QUOTED OR OTHERWISE UTILIZED

(The books and some other writings of Charles Darwin are listed first; then books, pamphlets and articles dealing specifically with him; finally other miscellaneous works. The order of the first two sections is chronological; that of the third alphabetical, some of the items, however, being placed under subject rather than author, all books on Cambridge or Erasmus Darwin, for example, appearing under C or D respectively.)

I

Letters to Professor Henslow. 1835. (In her *Beagle Diary* Bibliography, Mrs. Nora Barlow refers to Darwin's and FitzRoy's letter on the "Moral State of Tahiti," printed September 1836—see below—as "the first contribution to current periodical literature bearing Darwin's signature." It seems to have escaped attention that a series of twelve short extracts from this Henslow pamphlet were printed in the *Entomological Magazine* for April 1836, pp. 457-60.)

Narrative of the Surveying Voyages of Her Majesty's Ships *Adventure* and *Beagle* between the years 1826 and 1836, describing their examination of the Southern shores of South America, and the *Beagle*'s circumnavigation of the globe. Vol. III: *Journal and Remarks*, 1832-1836. 1839. (Issues of this third volume as a separate work appeared in 1839 and 1840, and the popular revised edition in John Murray's Home and Colonial Library in 1845, as: *Journal of Researches into the Natural History and Geology of the countries visited during the Voyage of H.M.S. "Beagle" round the world, under the command of Capt. FitzRoy, R.N.* Here again a point of minor bibliographical interest appears not to have been noticed—the first issue of the 1845 edition in three separate, presumably paper-covered parts at half-a-crown each. When Darwin referred to these parts in letters to Lyell—see the *Life and Letters*, Vol. I, pp. 337, 338, 339—it was supposed by Francis Darwin that he meant proof-sheets, but advertisements in the *Athenaeum* for 1845, at pp. 651, 754, 862, and 1004, make it clear that the parts appeared approximately on July 5th, August 2nd, and August 30th as numbers xxii to xxiv of the Library, and then as one bound volume, at seven-and-sixpence, in mid-October. The parts consisted of pp. 1-176 (Chapters I-VIII), 177-336 (IX-XV), and 337-520 (XVI-XXI). Rather oddly, while Murray advertised the series as the Home and Colonial Library, the final volume at least was stamped Colonial and Home Library. The 1860 edition of this work, which has a new postscript by the author, seems only less rare than the parts-issue; neither the British Museum Reading Room nor the Bodleian Library has a copy.)

Zoology of the Voyage of H.M.S. "Beagle." Edited and superintended by Charles Darwin. Part I: *Fossil Mammalia*, by Richard Owen. Geological Introduction by Charles Darwin. 1840. Part II: *Recent Mammalia*, by George R. Waterhouse. With a notice of their habits and ranges, by Charles Darwin. 1839. Part III: *Birds*, by John Gould and G. R. Gray. 1841. Part IV: *Fish*, by Leonard Jenyns. 1842. Part V: *Reptiles*, by Thomas Bell. 1843.

The Structure and Distribution of Coral Reefs. Being the First Part of the Geology of the Voyage of the *Beagle*. 1842. Second edition, 1874.

Geological Observations on the Volcanic Islands, visited during the Voyage of H.M.S. *Beagle*. Being the Second Part of the Geology of the Voyage of the *Beagle*. 1844. Second edition, with the *South America*, 1876.

Geological Observations on South America. Being the Third Part of the Geology of the Voyage of the *Beagle*. 1846. Second edition, with the *Volcanic Islands*, 1876.

A Monograph on the Fossil Lepadidae; or, Pedunculated Cirripedes of Great Britain. 1851.

A Monograph on the sub-class Cirripedia, with Figures of all the Species. The Lepadidae; or, Pedunculated Cirripedes. 1851.

A Monograph on the sub-class Cirripedia, with Figures of all the Species. The Balanidae (or Sessile Cirripedes); The Verrucidae, etc. 1854.

A Monograph on the Fossil Balanidae and Verrucidae of Great Britain. 1854.

On the Origin of Species by means of Natural Selection, or the Preservation of Favoured Races in the Struggle for Life. 1859. Revised editions 1860, 1861, 1866, 1869, 1872. (Also World's Classics reprint of 1860 edition, with comparison of first, second and sixth editions by Irene Manton. 1929.)

On the Various Contrivances by which Orchids are Fertilized by Insects. 1862. Second edition, 1877.

The Movements and Habits of Climbing Plants. 1865. (Appeared as part of Vol. IX of the *Journal* of the Linnean Society, but was also issued separately and sold by the Society and two other publishers.) Second enlarged edition, 1875.

The Variation of Animals and Plants under Domestication. Two volumes. 1868. Second revised edition, 1875.

The Descent of Man, and selection in relation to Sex. Two volumes. 1871. Second revised edition, 1874.

The Expression of the Emotions in Man and Animals. 1872.

Insectivorous Plants. 1875.
The Effects of Cross and Self Fertilization in the Vegetable Kingdom. 1876. Second edition, 1878.
The Different Forms of Flowers on Plants of the Same Species. 1877. Second edition, 1880.
The Power of Movement in Plants. 1880.
The Formation of Vegetable Mould, through the action of Worms, with observations on their habits. 1881.
The Life and Letters of Charles Darwin, including an autobiographical chapter. Edited by his son, Francis Darwin. Three volumes. 1887. Abridged edition in one volume, with some new material, issued 1892 as *Life of Charles Darwin.*
More Letters of Charles Darwin: a record of his work in a series of hitherto unpublished letters. Edited by Francis Darwin and A. C. Seward. Two volumes. 1903.
The Foundations of the Origin of Species: Two Essays written in 1842 and 1844 by Charles Darwin. Edited by his son, Francis Darwin. 1909.
Charles Darwin's Diary of the Voyage of H.M.S. "Beagle." Edited from the MS. by Nora Barlow. 1933.

A Manual of Scientific Enquiry; prepared for the use of Her Majesty's Navy; and adapted for travellers in General. Edited by Sir John Herschel. 1849. "Geology," by Charles Darwin, pp. 156-195.
Memoir of the Rev. John Stevens Henslow. By Leonard Jenyns. Chapter III has recollections by Darwin, pp. 51-55. 1862.
Erasmus Darwin. By Ernst Krause. Translated from the German by W. S. Dallas. With a preliminary notice by Charles Darwin (pp. 1-127). 1879.
Mental Evolution in Animals. By G. J. Romanes. With a posthumous essay on "Instinct" by Charles Darwin (pp. 355-384). (This would appear to be the only case of part of the original manuscript of the Great Work being put into print; that is, in fact, save for the specialist, its principal interest.) 1883.

"A letter containing remarks on the Moral State of Tahiti, New Zealand, etc." *South African Christian Recorder*, September 1836, pp. 221-238. (Signed by both FitzRoy and Darwin, and containing several paragraphs from the latter's *Beagle* diary.)
"Notes upon the Rhea Americana." *Zoological Soc. Proc.*, v., 1837, pp. 35-36.
"Geological Notes made during a survey of the East and West Coasts of South America in the years 1832-1835, etc." *Geological Society Proceedings*, ii., 1838, pp. 210-212.
"Observations of proofs of recent elevation on the coast of Chile, etc." *Ibid.*, ii., 1838, pp. 446-449.
"A sketch of the deposits containing extinct Mammalia in the neighbourhood of the Plata." *Ibid.*, ii., 1838, pp. 542-544.

"On certain areas of elevation and subsidence in the Pacific and Indian oceans, as deduced from the study of coral formations." *Ibid.*, ii., 1838, pp. 552-554.

"On the Formation of Mould." *Ibid.*, ii., 1838, pp. 574-576.

"On the Connexion of certain Volcanic Phenomena and on the Formation of Mountain-Chains and the effects of Continental Elevations." *Ibid.*, ii., 1838, pp. 601-632.

"Observations on the Parallel Roads of Glen Roy, etc." *Philosophical Transactions*, 1839, pp. 39-82.

"On the Tendency of Species to form Varieties; and on the Perpetuation of Varieties and Species by Natural Means of Selection." By Charles Darwin and Alfred Wallace. Communicated by Sir Charles Lyell and J. D. Hooker. *Journal* of Linnean Society, Zoology, iii., 1859, pp. 45-62.

"A Biographical Sketch of an Infant." *Mind*, July 1877, pp. 285-294.

II

Charles Darwin: Memorial Notices reprinted from "Nature." 1882.

Addresses delivered on the occasion of the Darwin Memorial Meeting of the Biological Society of Washington, on May 12th, 1882. 1882.

The Life and Works of Charles Darwin. A lecture by L. C. Miall. 1883. (Mentions in passing a forthcoming *Life of Darwin* by Miss Arabella Buckley, once secretary to Lyell and author of several other works.)

The Religious Views of Charles Darwin. By E. B. Aveling. 1883.

"Charles Darwin": a paper contributed to the *Transactions of the Shropshire Archaeological Society*. By Edward Woodall. 1884.

"A Reminiscence of Mr. Darwin." Anonymous. *Harper's Magazine*, October 1884.

Charles Darwin. By Grant Allen. 1885.

Charles Darwin, Naturalist. By J. T. Cunningham. 1886.

Life of Charles Darwin. By G. T. Bettany. 1887.

Pedigree of the Family of Darwin. Compiled by H. Farnham Burke. 1888.

Charles Darwin: his Life and Work. By Charles Frederick Holder. 1891.

Darwin: his Work and Influence. By E. A. Parkyn. 1894.

The Ill-Health of Charles Darwin: its Nature and its Relation to his Work. By W. W. Johnston. 1901.

The Darwin-Wallace Celebration held on July 1st, 1908, by the Linnean Society of London. 1908.

Catalogue of the Library of Charles Darwin. Compiled by H. W. Rutherford. Introduction by Francis Darwin. 1908.

Order of the Proceedings at the Darwin Celebration held at Cambridge, June 22nd-24th, 1909. With a Sketch of Darwin's Life. 1909.

Darwin Celebration, Cambridge, June 1909. Speeches delivered at the Banquet held on June 23rd. 1909.

Memorials of Charles Darwin: a collection of manuscripts, etc., to commemorate the centenary of his birth, etc. By W. G. Ridewood. 1909.
Charles Darwin and the Origin of Species. By Edward B. Poulton. 1909. (Contains new letters, etc.)
Darwin and Modern Science: Essays in commemoration of the centenary, etc. Edited by A. C. Seward. 1909.
Charles Darwin as Geologist. By Sir Archibald Geikie. 1909.
"Charles Darwin's Earliest Doubts Concerning the Immutability of Species." By J. W. Judd. *Nature*, November 2nd, 1911.
Charles Darwin. By Leonard Huxley. 1921.
Charles Darwin. By Karl Pearson. 1923.
Darwin. By Gamaliel Bradford. 1926.
The Evolution of Charles Darwin. By G. A. Dorsey. 1927.
Charles Darwin: The Man and his Warfare. By Henshaw Ward. 1927.
"Charles Robert Darwin." By G. P. Wells. Essay in *The Great Victorians*, pp. 151-60. 1932.
Darwin. By R. W. G. Hingston. 1934.
"Charles Darwin as a Student in Edinburgh, 1825-7." By J. H. Ashworth. *Nature*, December 28th, 1935, pp. 1011-1014.

III

Letters and Recollections of Alexander Agassiz. Edited by G. R. Agassiz. 1913.
Louis Agassiz: his Life and Correspondence. Edited by E. Cary Agassiz. 1885.
Annual Register. Various dates.
Athenaeum. Various dates.
John James Audubon: his Life and Adventures. Edited by Robert Buchanan. 1869.
Memoirs, Journal, etc., of Charles Cardale Babbington. 1897.
Nomogenesis: or Evolution Determined by Law. By Leo S. Berg. 1926.
Autobiographical Sketches. By Annie Besant. 1885.
Matthew Boulton. By H. W. Dickinson. 1937.
Galatea: or, The Future of Darwinism. By W. Russell Brain. 1925.
Life, Letters, etc., of Sir Charles J. F. Bunbury. Edited by his wife, F. J. Bunbury. Three volumes. 1894.
Life and Letters of Samuel Butler, D.D. By his grandson, Samuel Butler. Two volumes. 1896.
Dr. Butler of Shrewsbury School. By William Heitland. 1897.
Life and Habit. By Samuel Butler. 1878.
Evolution Old and New. By Samuel Butler. 1879.
Unconscious Memory. By Samuel Butler. 1880.
Luck, or Cunning. By Samuel Butler. 1887.
Charles Darwin and Samuel Butler: a step towards reconciliation. By Henry Festing Jones. 1911.
The Notebooks of Samuel Butler. Edited by H. F. Jones. 1912.

Samuel Butler: a Memoir. By H. F. Jones. Two volumes. 1919.
Further Extracts from the Notebooks of Samuel Butler. Edited by A. T. Bartholomew. 1934.
Letters between Samuel Butler and Miss E. M. Savage, 1871-1885. Edited by G. Keynes and B. Hill. 1935.
"The Centenary of Samuel Butler." By St. John Ervine. *Fortnightly Review,* December, 1935.
The Cambridge University Calendar, 1822-1837.
The University of Cambridge. By D. B. Walsh. 1837.
Biographical Register of Christ's College, Cambridge, 1505-1905. Compiled by John Peile. Two volumes. 1910-1913.
The Cambridge Undergraduate a Hundred Years Ago. By Oskar Teichman. 1926.
"Reminiscences of Christ's College." By E. T. V. *Christ's College Magazine,* Michaelmas Term, 1893.
New Letters, etc., of Jane Welsh Carlyle. Edited by Alexander Carlyle. Two volumes. 1903.
New Letters of Thomas Carlyle. Edited by Alexander Carlyle. Two volumes. 1904.
Life of Thomas Carlyle. By David Alec Wilson. Six volumes. 1923-1934.
Vestiges of the Natural History of Creation. By Robert Chambers. 1844.
Life of Frances Power Cobbe, as told by herself. 1904.
Biography of the late John Coldstream. By J. H. Balfour. 1865.
Autobiography, etc., of Moncure Daniel Conway. Two volumes. 1904.
Life of James Dwight Dana. By Daniel C. Gilman. 1899.
Green Memories. By Bernard Darwin. 1928.
Experiments Establishing a Criterion between Mucaginous and Purulent Matter, and An Account of the Retrograde Motions of the Absorbent Vessels of Animal Bodies in Some Diseases. (By Charles Darwin, son of Erasmus of Lichfield.) With a Life of the author. (By Erasmus Darwin.) 1780.
Emma Darwin, Wife of Charles Darwin: a Century of Family Letters. By her daughter, H. E. Litchfield. Two volumes. Privately printed. 1904. Subsequently published, with some minor omissions and amendments, as *Emma Darwin: a Century of Family Letters,* etc. Two volumes, 1915. (All my references to this work, with one exception duly noted, are to the 1915 edition.)
The System of Vegetables. Translated from Linneus (partly by Erasmus Darwin) and published by the Lichfield Botanical Society. 1781.
The Families of Vegetables. Translated from Linneus, etc., as previous item. 1787.
The Loves of the Plants. By Erasmus Darwin. 1789.
The Economy of Vegetation. By Erasmus Darwin. 1791.
Zoonomia. By Erasmus Darwin. 1794-1796. Two volumes. Third edition. Four volumes. 1801.
A Plan for the Conduct of Female Education in Boarding Schools. By Erasmus Darwin. 1797.

BIBLIOGRAPHY

Phytologia: or the Philosophy of Agriculture and Gardening. By Erasmus Darwin. 1800.
The Temple of Nature. By Erasmus Darwin. 1804.
Memoirs of the Life of Dr. Darwin. By Anna Seward. 1804.
Erasmus Darwin: Philosopher, Poet, and Physician. By John Dowson. 1861.
Dr. Darwin. By Hesketh Pearson. 1931.
The Poetry and Aesthetics of Erasmus Darwin. By James V. Logan. 1936.
The MSS. of Francis Darwin, Esq. of Creskeld, Co. York. Hist. MSS. Commission, 11th Report, Appendix, Part VII. 1888.
Scientific Papers. By Sir George Howard Darwin. Vol. V, containing biographical memoirs by Francis Darwin, etc. 1916.
Principia Botanica. By R. W. Darwin (brother of Erasmus of Lichfield). 1787.
Appeal to the Faculty, concerning the case of Mrs. Houlston. By R. W. Darwin (father of Charles Darwin). 1789.
Abstract of Papers communicated to the Royal Society of London, 1843-1850. (Obituary of R. W. Darwin, p. 883.) 1851.
The Mount, Shrewsbury: Sale by Auction, Nov. 19th-24th, 1866, of Furniture, etc., by the order of the executors of the late Miss Darwin. 1866.
Remarks on the Final Causes of the Sexuality of Plants, with particular reference to Mr. Darwin's work on the origin of species. By C. G. B. Daubeny. 1860.
Dictionary of National Biography.
Notes from a Diary, 1851-72. By Sir M. E. G. Duff. Two volumes. 1897.
Edinburgh University: a Sketch of its Life for 300 years. 1884.
King Edward VII. By Sir Sidney Lee. Two volumes. 1925.
Evolution in the Light of Modern Knowledge: a Collective Work. 1925.
Palaeontological Memoirs, etc., of Hugh Falconer. Edited by Charles Murchison. Two volumes. 1868.
Narrative of the Surveying Voyages, etc. Edited and in part written by Robert FitzRoy. Four volumes. 1839.
Captain FitzRoy's Statement. 1841.
The Conduct of Capt. R. FitzRoy. By William Sheppard. 1842.
Remarks on New Zealand. By Robert FitzRoy. 1846.
"FitzRoy and Darwin, 1831-36." By Francis Darwin. *Nature,* Feb. 22nd, 1912.
"Robert FitzRoy and Charles Darwin." By Nora Barlow. *Cornhill,* April 1932.
Life, Letters, etc., of Francis Galton. By Karl Pearson. Four volumes. 1914-1930.
Geddes, Patrick. See J. A. Thomson.
A Long Life's Work. By Sir Archibald Geikie. 1924.
Gentleman's Magazine. Various dates.
Life of Gladstone. By John Morley. Three volumes. 1903.

Biographic Clinics: the Origin of the Ill-Health of . . . Darwin (and others). By George M. Gould. 1903.
Darwiniana. By Asa Gray. 1876.
Letters of Asa Gray. Edited by Jane Loring Gray. Two volumes. 1893.
Memory's Harkback. By F. E. Gretton. 1889.
The Causes of Evolution. By J. B. S. Haldane. 1932.
The Early Life of Thomas Hardy, 1840-91. By Florence Emily Hardy. 1928.
The Sources of Civilization. By Gerald Heard. 1935.
The Third Morality. By Gerald Heard. 1937.
Preliminary Discourse on the Study of Natural Philosophy. By J. F. W. Herschel. 1830.
Cheerful Yesterdays. By T. W. Higginson. 1898.
Life and Letters of Joseph Dalton Hooker. By Leonard Huxley. Two volumes. 1918.
Memoir of Leonard Horner. Edited by his daughter, K. M. Lyell. Two volumes. 1890.
A History of Darwin's Parish: Downe, Kent. By O. J. R. and E. K. Howarth. 1933.
Personal Narrative of Travels, etc. By Alexander de Humboldt. Seven volumes. 1814-1829.
We Europeans. By Julian Huxley and A. C. Haddon. 1935.
Man's Place in Nature. By T. H. Huxley. 1863.
Darwiniana. By T. H. Huxley. 1893.
Life and Letters of T. H. Huxley. By Leonard Huxley. Two volumes. 1900.
Huxley: Prophet of Science. By Houston Peterson. 1932.
Prometheus: or, Biology and the Advancement of Man. By H. S. Jennings. 1925.
Reminiscences of Prideaux John Selby. By L. Blomefield (that is, Jenyns). 1885.
Chapters in my Life. By L. Blomefield (that is, Jenyns). 1889.
The Coming of Evolution. By J. W. Judd. 1910.
Concerning Man's Origin. By Sir Arthur Keith. 1927.
Darwinism and what it Implies. 1928.
Darwinism and its Critics. By Sir Arthur Keith. 1935.
Psychopathology. By Edward J. Kempf. 1921.
Charles Kingsley: his letters, etc. Edited by his wife. Two volumes. 1877.
The Psychology of Men of Genius. By Ernst Kretschmer. 1931.
Memoirs of a Revolutionist. By P. Kropotkin. Two volumes. 1899.
Life of Sir John Lubbock, Lord Avebury. By H. G. Hutchinson. Two volumes. 1914.
Principles of Geology: being an enquiry how far the former changes on the earth's surface are referable to causes now in operation. By Charles Lyell. Three volumes. 1830-1833.
The Geological Evidences of the Antiquity of Man. By Charles Lyell. 1863.

BIBLIOGRAPHY

Life, Letters, etc., of Sir Charles Lyell. Edited by his sister-in-law, Mrs. Lyell. Two volumes. 1881.
An Essay on the Principle of Population; etc. By T. R. Malthus. Sixth edition. Two volumes. 1826.
Karl Marx and Friedrich Engels: Correspondence, 1846-95. 1934.
Reminiscences chiefly of Towns, Villages, and Schools. By T. Mozeley. Two volumes. 1885.
John Murray III. By John Murray IV. 1919.
At John Murray's. By George Paston. 1932.
Nature. Various dates.
Letters of Charles Eliot Norton. Edited by Sara Norton and M. A. D. Howe. 1913.
Notes and Queries. Various dates.
Impressions of Great Naturalists. By H. F. Osborn. 1924.
From the Greeks to Darwin. By H. F. Osborn. Second revised edition. 1929.
Life of Richard Owen. By his grandson, Richard Owen. Two volumes. 1894.
The Principle of Moral and Political Philosophy. By William Paley. 1811 edition.
A View of the Evidences of Christianity. By William Paley. 1823 edition.
Natural Theology. By William Paley. 1826 edition.
Scientific Correspondence of Dr. Joseph Priestley. Edited by H. C. Bolton. 1892. (Appendix on the Lunar Society.)
The Richmond Papers. By A. M. W. Stirling. 1926.
A History of Freethought in the Nineteenth Century. By J. M. Robertson. Two volumes. 1929.
Salopian Shreds and Patches. 1884.
Life of Mary Anne Schimmelpennick. Edited by C. C. Hankin. Two volumes. 1884.
Life and Letters of Adam Sedgwick. By J. W. Clark and T. McK. Hughes. 1890.
The Singing Swan: Anna Seward. By Margaret Ashmun. 1931.
The Shrewsbury Guide. 1809.
A Guide through the town of Shrewsbury. 1836.
A History of Shrewsbury. By Hugh Owen and J. B. Blakeway. Two volumes. 1825.
The Old Houses of Shrewsbury. By H. E. Forrest. 1911 and later editions to 1935.
Shrewsbury School Register, 1798-1898. Compiled by J. E. Auden. 1898.
A Short History of Biology. By Charles Singer. 1931.
Proceedings and Transactions of the London Entomological, Geological, Linnean, Royal, and Zoological Societies. Various dates. Also Edinburgh Societies, 1825-1827.
The Life and Letters of Herbert Spencer. By David Duncan. 1908.
The Victorian Tragedy. By Esmé Wingfield Stratford. 1930.

Life and Letters of Admiral Sir B. J. Sulivan. Edited by H. N. Sulivan. 1896.
Life: Outlines of General Biology. By J. Arthur Thomson and Patrick Geddes. Two volumes. 1931.
Evolution. By Patrick Geddes and J. Arthur Thomson. 1911.
The Times. Various dates.
Reminiscences of Oxford. By William Tuckwell. 1900.
A Sketch of The Origin of Species. By Paul B. Victorius. *Colophon*, Part IX. 1932.
Natural Selection and Tropical Nature. By Alfred Russel Wallace. 1891.
Darwinism. By A. R. Wallace. 1889.
My Life. By A. R. Wallace. Two volumes. 1905.
Alfred Russel Wallace: Letters and Reminiscences. By James Marchant. Two volumes. Cassell. 1916. (Presents the complete extant Darwin-Wallace correspondence in chronological order, 1857-81.)
The Wedgwoods. By Llewellyn Jewitt. 1865.
The Life of Josiah Wedgwood. By Eliza Meteyard. Two volumes. 1865.
A Group of Englishmen, 1795-1815: Being records of the younger Wedgwoods and their friends. By Eliza Meteyard. 1871.
Wedgwood and his Works. By Eliza Meteyard. 1873.
Josiah Wedgwood: his personal history. By Samuel Smiles. 1894.
Tom Wedgwood: the first Photographer. By R. B. Litchfield. 1903.
A History of the Wedgwood Family. By Josiah C. Wedgwood. 1909.
The Personal Life of Josiah Wedgwood. By Julia Wedgwood. 1915.
Wedgwood Pedigrees. By Josiah C. Wedgwood and Joshua G. E. Wedgwood. 1925.
The Science of Life. By H. G. Wells, Julian Huxley, and G. P. Wells. Two volumes. 1930.
A History of the Warfare of Science with Theology in Christendom. By Andrew Dickson White. Two volumes. 1896.
Thomas Woolner, R.A.: his Life in Letters. By his daughter, Amy Woolner. 1917.
Celebrities at Home: second series. By Edmund H. Yates. "Mr. Darwin at Down," pp. 223-30. 1878.

INDEX

Acquired characters, inheritance of 26, 151, 180, 255, 320, 321
adaptation 27, 75, 122, 150, 156
Adventure 88, 116, 121, 127, 128
Agassiz, Alexander 243, 325
Agassiz, Louis 233-4, 243
Ainsworth, W. 62
Allen, Catherine (later Mackintosh) 33, 57
Allen, Elizabeth (wife of Josiah Wedgwood of Maer) 33, 44, 56-8, 147, 162, 163, 164, 178, 193
Allen, Emma 208, 265
Allen, Fanny 208, 262, 265, 291
Allen, Grant 310
Allen, Jessie (later Sismondi) 57, 65, 162, 163, 208
Allen, John Bartlett 57, 162
Allen, Louisa Jane (later Wedgwood) 57
American Civil War 253, 262
Andes, the 123, 126-7, 130-1
Aristotle 26, 292
Ashmun, Margaret 20 n.
Athenaeum 245, 251, 253, 270
Audubon, J. J. 64
Australia 136-8

Babbage, C. 149
Bacon, F. 151
Bahia 102, 140
Barlow, Nora 105, 149 n.
Bates, H. W. 219, 308
Bateson, W. 320
Beagle 84, *passim* 87-144, 145, 164, 186, 198, 202, 203, 214, 292, 307
Beaufort, F. 84, 86, 89, 90
Bell, Sir Charles 280
Bell, T. 146, 154, 196

Bentham, G. 228, 243
Bentley, T. 21, 23
Berg, L. 320
Bernhardi 318
Besant, Annie 303
Bilsborrow, Dewhurst 32
Biographical Sketch of an Infant 292
birth-control 303, 322, 328
Blyth, E. 275
Botanic Garden 24, 27
botany 17, 18, 45, 71, 158, 188, 260, 284-6
Boulton, Matthew 4, 14, 15, 31
Bradlaugh, C. 303
British Association 167, 194, 200, 202, 219, 249-53
Broderip, W. J. 146
Brodie 178, 208, 263
Brown, Robert 146, 165, 166
Browning, Robert 152
Buckland, W. 200
Buddha 324
Buffon, G. L. L. 26, 28, 151, 296
Bunbury, C. 173, 187, 221, 243
Burdon-Sanderson, J. 288, 303
Butler, Samuel, D.D. 38, *passim* 49-53
Butler, Samuel 29 n., 295-8

Cambridge 8, *passim* 66-78, 87, 147, 148, 230, 282, 292, 304, 310
Candolle, A. de 287
Carlyle, Jane 149, 185, 245, 310
Carlyle, Thomas 57, 63, 149, 162, 166, 185, 243, 267, 288, 304
Carpenter, W. B. 234, 242, 247, 308
Certain areas of elevation, etc., On 153

343

Chambers, Robert 188, 250
Christ's College, Cambridge 67, 72, 73, 147
Cirripedes monographs 198, 202, 212
Climbing Plants. See *Movements, etc.*
Cobbe, F. P. 274, 276-7, 294
Colburn, H. 155, 167
Coldstream, J. 62, 63
Coleridge, S. T. 33
Connexion of certain volcanic phenomena, etc., On the 158
Conway, Moncure 287, 308
coral islands 133, 138, 146, 153
Coral Reefs. See *Structure, etc.*
Corfield, Richard 126, 127, 132
Covington 120, 129, 164, 165
Cross and Self Fertilization. See *Effects, etc.*
Crystal Palace 211
Cuvier 26, 64, 151

Darwin, Anne Elizabeth (daughter) 168, 208-9
Darwin, Bernard (grandson) 291, 294, 307, 310
Darwin, Caroline (sister) 39, 43, 54, 57, 58, 61, 64, 65, 81, 153, 208, 262, 311
Darwin, Catherine (sister) 39, 44, 45, 54, 58, 148, 153, 158, 161, 162, 184, 201, 219, 265
Darwin, Charles (uncle) 12, 19-20, 60
DARWIN, CHARLES ROBERT 3, 7, 38, 39, *passim* 43-331. *Pre-Voyage*: childhood memories 43-5; traits 44-5, 48, 51-2; influences 45-6, 47; collecting habits 47-8, 51, 55, 63, 69, 70, 71-2; schooling 48-53; friendships 48, 52, 62-4, 66-9, 71, 74, 77-8, 85, 87; reading 51-2, 53, 54, 72, 75, 76-7; home and family life 54-8; love of sport 55-6, 58, 59, 61, 66, 70, 74, 80; university years 59-78; tedium of study 60-1, 62, 64, 69, 73, 75; scientific interests 62-4, 69, 70, 71-2, 77-80; religious views 65, 72-3. *Voyage*: voyage proposed 81-6; preparations 86, 90-1; waiting to sail 91-5; ill-health 93, 111, 127; voyage, *passim* 96-141; seasickness 94, 95, 96, 100, 109, 111, 114, 124, 136, 138, 140, 141; reading 96, 97-9, 116, 119, 124; geology 97-9, 100, 116-8, 121, 123, 124-5, 128, 130-1, 139; scientific ambitions 113-4, 120-1, 140; species views 116, 118, 134-5, 137-8, 149; doubts of religious vocation 133; home-coming 141-2; view of voyage 142-3. *Post-Voyage*: diary publication 139, 147-8, 149, see *Journal and Remarks*; friendships 146, 149, 165, 166, 169, 184-7, 210-1; work on voyage specimens 147, 149-50, 153, 154, 155, 158, 164, 166, 170-171, 179, 180, 188, 189, 193, 195-6, 197; species work 151-2, 155-8, 159-60, 171-2, 180-3, 187-8, 197-198, 199, 212-3, 214-5, 217-8, 221-4, 229-30; rumoured engagement 153; on marriage 153; ill-health 154, 166, 167-8, 169, 174, 179, 189, 193, 194, 200-4, 209, 211, 222, 230, 232, 241; reading 154, 159, 188, 191, 200, 213, 232; holidays 158, 167, 172-3, 189, 193, 200, 210, 216-7; marriage 160-1, 164-5; appearance 174, 211; range of interests 188; daily routine 204-6, 290; as father, 209, 267, 291; mental qualities 213-4; wish for priority 225-6, 259. *Post-Origin*: ill-health 257, 260, 263-4, 275, 282, 283, 304-5; zoological work 259, 261, 263, 264, 267-71, 274-5,

INDEX 345

Darwin, Charles Robert—*contd.* 277, 278, 283, 286, 299; friendships 259, 274, 275, 287-9, 293, 294; botanical work 260, 261-2, 263-4, 283, 284-6; reading 261, 277, 290-1, 300, 303, 327; holidays 263, 265, 275, 276-7, 282, 287, 293-4, 304; honours 265, 292; need to work 284; appearance 290; autobiography 292-3, 300; attitude to controversy 294; loss of aesthetic appreciation 300; mental qualities 300-1; religious views 301-3; social views 303; last days 303-6; death 306-7; funeral 308-9; estimate of achievement 315-31

Darwin, Charles Waring (son) 222, 227

Darwin, Elizabeth (daughter) 208, 292, 294, 307

Darwin, Emma (wife). See Wedgwood

Darwin, Erasmus (grandfather) *passim* 3-32, 39, 60, 63, 68, 75, 151, 152, 192, 259, 293, 296, 299, 328

Darwin, Erasmus 283, 293, 294-7

Darwin, Erasmus (uncle) 12, 20, 29, 30-1

Darwin, Erasmus Alvey (brother) 39, 47, 48, 51, 52, 55, 59, 60, 61, 65, 68, 71, 86, 93, 94, 142, 145, 148, 149, 162, 164, 165, 166, 178, 184-5, 186, 193, 203, 210, 234, 242, 259, 261, 267, 275, 286, 302, 304

Darwin, Francis (son) 68, 202, 208, 213, 266, 286, 289, 291, 292, 294, 296, 297, 305, 307, 310

Darwin, George (son) 193, 208, 266, 282, 292, 302, 303, 305, 310

Darwin, Henrietta (daughter) 178, 208, 211, 227, 259, 260-1, 266, 276, 282, 294, 302, 307

Darwin, Horace (son) 208, 267, 282, 292, 304, 305, 310

Darwin, Leonard (son) 208, 262-3, 267, 292, 305, 309, 310

Darwin, Marianne (sister) 39, 54, 58, 229, 265

Darwin, Mary Eleanor (daughter) 173, 178

Darwin, Robert Waring (father) 12, 20, 23, 29, 30, 32, 34, 35-9, 44, 46-7, 50, 52, 54, 55, 56, 57, 59, 65, 66, 80, 82-4, 86, 102, 108, 120-1, 127, 133, 141, 144, 145, 162, 163, 168, 173, 178-9, 184, 189, 193, 200-1, 202, 210, 276, 292, 328

Darwin, Susan (sister) 39, 54, 56, 58, 65, 153, 185, 201, 219, 265-6

Darwin, Susanna (great-aunt) 8, 36

Darwin, Violetta (step-aunt) 29, 30

Darwin, William Erasmus (son) 167, 208, 209, 230, 263, 266, 267, 287, 291, 292, 305, 309

Darwin ancestors 6-7

Darwinism 315-29

Daubeny, C. 249, 253

Day, Thomas 15, 16, 31

Deposits containing extinct mammalia, Sketch of the 153

Derby Philosophical Society 21

Descartes 151

Descent of Man, The 277-80, 281, 283, 286

Diderot 151

Different Forms of Flowers, etc., The 283, 285

divergence 172, 182, 214-5, 223, 227, 320

Downe 173, 177

Down House 173, 177-8, 184, 218, 267, 289-90

Duncan, A. 60, 63

Earle, A. 100, 108, 109

Economy of Vegetation, The 24, 27

Edgeworth, R. L. 15, 16, 18, 31, 32

Edinburgh University *passim* 60-6
Effects of Cross and Self Fertilization, etc., The 283, 285
Eliot, George 302
Elwin, Whitwell 231, 277-8
Emerson, R. W. 152, 245, 310
Empedocles 26
Ervine, St. J. 298 n.
Etruria Hall 21, 33
Euclid 51
evolution 9, 25-8, 63, 75, 116-8, 151-2, 254, 272, 284, 315-6, 319-320; and improvement 27, 63, 183, 237-8
Expression of the Emotions, etc., The 278, 280-1
Eyre, E. J. 267

Falconer, H. 234, 243
Falkland Islands 115-6, 118, 122
Farrer, Sir Thomas 287, 304
fertilization, cross and self 28, 158, 188, 285
Fitton, W. H. 165
FitzRoy, Robert *passim* 81-139, 148, 149, 165-6, 252, 293
Formation of Vegetable Mould, etc., The 154, 283, 299
Fox, W. Darwin 68-9, 70, 72, 74, 87, 133, 154, 209, 217, 234, 288, 327
Franklin, B. 16
Freud 317
Fuegians, FitzRoy's 88-9, 114, 115, 122

Galapagos Islands 133-5, 149 n., 150, 191-2, 220
Galileo, 317, 323
Galton, Francis 29, 158, 288, 302, 308, 325
Galton, Samuel 15, 16; the third 29, 55
Geddes, P. 321-2, 323-4
Geikie, A. 242

geographical distribution 155, 181, 191-2, 218, 237
Geological Observations on South America 183, 188-9, 193, 194-6, 283
Geological Observations on the Volcanic Islands 166, 179-80, 283
geological record, imperfection of the 181, 195, 236
Geological Society 146, 148, 149, 154, 166, 168, 184
George III 4, 13
Gladstone 288, 303
Glen Roy 158
Goethe 63, 151
Gosse, P. H. 243
Gould, G. M. 203, 204
Gould, John 154
Gourmet Club 67-8
Graham (of Christ's College) 73, 75
Grant, R. E. 62, 63-4, 147, 152, 293 327
Gray, Asa 155, 210, 217, 222, 226, 234, 242, 243, 253, 262, 263, 270, 271, 275, 288
Gray, J. E. 197, 243
Great Exhibition 210

Haeckel 253, 281, 287
Haldane, J. B. S. 320
Hardy, Thomas 308
Harvey, W. H. 243
Heard, Gerald 317, 323
Helvetius 28
Henslow, J. S. *passim* 69-90, 97, 112, 120, 126, 143, 145, 147, 154, 158, 165, 183, 186, 196, 199, 200, 223, 234, 243, 249, 250, 251, 259, 261, 282, 289, 293, 327
Herbert, J. M. 68, 69-70, 72, 74, 276
Herder 151
Herschel, Sir J. 77, 78, 79, 139, 166, 212, 234, 243, 327
Holland, H. 64, 72, 148, 165, 167, 243

INDEX

Hooker, J. D. 155, 183, 186-7, 192, 196, 198, 200, 210, 213, 217, 221, 224, 225, 226, 227, 228, 230, 231, 233, 242, 247, 250-3, 259, 262, 263, 270, 275, 288, 308, 309
Hope, F. W. 71, 73, 74
Horner, Leonard 64, 148, 155, 186, 202
Howard, Mary (first wife of Dr. Erasmus Darwin) 11-12, 16, 36
Humboldt, A. von 76-7, 87, 90, 92, 96, 169, 327
Hume 27
Hunter, J. 9
Hunter, W. 8
Hutton, J. 16, 78, 98
Huxley, J. 255 n.
Huxley, T. H. 210, 213, 233, 242, *passim* 245-53, 259, 264, 267, 270, 274, 288, 292, 294, 297, 302, 303, 308, 309, 329

Insectivorous Plants 283, 286
International Congress of Medicine 304

Jameson, R. 62, 78
Jenyns, L. 74, 77, 89, 93, 154, 158, 171, 187, 234, 243, 293
Jesus 324
Johnson, Samuel 13
Johnston, W. W. 203-4
Journal and Remarks 148, 149, 150-151, 152, 153, 154-5, 164, 167
Journal of Researches (second edition of foregoing) 186, 189-92, 195, 196, 213, 220, 333-4
Jukes, J. B. 242

Kant 151, 277
Keir, James 8, 9, 15, 16
Keith, Sir A. 316, 323
Kempf, E. J. 46, 107 n., 168, 203
King, P. G. 95, 100, 103, 107, 108, 113

King, P. P. 87, 147, 166
Kingsley, C. 234, 241, 243, 267
Krause, E. 293, 296, 297, 298 n.
Kretschmer, E. 325-6
Kropotkin, P. 156-7, 322

Lamarck, J. B. 28, 29 n., 63, 116, 151, 152, 180, 187, 227, 246, 254, 259, 273, 295, 296, 320, 321
Langton, C. 265
Leibnitz 151
Lessing 151
Letter containing remarks on . . . Tahiti, A 139, 333
Letters to Professor Henslow, 143, 333
Lewes, G. H. 302
Lichfield Botanical Society 17-18
Life and Letters of C. Darwin, The 310
Lincoln, A. 329
Linnean Society, 72, 146, 211, 227, 229, 262, 263
Linneus 17, 26
Litchfield, R. B. 282, 287, 297, 302
Longfellow, H. W. 275
Lonsdale, W. 146
Loves of the Plants, The 24, 25, 27
Loves of the Triangles, The 24
Lowell, J. R. 308
Lubbock, J. 211, 234, 242, 252, 287-8, 308, 309
Lucretius 26
Lunar Society 15, 16, 21, 37
Lyell, Charles 78, 90, 92, 96, 97-9, 110, 116-8, 123, 124-5, 133, 146, 147, 148, 149, 152, 156, 157, 159, 162, 165, 183, 186, 187, 191, 192, 202, 203, 220, 221, 224, 226, 227, 228, 230, 231, 233, 241, 242, 260, 262, 268, 271, 273, 275, 291, 294

Macgillivray, W. 62-3
Mackintosh, Fanny (later Wedgwood) 145-6, 153, 161, 165, 208

Mackintosh, Sir James 33, 34, 57, 58, 65, 86
Maer Hall 34, 56-8, 61, 147, 164, 193-4
Mahon, Lord 211
Malibran, M. F. 73, 163
Malthus, T. R. 159, 171, 190, 328
man 28, 155, 223, 237, 244, 245, 251-2, 270, 272-4, 275, 278-81
Manual of Scientific Inquiry 212
Martineau, Harriet 185
Marx, K. 316
Matthew, Patrick 152, 238, 259
Matthews (missionary) 101, 114, 115
Mauritius 138
Mellersh, A. 100, 113
Melville, H. 225, 323, 330-1
Mendel 320
Mill, J. S. 247, 267, 277
Milman, Dean 166
Milton 72, 92, 300
Mivart, St. G. 281
Monro, A. 60
Monte Video 111-2, 113
Morley, J. 281, 288, 308
Mount, The 37-8, 43, 54-5, 178, 265-6, 276
Movements and Habits of Climbing Plants, The 263-4, 283, 285-286
Müller, Fritz, 253
Murchison, R. 79, 166, 169
Murray, A. 243
Murray, John (the third) 189, 231-2, 241, 261, 267-8, 277, 308
Mussolini 318
mutation 254-5, 320-1

Napoleon 22, 44, 139
Narrative of the Surveying Voyages, etc. 167
natural selection 28, 152, 156, 171, 222, 227-8, 231, 235, 237, 238, 246, 248, 250, 253, 255, 259, 269, 270, 279, 281, 284, 315, 317, 319, 320, 322-3
Naudin, C. 151
Newton 317, 323
New Zealand 136
Nietzsche, 316
Norton, C. E. 275, 287
Novello, Clara 163, 211

Oken, L. 151
Origin of Species, The: first sketch 170-1; second sketch 180-3, 195; writing of Great Work 221-4; writing of *Origin* 229-32; described 234-40; reception 241-53; second edition 241; third 260; fourth 264-5; fifth 275; last 254, 278, 281; other references 254-6, 257, 258, 268, 273, 278, 283-4, 293, 300, 302, 318, 325; Historical Sketch 242, 259, 260, 264, 296
Osborn, H. F. 330
Owen, Fanny 54, 56
Owen, Mr. 56
Owen, Richard 146, 147, 149, 154, 155, 164, 171, 183, 187, 200, 234, 243, 245, 247, 248, 249, 250, 259, 264-5, 270
Owen, Sarah 54, 56, 70

Paley, W. 75, 327
pangenesis 258, 264, 267, 268-71, 320, 325
Parkers, The Miss 17, 29, 31
Parslow 178
Peacock, G. 81, 82, 84, 89, 90
Peterson, Houston 325
Philosophical Club 211
Phytologia 27, 28, 31
Plan for the Conduct of Female Education 29
Playfair, Lord 288
Plinian Society 63, 293, 327
Pole, Colonel 19, 20

Pole, Elizabeth Chandos (second wife of Dr. Erasmus Darwin) 19, 20, 23, 29, 31, 68, 108
Poole, T. 33
Power of Movement in Plants 283, 285-6, 293
Priestley, J. 15, 16
Prince of Wales (later Edward VII) 265, 304
Principia Botanica 7
Proofs of recent elevation, etc., Observations of 141
Punta Alta 113, 118, 119, 150

Quatrefages, J. L. A. de 234

Ramsay, A. 242
Ray Society 212
Rhea Americana, Notes upon the 149
Rio de Janeiro 108
Robinet 151
Romanes, G. J. 288, 302, 306
Rousseau 15, 16
Rowlett, G. 99, 100, 103, 124
Royal Institution 72, 303-4
Royal Society of Edinburgh 64, 249, 265
Royal Society of London 35, 36, 53, 146, 184, 211, 265
Ruskin, J. 267, 294

St. Hilaire, I. G. 63, 151, 180
St. Jago 99-100
Schelling 151
science and reality 323-4, 329
Scott, Sir Walter 64
Sedgwick, Adam 77-8, 79-80, 110, 139, 140, 143-4, 154, 158, 165, 173, 212, 234, 243-5, 248, 249, 282, 293, 317
Seward, Anna 10, 12, 17-19, 20 n., 21, 31, 32, 293
sexual selection 171, 180, 227, 235, 258, 274, 275, 278, 279, 320, 325

Shakespeare 51, 72, 164, 207, 265, 300
Shaw (of Christ's College) 67, 73
Shrewsbury 34-5, 36, 44, 265-6, 276
Shrewsbury Grammar School 48-9, 51
Singer, Charles 315, 319
Sismondi, J. C. L. de 57, 185
Sismondi, Jessie. See Allen
Small, W. 15, 16
Smith, Andrew 139
Smith, Sydney 33, 56, 57, 74, 166
Sombart, Werner 329
South America. See *Geological Observations, etc.*
Spencer, Herbert 238, 242, 253, 261, 267, 280, 308
Spengler, O. 323
Stephens, J. S. 71-2
Stokes, J. L. 91, 100, 108, 113
Stratford, E. Wingfield 206
Strickland, H. E. 183
Structure and Distribution of Coral Reefs, The 166, 167, 169-70, 171, 283
struggle for existence 28, 151, 235, 254, 270, *passim* 316-24
Sulivan, B. J. 95, 100, 108, 128, 203, 210, 288
survival of the fittest 25-6, 28, 151, 171, 180, 254, 255, 270, 319, 320

Tahiti 135
Temple of Nature 24, 25, 27-8, 32
Tendency of species to form varieties, On the 227-8
Teneriffe 76, 81, 87, 96-7
Tennyson, A. 152, 207, 267, 275
Thomson, J. A. 321-2, 323, 324
Tierra del Fuego 81, 82, 113-5, 120-3
Times 246, 251, 281, 306, 307, 308

Treviranus 151
Twain, Mark 290, 294
Tylor, E. B. 302, 308
Tyndall, J. 275

Unity of organic life 26, 28, 182, 237, 270

Valparaiso 125-6
variation 75, 150, 156, 171, 180-1, 234-5, 254, 255, 256, 268, 269, 270, 316, 320-1
Variation of Animals and Plants under Domestication, The 261, 263, 264, 268-71, 272, 283, 286
Various Contrivances by which Orchids are Fertilized, etc., The 261-2, 283, 284, 285
Vestiges of Creation 188, 215, 219, 246
Victoria, Queen 154, 158, 211
Volcanic Islands. See *Geological Observations, etc.*
Vries, H. de 320

Wallace, A. R. 192, 211, 219-20, 222, 223, 224-8, 230, 234, 238, 242, 255, 262, 271, 273-4, 275, 278, 279, 285, 288, 302, 303, 308
Ward, Henshaw 79, 330
Waterhouse, G. R. 154, 170, 182 n., 196, 212
Waterton, C. 193
Watt, James 4, 15, 16
Wedgwood, Catharine (aunt) 30, 33, 34, 44, 56
Wedgwood, Charlotte (later Langton) (cousin) 54, 58, 91, 142, 147, 162, 194, 208, 260, 262
Wedgwood, Elizabeth (cousin) 54, 142, 147, 161, 162, 163, 167, 193, 208, 229, 299

Wedgwood, Emma (later Darwin) (wife) 55, 58, 64, 91, 147, 148, 153, 158, 159, 161-5, 166, 167, 168, 171, 173, 178, 183, 184, 193, 194, 202, 205, 206-7, 208, 210, 211, 216, 219, 222, 262, 263, 264, 265, 268, 275, 276, 280, 282, 286, 289, 290, 292, 294, 299, 300, 304, 305, 306, 307, 309-11
Wedgwood, Frances (cousin) 64, 91, 147, 162, 163
Wedgwood, Francis (cousin) 91, 147, 193, 208, 311
Wedgwood, Harry (cousin) 55, 91, 147, 193, 208, 311
Wedgwood, Hensleigh (cousin) 91, 145-6, 149, 161, 165, 178, 193, 208, 274, 292, 302, 311
Wedgwood, John (uncle) 23, 33, 34
Wedgwood, Josiah, of Etruria (grandfather) 3, 4-5, 11, 14, 15, 17, 21-3, 30, 33, 221, 328
Wedgwood, Josiah, of Leith Hill (cousin) 91, 147, 153, 193, 208, 262, 299
Wedgwood, Josiah, of Maer (uncle) 23, 33, 34, 56-7, 64, 83-4, 87, 91, 142, 147, 154, 161, 163, 164, 165, 167, 174, 178, 292
Wedgwood, Sarah (aunt) 30, 33, 34, 44, 56, 194, 208, 221-2
Wedgwood, Sarah (grandmother) 22-3, 33, 56
Wedgwood, Susannah (later Darwin) (mother) 23, 30, 33, 34, 37, 38, 39-40, 45-6, 47, 48, 201, 293
Wedgwood, Thomas (uncle) 23, 30, 33, 34
Weir, Jenner 275
Weismann, A. 320
Wells, W. C. 151-2, 238, 259
Wernerian Society 64

INDEX

Whewell, W. 74, 154, 243
Whieldon, T. 5
Whitley, C. 68, 72
Wickham, J. C. 100, 101, 104, 113, 126, 127, 132
Wilberforce, S., Bishop of Oxford 248, 250-3
Wilcox, 58, 61
William IV 86-7, 154, 158
Withering, Dr. 37

Woodhouse 56
Wordsworth 33, 159

Yarrell, W. 86, 146, 218

Zoological Society 72, 86, 149, 184
Zoology of Voyage, etc. 154, 164, 166, 170-1, 196
Zoonomia 18, 24, 25-7, 29, 53, 63, 75, 151, 243, 299, 327

For Product Safety Concerns and Information please contact our EU representative GPSR@taylorandfrancis.com
Taylor & Francis Verlag GmbH, Kaufingerstraße 24, 80331 München, Germany

www.ingramcontent.com/pod-product-compliance
Lightning Source LLC
Chambersburg PA
CBHW071232290426
44108CB00013B/1386